COMMON CORE | ACHIEVE

Mastering Essential Test Readiness Skills

MATHEMATICS

Mc Graw Hill Education

Bothell, WA • Chicago, IL • Columbus, OH • New York, NY

MHEonline.com

Copyright © 2015 McGraw-Hill Education

All rights reserved. No part of this publication may be reproduced or distributed in any form or by any means, or stored in a database or retrieval system, without the prior written consent of McGraw-Hill Education, including, but not limited to, network storage or transmission, or broadcast for distance learning.

Send all inquiries to:
McGraw-Hill Education
8787 Orion Place
Columbus, OH 43240

ISBN: 978-0-02-143257-8
MHID: 0-02-143257-0

Printed in the United States of America.

4 5 6 7 8 9 QLM 17 16 15 14

Table of Contents

Congratulations! If you are using this book, it means that you are taking a key step toward achieving an important new goal for yourself. You are preparing to take your high school equivalency test, an important step in the pathway toward career, educational, and lifelong well-being and success.

Common Core Achieve: Mastering Essential Test Readiness Skills is designed to help you learn or strengthen the skills you will need when you take your high school equivalency test. The program includes four core student modules – *Reading & Writing, Mathematics, Science,* and *Social Studies.* Each of these modules provides subject-level pre- and posttests, in-depth instruction and practice of the core skills and practices required for high school equivalency tests, and a number of additional helpful features to help you master all the skills you need for success on test day and beyond.

How to Use This Book

Before beginning the lessons in each module, take the **Pretest**. This will give you a preview of the types of questions you will be answering on the high school equivalency test. More important, it will help you identify which skill areas you need to concentrate on most. Use the Check Your Understanding chart at the end of the Pretest to pinpoint the types of questions you have answered incorrectly and to determine which skills you need to work on. The chart will also help you identify where to go within the text for additional instruction and practice. You may decide to concentrate on specific areas of study or to work through the entire module. It is highly recommended that you do work through the whole module to build a strong foundation in the core areas in which you will be tested.

Common Core Achieve: Mastering Essential Test Readiness Skills includes a number of features designed to familiarize you with high school equivalency tests and to prepare you for test taking. At the start of each chapter, the **Chapter Opener** provides an overview of the chapter content and a goal-setting activity. The lessons that follow include the following to help guide and enhance your learning.

- **Lesson Objectives** state what you will be able to accomplish after completing the lesson.

- **Key Terms and Vocabulary** critical for understanding lesson content is listed at the start of every lesson. All boldfaced words in the text can be found in the Glossary.

- The **Key Concept** summarizes the content that is the focus of the lesson.

- **Core Skills** are emphasized with direct instruction and practice in the context of the lesson. Each of the Core Skills aligns to the Common Core State Standards.

- **Core Practices** build important reasoning skills in Math and Science and align to key skills specified in the Common Core State Standards and other standards.

- Special features within each lesson include **21st Century Skills, Technology Connections, Workplace Skills,** and **Test-Taking Skills** to help you activate high-level thinking skills by using real-word application of these skills.

- The **Calculator Skills** feature will help you learn important tips to simplify your work with mathematical concepts or numerical data.

- **Think about Math** questions check your understanding of the content throughout each lesson.

- End-of-lesson **Vocabulary Review** checks your understanding of important lesson vocabulary, while the **Skill Review** checks your understanding of the content and skills presented in the lesson.

- **Skill Practice** exercises appear at the end of every lesson to help you apply your content knowledge and skill fundamentals.

In addition to the above lesson-level features, each *Common Core Achieve* module also includes the features to help you check your understanding as you prepare for the test.

- The end-of-chapter **Review** tests your understanding of the chapter content and skills.

- **Check Your Understanding** charts allow you to check your knowledge of the skills you have practiced, and references where you can go to review skills that you should revisit.

- The **Answer Key** explains the answers for the questions in the book.

- After you have worked through the book, take the **Posttest** to see how well you have learned the skills presented in this book.

Good luck with your studies, and remember: you are here because you have chosen to achieve important and exciting new goals for yourself. Every time you begin working within the materials, keep in mind that the skills you develop in *Common Core Achieve: Mastering Essential Test Readiness Skills* are not just important for passing the high school equivalency test; they are keys to lifelong success.

1. Order the following numbers from largest to smallest.

$$2.3, 2\tfrac{1}{3}, 2.03$$

   A. $2.3 > 2\tfrac{1}{3} > 2.03$

   B. $2\tfrac{1}{3} > 2.3 > 2.03$

   C. $2.3 > 2.03 > 2\tfrac{1}{3}$

   D. $2\tfrac{1}{3} > 2.03 > 2.3$

2. What is the prime factorization of 378?

   $2 \times \underline{} \times \underline{}$

3. The approximate volume of a beach volleyball court is 64,000,000,000 cubic millimeters. If there are 8 grains of sand in each cubic millimeter, how many grains of sand are contained in a beach volleyball court, expressed in scientific notation?

   A. 5.12×10^{10}
   B. 6.4×10^{10}
   C. 5.12×10^{11}
   D. 6.4×10^{11}

4. Which property is illustrated in the following equation?

   $4^3 \times 4^{-2} = 4^{3+(-2)} = 4^1 = 4$

   A. Product of Powers Property
   B. Quotient of Powers Property
   C. Power of a Power Property
   D. Power of a Product Property

5. The bottom of a square pizza box has an area of 196 square inches. What is the length of a side of the pizza box, rounded to the nearest inch?

   A. 6
   B. 14
   C. 49
   D. 98

6. You want to get an old family photograph enlarged and framed as a present. The original picture was 3 inches wide and 5 inches long, and the enlarged photograph is 20 inches long. How wide is the enlarged photograph?

   _____ inches

7. If your restaurant bill comes to $13 and you want to leave a 15% tip, how much is the total cost of your meal?

   A. $1.95
   B. $11.05
   C. $14.95
   D. $15.00

8. A student borrows $5,000.00 at a fixed 4.8% simple interest rate. The loan period is 3 years. How much interest will the student pay on the loan?

9. Sam's Salads offers 4 different types of specialty vegetables, 2 different types of cheese, and 5 different dressings. If you can only choose one from each category, how many unique salads can be formed?

10. A grocery store offers raffle tickets to all customers who pick the "lucky" checkout line. If the chance of winning a prize in the raffle is $\tfrac{1}{6}$ and the chance of picking the "lucky" checkout line is $\tfrac{1}{3}$, what is the probability of both getting a raffle ticket and then winning a prize?

   A. $\tfrac{1}{18}$

   B. $\tfrac{1}{9}$

   C. $\tfrac{1}{6}$

   D. $\tfrac{1}{2}$

11. Evaluate the expression $7x + 8y$ for $x = 3$ and $y = -2$.

12. If $7(n - 3) = 35$, what is the value of $2n$?

A. 2
B. 4
C. 8
D. 16

13. A highway has a minimum speed limit of 45 mph, and a maximum speed limit of 65 mph. If a car's speed is represented by the variable x, which of the following inequalities describes the legal speeds that a car can drive on the highway?

A. $45 \geq x \leq 65$
B. $45 \leq x \geq 65$
C. $45 \leq x \leq 65$
D. $45 \geq x \geq 65$

14. A nursery sells perennial flowers for $7 and annual flowers for $5.50. If you bought 3 perennials and some annuals and spent a total of $43, how many annuals did you buy?

_____ annuals

15. Jack invested $10,000 in an account earning simple interest. After 7 years, the account is worth $11,960. What is the rate of simple interest that Jack earns in this account?

A. 1.196%
B. 2.3%
C. 2.8%
D. 17%

16. What is the degree of the product of these two polynomials?

$(4x^2 - 15 + x)(x^3 - 19x)$

A. 3
B. 5
C. 6
D. 8

17. Which factor pair can you use to find m and n in the equation $(x + m)(x + n) = x^2 - 3x - 18$?

A. -3 and 6
B. 3 and -6
C. -2 and 9
D. 2 and -9

18. Solve the quadratic equation $(x + 7)^2 = 81$. Which of the following are solutions to the equation?

A. 2 and -16
B. 7 and -7
C. 9 and -9
D. 16 and -2

19. Which shows the first step in using the Quadratic Formula to solve the quadratic equation $3x^2 - 5x + 17 = 0$?

A. $x = \dfrac{-5 \pm \sqrt{5^2 - 4(3)(17)}}{2(3)}$

B. $x = \dfrac{5 \pm \sqrt{(-5)^2 - 4(3)(17)}}{2(3)}$

C. $x = \dfrac{-3 \pm \sqrt{(3)^2 - 4(-5)(17)}}{2(3)}$

D. $x = \dfrac{3 \pm \sqrt{(3)^2 - 4(-5)(17)}}{2(3)}$

20. What are the restricted values of $\dfrac{x - 7}{x^2 - 49}$?

$x \neq$ _____

21. What is the slope of a line that passes through the points $(2, 7)$ and $(3, 1)$?

$m =$ _____

22. Janet can stuff 35 envelopes in 20 minutes. What is her unit rate of envelopes per minute?

   A. $\frac{3}{5}$

   B. $\frac{3}{4}$

   C. $1\frac{3}{5}$

   D. $1\frac{3}{4}$

23. A line has slope 8 and passes through $(0, 0)$. What is the equation of the line in slope-intercept form?

   A. $x + y = 8$
   B. $y = 8x + 8$
   C. $y = 8x$
   D. $8x + 8y = 0$

24. Fill in the values of y in the following table of points that lie on the line $y = -3x + 4$.

x	0	1	2	3
y				

25. Which statement or statements are true of the system of linear equations?

$3x - 5y = 13$
$8x + 7y = -6$

   A. The system is independent.
   B. The system is dependent.
   C. The solution of the system is $(1, -2)$.
   D. The solution of the system is $(2, -3)$.

26. For every element in the domain, how many values can a function have in the range?

27. Which of the following is not an example of a quadratic function?

   A. $f(x) = 3x^2 - 15x + 2$
   B. $f(x) = (x - 1)(x + 2)$
   C. $f(x) = -1 + x^2 + 7x$
   D. $f(x) = \sqrt{x^2 + 7}$

28. Which of the following functions when graphed does not have an x-intercept?

   A. $f(x) = 3x$
   B. $f(x) = -3x$
   C. $f(x) = x^2 + 3$
   D. $f(x) = x^2 - 3$

29. If a car drives at a constant speed, the function comparing distance versus time is a proportional relationship. If after 3 hours, the car has traveled 96 miles, what is the slope of the graph of that proportional relationship?

30. A pool needs to be fenced in for safety. The rectangular deck area containing the pool is 18 feet wide by 25 feet long. How many feet of fencing will need to be bought?

31. The area of a trapezoid is 64 square centimeters. Its height is 8 centimeters and one of its bases measures 10 centimeters. What is the length of the other base?

   A. 6 centimeters
   B. 8 centimeters
   C. 10 centimeters
   D. 12 centimeters

32. An Indy-style racecar has to have certain engine, body, and wheel specifications in order to compete in a race. The diameter of the wheel must be 15 inches. What is the circumference of the wheel?

33. What is the approximate area of a circle with a circumference of 18π centimeters?

   A. 9π cm^2

   B. 18π cm^2

   C. 81π cm^2

   D. 324π cm^2

34. A standard shoe box is 11.5 inches by 7 inches by 3.75 inches. What is the volume of the shoe box in cubic inches?

35. A farmer has a corn silo in the shape of a cylinder with a radius of 3 meters and a height of 8 meters. The top of the silo is a hemisphere. What is the approximate volume of the silo?

   A. 74π m^3

   B. 78π m^3

   C. 90π m^3

   D. 108π m^3

36. Which of the following measures of central tendency could have a negative value? Select all that apply.

   A. Mean

   B. Median

   C. Mode

   D. Range

37. Which is greater, the mean or the mode of the following data set?

12, 20, 3, 8, 12

38. A sector in a circle graph with a central angle of 90° corresponds to what percent?

   A. 9%

   B. 25%

   C. 45%

   D. 90%

39. Identify the outlier in the following set of data points:

9, 12, 15, 13, 40, 10, 14, 14, 11, 10

40. If a scatter plot shows a steep negative trend, which would best describe a line modeling the data?

   A. Positive slope

   B. Negative slope

   C. Horizontal line

   D. Vertical line

1. **B** To compare the numbers convert the fraction to a decimal. $2\frac{1}{3} \approx 2.33$ so it is greater than 2.3, and 2.3 is greater than 2.03. Incorrect responses may result from errors when converting from a fraction to a decimal, or comparing the decimals incorrectly.

2. $2 \times 3^3 \times 7$ Finding the prime factorization involves breaking a number down into the product of prime numbers. Incorrect responses may include composite numbers that would need to be broken down further into their prime factors. Other incorrect responses may result from computational errors while dividing.

3. **C** The volume of the court can be expressed in scientific notation as 6.4×10^{10} cubic millimeters and then multiplied by 8 grains of sand. $6.4 \times 10^{10} \times 8 = 6.4 \times 8 \times 10^{10} = 51.2 \times 10^{10} = 5.12 \times 10^{11}$. Answer choices B and D, express the volume in scientific notation, but do not calculate the number of grains of sand and answer choice D incorrectly applies scientific notation. Answer choice A follows the right steps but incorrectly uses scientific notation and makes an error in the power of 10.

4. **A** You are multiplying two powers with the same base. This is the same as adding the exponents according to the Product of Powers Property. The other properties describe different situations involving quotients of powers, powers of a power, or powers of a product.

5. **B** To find the length of the side, you have to take the square root of $\sqrt{196} = 14$. Answer choice A is the cube root of the area. Answer choice D divides the area by 2 and answer choice C divides the area by 4, instead of taking the square root.

6. **12 inches** To solve this problem you can set up a proportion $\frac{3}{5} = \frac{x}{20}$ or solve for the scale factor. The original photo is 5 inches long and gets enlarged to 20 inches; this is 4 times as large. The width of the picture will be 4 times larger, or $3 \times 4 = 12$ inches. Incorrect responses may occur by setting up the wrong proportion, or multiplying by the wrong scale factor.

7. **C** To find the amount of the tip, multiply the amount of the bill by 0.15, $13 \times 0.15 = 1.95$. The total cost of the meal is $13 plus the amount of the tip $1.95, or $14.95. The incorrect responses may mean you calculated only the amount of the tip, or subtracted the tip from the cost instead of adding.

8. **$720** Simple interest is calculated by the formula $I = P \times r \times t$. Substitute the values given in the problem and solve for the amount of interest: $I = 5,000 \times 0.0048 \times 3 = 720$. Incorrect responses may make a mistake converting the percentage interest rate to a decimal, or may use the wrong formula to calculate simple interest.

9. **40 salads** If you can choose one from each category, you can multiply the number in each category to determine how many unique combinations there are: $4 \times 2 \times 5 = 40$. Incorrect responses may result from adding the categories instead of multiplying, or errors in computation.

10. **A** The probability of both getting a ticket and winning a prize, is the product of the probability of each event happening: $\frac{1}{3} \times \frac{1}{6} = \frac{1}{18}$. Answer choice B could result from adding the denominators instead of multiplying. Answer choice C could result from subtracting the probabilities, and answer choice D could result from adding the probabilities.

11. **5** To evaluate the expression substitute the values for each variable and perform the calculations: $7x + 8y = 7(3) + 8(-2) = 21 - 16 = 5$. Incorrect responses may switch the variables when substituting or make an error when performing the calculations.

12. **D** To solve this equation use inverse operations to isolate the variable, and find that $n = 8$. Then evaluate the expression $2n$ for the value of n: $2n = 2(8) = 16$. Incorrect responses may just solve for n, or incorrectly apply inverse operations in order to find n.

13. **C** A car can legally drive between 45 and 65 mph, so the value of x must be greater than or equal to 45 and less than or equal to 65. This looks like $45 \le x \le 65$. Incorrect responses use the incorrect inequality symbols to compare x to 45 and 65 mph.

14. **4 annuals** You can represent the problem using a variable to represent the number of annuals: $7(3) + 5.50A = 43$. Use inverse operations to solve for the variable, and find the number of annuals purchased, $A = 4$. Incorrect responses may result from mixing up the number of perennials and annuals, or incorrectly setting up or solving the equation.

15. **C** In order to solve this problem use the simple interest formula: $I = P \times r \times t$. The amount of interest Jack earns is equal to the final value of his investment minus his original investment: $11,960 - 10,000 = 1,960$. Substitute all the known values in the formula and solve for the rate: $1,960 = 10,000 \times r \times 7$, $r = 0.028 = 2.8\%$. Incorrect responses may result from substituting the ending value instead of the initial value for the principal, or incorrectly using the simple interest formula.

16. **B** The degree of a polynomial is the highest exponent of the variable. To find the degree of a product, look at the degree of each factor, and add. The first factor has degree 2, and the second factor has degree 3, so the degree of the product will be 5. You can see this by a simple example: $x^2x^3 = x^{2+3} = x^5$. Incorrect responses may multiply the degrees, or just choose the factor with the higher degree.

17. **B** The factor pair must satisfy the following conditions: $m + n = -3$ and $m \times n = -18$. 3 and -6 are the only factor pair that satisfy both conditions. The incorrect responses make a sign error or do not satisfy both conditions.

18. **A** To solve, take the square root of both sides of the equation and get the resulting equation $x + 7 = \pm 9$. Set the left side of the equation to both values 9 and -9 and solve. The solutions of the equation are 2 and -16. Incorrect responses may result from only taking the square root of the right side, or incorrectly applying inverse operations to solve.

19. **B** The Quadratic Formula is
$$x = \frac{5 \pm \sqrt{(-5)^2 - 4(3)(17)}}{2(3)}.$$ In this problem $a = 3$, $b = -5$, and $c = 17$. Substitute the values of a, b, and c to set up the Quadratic Formula. Incorrect responses mistake the values of a and b, or make a sign error in setting up the Quadratic Formula.

20. $x \ne 7$ The restricted values can be found by factoring the denominator:
$$\frac{x - 7}{x^2 - 49} = \frac{x - 7}{(x - 7)(x + 7)}$$
The denominator of an expression cannot equal 0, so the restricted values are any values of x that make the denominator 0, or $x = -7$ or 7. Incorrect responses may make an error when finding values that make the denominator 0.

21. **−6** The slope of a line between two points is the quotient of the change in y over the change in x. You can substitute the points into the equation: $m = \frac{y_2 - y_1}{x_2 - x_1} = \frac{1 - 7}{3 - 2} = \frac{-6}{1} = -6$. Incorrect responses may result from mixing up the x and y values, or incorrectly setting up the slope formula.

22. **D** The unit rate is the number of envelopes Janet can stuff in one minute. The unit rate can be found by dividing 35 by 20, to get $1\frac{3}{4}$ envelopes per minute. Incorrect responses come from dividing the number of minutes by the number of envelopes, or incorrectly leaving off, or adding, the whole number part.

23. **C** Slope-intercept form is written as $y = mx + b$. The slope is 8, and the y-intercept is 0, because the line passes through the point $(0, 0)$. By substituting the values of m and b, the equation is $y = 8x$. Incorrect responses may be written in standard form, and incorrectly calculate the value of m and b.

24. **(0, 4), (1, 1), (2, −2), and (3, −5)** To fill in missing values of the table, substitute the value of x into the equation and solve for y. Incorrect responses may result from substituting for y instead of x, or errors in calculation.

25. **A and C** The point that satisfies both equations is $(1, -2)$, so it is a solution of the system. There is only one unique solution to the system, so the system of equations is independent. Incorrect responses may only identify one of the correct responses, or incorrectly identify the system as dependent, or make a calculation error when finding the solution.

26. **one** For every element in the domain, a function can only have one value in the range. If a function has more than one value of $f(x)$ for a value of x then the graph of the equation will not pass the vertical line test, and the equation is not a function. Incorrect responses come from a misunderstanding of the rules of functions.

27. **D** A quadratic function's highest degree variable is raised to the second power. Answer choice D includes x to the second power, but it is underneath a radical sign, so it does not count as a quadratic function. Incorrect responses result from misunderstanding the definition of a quadratic function.

28. **C** This problem can be solved by graphing all four functions and determining which graph does not intersect the x-axis, or you can set each function equal to 0, and find the function that cannot be solved for 0.

29. **32** A proportional relationship is defined as $y = kx$, where k is the slope of the line. Substitute the known values for x and y and solve for k: $96 = 3k$, $k = 32$. Incorrect responses may make errors in calculation, or switch the values of y and x.

30. **86 feet of fencing** The fence will define the perimeter of the pool area and is rectangular in shape. The perimeter of a rectangle is 2 times the width plus 2 times the length: $P = 2(18) + 2(25) = 86$. Incorrect responses may find the area of the rectangle instead of the perimeter, or use the incorrect formula for perimeter.

31. **A** The formula for the area of a trapezoid is $A = \frac{1}{2}h(b_1 + b_2)$. Substitute all the known values in the problem and solve for the missing base length: $64 = \frac{1}{2}(8)(10 + b_2)$, $b_2 = 6$. Incorrect responses result from not knowing the area formula for a trapezoid or making errors in the calculation.

32. **15π or approximately 47.12** The circumference of a circle is found using the formula $C = \pi d$, or π times the diameter. By multiplying the diameter of the wheel, 15 inches, by π, you find the circumference of the Indy car wheel. Incorrect responses may confuse radius and diameter, or use the area formula of a circle instead of the circumference.

33. **C** The circumference of a circle is π times the diameter, therefore this circle has a diameter of 18, so the radius of the circle is 9. Plug the value of the radius into the area formula, the area of the circle is 81π cm^2. Incorrect responses may confuse the radius and the diameter, or use the incorrect area and circumference formulas.

34. **301.875 cubic inches** A shoebox is a rectangular prism, so the volume of the shoebox is the product of the dimensions, $11.5 \times 7 \times 3.75 = 301.875$. Incorrect responses may use the incorrect formula or make an error when multiplying the decimals.

35. **C** The volume of the silo is the sum of the volume of the cylinder and the hemisphere, $V = \pi r^2 h + \frac{1}{2}\left(\frac{4}{3}\pi r^3\right)$. Substitute all of the known values and simplify, $V = \pi(3)^2(8) + \frac{1}{2}\left(\frac{4}{3}\pi(3)^3\right) = 72\pi + 18\pi = 90\pi$. The incorrect responses make an error in calculating the volume formulas.

36. **A, B, and C** If a set of data values includes negative numbers, than the mean, median and mode can all be negative. However the range is always the positive difference between the largest and smallest number and cannot be negative. Incorrect responses may not understand the definitions of each term.

37. **the mode** The mean of the data set is the sum of the numbers (55) divided by the number of data points (5), $55 \div 5 = 11$. The mode of the data set is the number that appears most often, or 12. Therefore the mode is greater than the mean of the data set. Incorrect responses make an error in calculating mean.

38. **B** The central angle 90° is $\frac{1}{4}$ of the whole circle, so the percentage is $\frac{1}{4}$ of 100% or 25%. The other answers confuse angles and percentages.

39. **40** The outlier of a data set is a number that is extremely different from the other values in the data set. In this set 40 is 25 away from the next closest data point. Incorrect responses may find the mode of the data set, or choose the one digit number as the outlier.

40. **B** A scatter plot with a negative trend can be modeled with a line with a steep negative slope. All of the other lines model situations with positive correlation, no correlation, or no change.

Check Your Understanding

On the following chart, circle any items you missed. This helps you determine which areas you need to study the most. If you missed many of the questions that correspond to a certain skill, you should pay special attention to that skill as you work through this book.

Item #	Reference	Item #	Reference	Item #	Reference	Item #	Reference
1	1.1	11	3.1	21	5.1	31	7.1
2	1.2	12	3.2	22	5.1	32	7.2
3	1.3	13	3.3	23	5.2	33	7.2
4	1.3	14	3.4	24	5.3	34	7.3
5	1.4	15	3.4	25	5.4	35	7.4
6	2.1	16	4.1	26	6.1	36	8.1
7	2.2	17	4.2	27	6.2	37	8.1
8	2.2	18	4.3	28	6.3	38	8.2
9	2.3	19	4.3	29	6.4	39	8.3
10	2.4	20	4.4	30	7.1	40	8.4

Chapter 1

Number Sense and Operations

Numbers are everywhere in your daily life. From the time
you wake up until you go to bed, you will encounter numbers
in a variety of forms. You will use numbers and math to help
understand situations, solve problems, and make decisions.
During your morning commute you may have to use a toll booth
or take a bus or subway. It is important to make sure you have the
correct change. At work, your boss may ask you to calculate the
number of sales for the month. On your way home, you might stop
to pick up food for dinner and use coupons to get the best deal.
These scenarios all involve numbers, likely written as fractions
and decimals. They will show up everywhere in your day, and it
is important to understand what the numbers mean and how to
calculate and use them.

©Floresco Productions/age fotostock

Lesson 1.1
Order Rational Numbers

When you are handed a memo at work, you may see numbers written as fractions or decimals. How do you compare the numbers and understand what the memo is trying to communicate? Learn how to identify and compare different types of numbers using a number line.

Lesson 1.2
Apply Number Properties

Numerical expressions can represent situations you encounter in your daily life, such as calculating a tip on a restaurant bill. You can use properties of numbers to quickly and accurately evaluate expressions. Learn how to apply the order of operations and such properties as the Distributive Property.

Lesson 1.3
Compute with Exponents

How can you find the area of the floor you need to tile or the volume of a container? Exponents are useful to calculate volume and area, as well as to solve other real-world situations. Learn how to apply the rules of exponents to rewrite and calculate exponent expressions.

Lesson 1.4
Compute with Roots

If addition is the inverse operation of subtraction, what is the inverse operation of exponents? Roots are operations that can "undo" the process of applying exponents. Learn how to calculate with roots and use roots to work backwards and solve problems involving exponents.

Goal Setting

Think about the last time you were in the grocery store. How did you decide what to buy? Do you always buy the bulk size or do you choose the cheapest option? If you have coupons, how do you figure out the reduced price of the item? How are prices labeled at your store? Where do you see fractions and decimals used in labels and packaging?

How could the lessons in this chapter help you make decisions while shopping? How could understanding how to compare rational numbers help compare prices and options?

LESSON 1.1　Order Rational Numbers

LESSON OBJECTIVES

- Identify rational numbers
- Order fractions and decimals on a number line
- Calculate absolute value

CORE SKILLS & PRACTICES

- Use Math Tools Appropriately
- Apply Number Sense

Key Terms

absolute value
the distance a number is from zero

integers
the set of whole numbers and their opposites

rational number
the set of numbers that can be expressed as the ratio of two integers

Vocabulary

denominator
the bottom number of a fraction that represents the total number of parts contained in the whole of a fraction

numerator
the top number in a fraction that represents the part of the whole the fraction is describing

order
to place in the proper sequence

Key Concept

Rational numbers include whole numbers, fractions, decimals, and their opposites. A number line is a useful math tool for comparing and ordering rational numbers.

Rational Numbers

Rational numbers are part of the set of real numbers. A real number is any number you would find on a number line, and there are many different types. The numbers you use every day are examples of rational numbers. A number identifies the subway line that you need. Other numbers tell you the cost of the fare, the time your train arrives at the station, and how many stops you will pass before reaching your destination.

Types of Numbers

When we count, we use the numbers 1, 2, 3, 4, 5… These are called natural numbers. If there are no objects to count, the number 0 is included. The set of natural numbers and 0 are the whole numbers.

Whole Numbers

In some instances, we need more than whole numbers to describe or measure a quantity. Think about temperature. Negative numbers describe temperatures below zero. **Integers** are the whole numbers and their opposites.

Integers

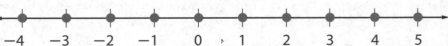

Order Rational Numbers

We often use numbers that fall between integers, like a 26.2 mile-long race or a −22.5°F temperature. Most of the numbers you encounter can be expressed as fractions or terminating decimals (decimals that have a finite number of figures). These form a larger set of numbers, called the **rational numbers,** or all numbers that can be expressed as the ratio of two integers, $\frac{a}{b}$, where $b \neq 0$. The rational numbers include the natural numbers, whole numbers, integers, fractions, and terminating or repeating (continuing a pattern forever) decimals. When writing repeating decimals, a bar is written over the number or numbers that repeat. For example, the number 0.416666..., where 6 repeats forever, would be written as $0.41\overline{6}$.

Example 1: Examples of Numbers

Natural Numbers	1	2	30	127
Whole Numbers	0	7	64	591
Integers	−27	−4	0	28
Fractions	$\frac{1}{2}$	$\frac{4}{9}$	$7\frac{3}{8}$	$\frac{12}{7}$
Terminating Decimals	−0.5	3.2	27.704	
Repeating Decimals	$-2.\overline{3}$	$0.\overline{12}$	$7.4\overline{63}$	$12.71\overline{4}$

Unlike rational numbers, **irrational numbers** cannot be expressed as the ratio of two integers. They are non–terminating decimals that do not repeat. They include the square roots of many whole numbers, such as $\sqrt{2} = 1.41421....$ Another example is the number pi, the ratio of the circumference of a circle by its diameter. Pi is represented by the symbol π, which is 3.14159.... When calculating using pi, most people use the estimation 3.14.

Fractions and Decimals

Whole numbers are not always as common as rational numbers in daily life. A kitchen is an example of a place where whole–number measurements are rare. For example, a recipe may call for $\frac{3}{4}$ cup of sugar.

Fractions

Fractions represent equal parts of a whole. The top number, or **numerator,** identifies the number of parts of the whole you are describing. The bottom number, or **denominator,** identifies the total number of parts contained in the whole. Together, whole numbers and fractions form mixed numbers like $3\frac{1}{5}$.

$$\frac{5}{8}$$ ← numerator—parts of the whole you have
← denominator—total number of parts in the whole

CALCULATOR SKILL

Many calculators are able to convert numbers between fractions and decimals. To convert a decimal to a fraction using the TI-30XS MultiView™, press the (2nd) key to access the second function, then the (table) key, whose second function allows you to "toggle" the number shown on the display back and forth from a decimal to a fraction.

Decimals

Terminating and repeating decimals are types of rational numbers. You rely on ten digits (0, 1, 2, 3, 4, 5, 6, 7, 8, 9) to write every number in our number system. Each digit in a number has a specific place value, or value based on its position in the number.

In a place–value chart like the one shown, a decimal point separates whole numbers from parts of a whole, or decimals. Whole numbers are to the left of the decimal point, and decimals are to the right. As you move to the right, each place value is one–tenth the value. The opposite is true as you move to the left.

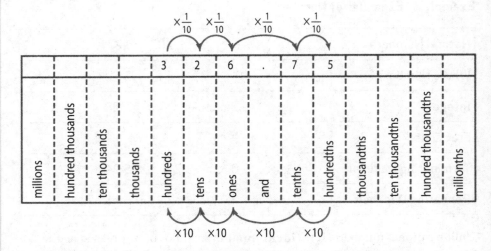

Since decimals represent fractional values, we read them as tenths, hundredths, thousandths, and so on. When reading decimals aloud, the word *and* represents the decimal point and only the last decimal place is named.

For example, read 326.75 as "three hundred, twenty–six and seventy–five hundredths." As a fraction, this would be written $326 \frac{75}{100}$ or $326 \frac{3}{4}$.

Think about Math

Directions: Answer the following questions.

1. Which number has a 3 in the tens place?

   A. 317.426
   B. 623.109
   C. 8,234.67
   D. 1,970.32

2. Which of the following categories apply to the number 7? Select all that apply.

   A. Whole number
   B. Integer
   C. Rational number
   D. Irrational number

Working With Fractions and Decimals

Fractions and decimals are common rational numbers you see and use each day. At a post office, for example, guidelines show the dimensions of letters, postcards, and boxes in fractions. Price charts show the cost of stamps and delivery options in decimals.

Compare and Order Fractions

To compare two fractions, you first want to take note if they have the same denominators or numerators.

Example 1: Same Denominators

Compare $\frac{6}{8}$ and $\frac{4}{8}$.

Step 1 Observe that both fractions have the same denominator. To compare them, read the numerators.

Step 2 The fraction with the greater numerator is the greater fraction. You can see this by comparing two fraction bars. Each bar is split into the same number of sections, but one has more filled in, and is therefore the greater fraction.

$$\frac{6}{8} > \frac{4}{8}$$

When fractions have different denominators and numerators, you can rewrite or rename one or more fractions in a set before you compare them. Once all of the fractions you are comparing share the same denominator, you can compare numerators.

Example 2: Different Numerators and Denominators

Compare $\frac{3}{4}$ and $\frac{5}{8}$. The fractions do not share the same denominator.

Step 1 Rewrite one or both fractions so that they have the same denominator by multiplying each fraction's numerator and denominator by the same number. Since $8 = 2 \times 4$, multiply $\frac{3}{4}$ by $\frac{2}{2}$ to get a fraction in eighths. As you can see in the fraction bars, it does not change the value of the fraction.

$$\frac{3}{4} = \frac{2 \times 3}{2 \times 4} = \frac{6}{8}$$

Step 2 Compare the fractions by comparing their numerators.

$\frac{6}{8} > \frac{5}{8}$, so $\frac{3}{4} > \frac{5}{8}$

Apply Number Sense

When you are given a set of numbers to compare, it is usually easiest to make sure they are of the same kind, either fractions or decimals. Since not all numbers can be written as fractions, decimals are an easier way to compare numbers.

For example, suppose a factory manager wants to know which of three products uses the most feet of plastic wrapping. Three employees report the average length of wrapping they use. The manager records the information in a data table.

Product A	Product B	Product C
5.25 feet	4.8 feet	$5\frac{3}{8}$ feet

Two of the values in the data table are decimals, and one is a fraction. To convert a fraction into a decimal, divide the numerator by the denominator.

$$5\frac{3}{8} = \frac{43}{8}$$
$$43 \div 8 = 5.375$$
$$5\frac{3}{8} = 5.375$$

Order the three numbers by comparing each digit from left to right. In this case, $4 < 5$ and so 4.8 is the smallest number. Since $0.2 < 0.3$, the next smallest number is 5.25. Therefore, the largest number is 5.375.

Use Math Tools Appropriately

To use math tools appropriately, first think about the problem you are trying to solve and which math tools can aid you in finding the answer. One such tool is a number line. A number line shows numbers from least to greatest. By plotting all the numbers on the number line, you can determine the order of the numbers from least to greatest by reading the numbers from left to right.

To compare and order decimals, it is helpful to use a number line marked off by tenths. This divides the space between each integer into 10 equal sections. To plot a number like 6.25, find the marks for 6.2 and 6.3 and plot the number halfway between.

Use the number line below to plot the decimals 6.75, 6.25, 6.4, and 7.1, and order them from least to greatest.

Compare and Order Decimals

To compare and **order** two decimals, you need to make sure you compare digits with the same place value. Suppose you want to compare 1.21 and 1.213.

Example 3: Compare Decimals

Step 1 To give both decimals the same number of digits before you compare them, add a zero to the end of 1.21. Adding a zero to the end of a decimal does not change its value.

1.210   1.213

Step 2 Compare the digits in the ones place, or whole number position.

1 = 1. The ones digits have the same value.

1.210   1.213

Step 3 Compare the digits in the tenths place.

2 tenths = 2 tenths. The tenths digits have the same value.

1.210   1.213

Step 4 Compare the digits in the hundredths place.

1 hundredth = 1 hundredth. The hundredths digits have the same value.

1.210   1.213

Step 5 Compare the digits in the thousandths place.

0 thousandth < 3 thousandths. Therefore 1.210 is less than 1.213.

1.210   1.213

1.21   <   1.213

◢ Think about Math

Directions: Compare each number to 4.65. Check the line that each number corresponds to.

1. $4\frac{3}{4}$ ___ Less than 4.65   ___ Greater than 4.65
2. 4.37   ___ Less than 4.65   ___ Greater than 4.65
3. 4.72   ___ Less than 4.65   ___ Greater than 4.65
4. $4\frac{1}{5}$ ___ Less than 4.65   ___ Greater than 4.65

Absolute Value

Positive and negative numbers express opposite amounts. Every integer has an opposite. For example, the opposite of 3 is −3.

On a number line, opposite numbers are always the same distance from zero. The distance from zero is called the absolute value of the number. The symbol for **absolute value** is | |. Because absolute value is the distance to 0, it is always a positive amount or 0.

The absolute value of 2 (written $|2|$) is 2, and the absolute value of –2 (written $|-2|$) is also 2 because both numbers are a distance of 2 units from 0. Because 0 is zero distance from itself, the absolute value of 0 is 0.

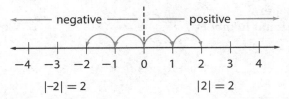

$$|-2| = 2 \qquad |2| = 2$$

Adding and Subtracting Integers Using Absolute Value

When adding two integers, look at the signs of the integers. If the integers have like signs, find the sum of the integers' absolute values. Then give the sum the same sign as both integers. For example, $-6 + -12 = -18$. If the integers have unlike signs, subtract the integers' absolute value as shown.

Example 4: Unlike Signs

Add $-8 + 6$.

Step 1 Subtract the integers' absolute values.

$$|-8| - |6| = 8 - 6 = 2$$

Step 2 Give the difference the sign of the integer with the greater absolute value.

$|-8| > |6|$, so make the difference negative. $-8 + 6 = -2$

Subtracting an integer is the same as adding the opposite of that integer. Change the number that is being subtracted to its opposite. Then add the integers. Once you know how to subtract integers, you can find the distance between two points.

Example 5: Finding Distance on a Number Line

The distance between two integers on a number line is the absolute value of their difference. Find the distance between –4 and –9.

Step 1 Find the difference of the two numbers. It does not matter in which order you subtract them because you will be taking the absolute value of the difference.

$$|-4 - (-9)| = |-4 + 9| = |5|$$

Step 2 Take the absolute value of the difference. $|5| = 5$

Think about Math

Directions: Choose the best answer to each question.

1. What is the distance between the numbers –1 and 5?
   A. –4
   B. –6
   C. 6
   D. 4

2. What is the sum of $-7 + 3$?
   A. –10
   B. 4
   C. –4
   D. 10

Environmental Literacy

In chemistry, a pH level indicates whether a solution is acidic, basic, or neutral. On a pH scale from 0 to 14, pure water has a pH of 7. Chlorine is added to swimming pools to destroy harmful organisms that may be in the water. For chlorine to be effective, a water pH of 7.3 is ideal. However, a pH level that is more than 0.3 away from ideal is considered unacceptable.

You can use absolute value to identify which pools in the table below have acceptable or unacceptable pH levels. For example, if a pool had a pH value of 7.8, you can find the distance from the ideal using absolute value and compare to 0.3.

$$|7.3 - 7.8| = |-0.5| = 0.5$$
$$0.5 > 0.3$$

The pH level is more than 0.3 away from the ideal. Therefore, the pool is unacceptable. Using the values in the table, determine which pools have acceptable pH levels and which do not.

pH Level	
Pool A	7.4
Pool B	7.7
Pool C	7.9
Pool D	7.1

Vocabulary Review

Directions: Write the missing term in the blank.

absolute value	**denominator**	**integers**
order	**numerator**	**rational number**

1. Rational numbers can be placed in _____ from least to greatest.

2. In the fraction $\frac{3}{4}$, the number 3 is the _____ .

3. A(n) _____ is any number that can be expressed as a ratio of two numbers.

4. The _____ is the total number of parts in a whole.

5. The _____ of −4 is 4.

6. The set of natural numbers, their opposites, and the number zero form the set

of _____ .

Skill Review

Directions: Read each problem and complete the task.

1. A lab technician measured the temperature of four different substances and recorded the temperatures in a data table. Now she wants to compare them.

Substance	X	Y	Z	W
Temperature	3.3°	3.15°	3.9°	3.55°

Order the temperatures on the number line. Then choose the appropriate ordering from least to greatest from the choices below.

A. 3.3°, 3.15°, 3.9°, 3.55°

B. 3.15°, 3.55°, 3.3°, 3.9°

C. 3.3°, 3.55°, 3.9°, 3.15°

D. 3.15°, 3.3°, 3.55°, 3.9°

2. The factory manager asks employees to use a new kind of transparent wrapping for their three top–selling items. The lengths recorded in the data table indicate how much wrapping each item requires. Compare the values and order them from least to greatest.

Item A	Item B	Item C
3.65 feet	4.1 feet	$3\frac{11}{16}$ feet

3. Explain the difference between rational and irrational numbers. Give examples of both in your explanation.

4. Determine which number has the greatest distance from the number 3.
   A. −6
   B. −2
   C. 7
   D. 11

5. A foot contains 12 inches. 5 inches is what fraction of a foot?
   A. $\frac{5}{12}$
   B. $\frac{1}{5}$
   C. $\frac{1}{12}$
   D. $\frac{5}{1}$

6. A wooden crate weighing $2\frac{5}{16}$ pounds contains grapefruit weighing $24\frac{1}{2}$ pounds. What is the combined weight of the crate and the grapefruit?
   A. $26\frac{6}{18}$ pounds
   B. 27 pounds
   C. $26\frac{13}{16}$ pounds
   D. $26\frac{3}{8}$ pounds

Skill Practice

Directions: Read each problem and complete the task.

1. Determine which rational number is represented on the number line shown.

   A. 0.5
   B. 0.6
   C. 1.5
   D. 1.6

2. Which rational numbers are within 1 unit of the rational number represented on the number line shown above?
   A. 1.5 and −1.5
   B. 2.5 and 0.5
   C. 2.6 and 0.6
   D. 1.6 and −1.6

3. Use a number line to compare the fractions $\frac{9}{5}$ and $\frac{5}{2}$.

4. Which of the following numbers have a distance of 3 from the number 8? Select all that apply.
   A. −5
   B. 2
   C. 5
   D. 11

5 Explain why absolute value is always positive or zero.

6. Find the sum of $3\frac{1}{3} + 2\frac{3}{4} + 5\frac{5}{6}$.
   A. $10\frac{9}{13}$
   B. $11\frac{11}{12}$
   C. $11\frac{8}{13}$
   D. $10\frac{11}{12}$

LESSON 1.2   Apply Number Properties

LESSON OBJECTIVES

- Determine LCM and GCF of two positive numbers (not necessarily different)
- Apply number properties (Distributive, Commutative, and Associative Properties) to rewrite numerical expressions
- Determine when a numerical expression is undefined

CORE SKILLS & PRACTICES

- Apply Number Sense Concepts
- Perform Operations

Key Terms

greatest common factor (GCF)
the greatest factor that is shared between the numbers

least common multiple (LCM)
the least multiple that is shared between the numbers

order of operations
the rules for the order that calculations should be done when evaluating an expression

Vocabulary

addend
a number that is added to another number

factor
a number that is multiplied by another number

undefined
an expression that cannot be evaluated

Key Concept

The least common multiple and greatest common factor of a pair of numbers can be used to solve problems. Awareness of number properties can be helpful in evaluating numerical expressions, although some expressions are undefined.

Factors and Multiples

Suppose you and a friend start jogging around a track at different speeds. You may meet up with each other at different points around the track. You can use the mathematical concepts of factors and multiples to find out how long it will take to meet up again at the starting point.

Prime Factorization

Whole numbers greater than 1 are considered either prime or composite. A prime number has only itself and the number 1 as its factors. A **factor** is any whole number that can be multiplied by another number. The result is the product. A composite number has itself, 1, and at least one other whole number as its factors. The number 1 is neither prime nor composite.

To list the factors of a composite number, identify whole numbers that divide evenly into the number. The number 48 can be divided evenly by 2, so both 2 and 24 are factors of 48.

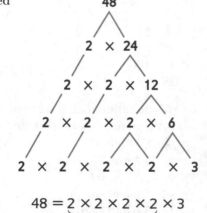

$$48 = 2 \times 2 \times 2 \times 2 \times 3$$
$$48 = \qquad 2^4 \qquad \times 3$$

The prime factorization of a composite number shows the number written as the unique product of its prime factors. Tree diagrams like the one shown are often used to break apart the number into its factors.

You can use powers to simplify a number's prime factorization by using repeated factors as the base and the number of times the factor appears as the exponent.

Greatest Common Factor

A pair of whole numbers can have many factors in common. To find common factors of a pair of numbers, list the factors of each number and identify the shared factors. There are four common factors for 24 and 30.

Factors of 24: 1, 2, 3, 4, 6, 8, 12, 24
Factors of 30: 1, 2, 3, 5, 6, 10, 15, 30
common factors

The **greatest common factor (GCF)** is the greatest factor that is shared between two composite numbers. You can find the GCF by finding the largest common factor from the list of the factors for each number. The GCF for 24 and 30 is 6.

$$189 = 3^3 \times 7$$
$$440 = 2^3 \times 5 \times 11$$
$$GCF(189, 440) = 1$$

189 and 440 are relatively prime.

Two numbers for which the GCF is 1 are said to be relatively prime. The numbers 189 and 440 are relatively prime because they have no common factors other than 1.

Least Common Multiple

A multiple of a number is the product of the number and any natural number. Just as with factors, pairs of numbers can have common multiples. Common multiples of 4 and 5 are 20, 40, 60, etc.

Multiples of 4: 4, 8, 12, 16, 20, 24, 28, 32, 36, 40, ...
Multiples of 5: 5, 10, 15, 20, 25, 30, 35, 40, 45, 50, ...
common multiples

The **least common multiple (LCM)** is the least multiple that is shared between the numbers. To find the LCM, you can write the first several multiples of each number and identify the least number in both lists. The LCM for 4 and 5 is 20.

You can also find the LCM of a pair of numbers by examining their prime factorizations. The least common multiple is the product of the highest power of each factor. Notice that the factors do not need to be shared by the numbers to be included in the LCM calculation.

$$120 = 2^3 \times 3 \times 5$$
$$252 = 2^2 \times 3^2 \times 7$$
$$LCM(120, 252) = 2^3 \times 3^2 \times 5 \times 7 = 2{,}520$$

Think about Math

Directions: Answer the following questions.

1. Which is the GCF of 36 and 90?
   A. 2
   B. 3
   C. 6
   D. 18

2. Which statement is true?
   A. The GCF of 5 and 15 is 15.
   B. The LCM of 7 and 21 is 21.
   C. The GCF of 60 and 126 is 36.
   D. The LCM of 24 and 27 is 648.

Apply Number Sense Concepts

The greatest common factor of two numbers is useful when reducing fractions, and the least common factor is used when adding and subtracting fractions. For instance, to reduce $\frac{8}{12}$, you find the GCF of 8 and 12, which is 4. Therefore, you can divide both 8 and 12 by 4 to get the reduced fraction $\frac{2}{3}$. When adding or subtracting fractions, you find the least common factor of the denominators (also known as least common denominator) instead of finding the GCF. Knowing whether the GCF or LCM is being asked in a problem is an important problem-solving skill.

For example, suppose a jeweler has 60 lengths of wire and 48 charms to use to make bracelets. He will use the same number of lengths of wire and the same number of charms on each bracelet, and he wants to make as many bracelets as possible, using all of his materials. How many bracelets will he be able to make?

Properties of Numbers

Construction workers rely on using the right tools in order to perform each job properly. Often times, different tools might be used for the same job, but using a specific tool makes the job less challenging. In mathematics, there are certain properties of numbers that you can use as tools to help make your calculations easier.

Commutative Property

The Commutative Properties deal with the order of numbers. The Commutative Property of Addition states that you can add two numbers in either order without affecting the sum. Think about a person training for a race. If they run 4 miles then 3 miles, or 3 miles then 4 miles, they still have run 7 miles total.

Similarly, the Commutative Property of Multiplication allows you to switch the order of two factors without changing the product. Both of these properties hold true for whole numbers, integers, and rational numbers, including fractions and decimals.

Commutative Property of Addition
$a + b = b + a$

Example:
$4 + 5 = 5 + 4$

$0.5 + (-1) = (-1) + 0.5$

$-\frac{1}{3} + \frac{2}{3} = \frac{2}{3} + \left(-\frac{1}{3}\right)$

Commutative Property of Multiplication
$a \times b = b \times a$

Example:
$2 \times 3 = 3 \times 2$

$-0.25 \times (-8) = -8 \times (-0.25)$

$\frac{4}{5} \times \frac{1}{4} = \frac{1}{4} \times \frac{4}{5}$

The Commutative Property does not hold for the operations of subtraction or division. For these two operations, the order in which the numbers are written have an effect on the difference and quotient.

Associative Property

The Associative Property of Addition states that you can group **addends**, or numbers added to get another number, in different ways without affecting the sum.

Similarly, the Associative Property of Multiplication allows you to change the grouping of factors without changing the product.

Associative Property of Addition
$(a + b) + c = a + (b + c)$

Example:
$(1 + 2) + 3 = 1 + (2 + 3)$

$3 + 3 = 1 + 5$

$6 = 6$

Associative Property of Multiplication
$(a \times b) \times c = a \times (b \times c)$

Example:
$(2 \times 4) \times 6 = 2 \times (4 \times 6)$

$8 \times 6 = 2 \times 24$

$48 = 48$

Using the Associative and Commutative Properties together, you can reorder and change the grouping of addends or factors to make computing with them easier. This can be helpful in evaluating complicated expressions using mental math.

Examples:

$$\left(-\frac{1}{4} + 3.9\right) + \frac{5}{4} \quad \text{and} \quad \left(-0.25 \times \frac{7}{9}\right) \times (-4)$$

Step 1

$$\left(-\frac{1}{4} + 3.9\right) + \frac{5}{4} \qquad \left(-0.25 \times \frac{7}{9}\right) \times (-4)$$

$$= \left(3.9 + \left(-\frac{1}{4}\right)\right) + \frac{5}{4} \qquad = \left(\frac{7}{9} \times (-0.25)\right) \times (-4) \;\leftarrow \text{Commutative Property}$$

Step 2

$$= 3.9 + \left(-\frac{1}{4} + \frac{5}{4}\right) \qquad = \frac{7}{9} \times (-0.25 \times (-4)) \;\leftarrow \text{Associative Property}$$

Step 3

$$= 3.9 + 1 = 4.9 \qquad = \frac{7}{9} \times 1 = \frac{7}{9} \qquad\quad \leftarrow \text{Simplify}$$

Distributive Property

The Distributive Property can be illustrated with the area of adjoining rectangles like the ones shown. On the left, the area of the entire rectangle is found by first finding the sum of the partial side lengths and then multiplying by the common side. On the right, the area is found by adding the areas of each of the smaller rectangles together. The result is the same.

Stated mathematically, you can see how the Distributive Property gets its name. The factor outside the parentheses is distributed to each of the addends inside the parentheses.

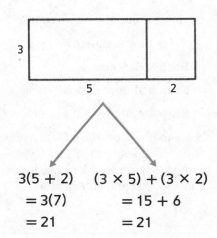

$$3(5 + 2) \qquad (3 \times 5) + (3 \times 2)$$
$$= 3(7) \qquad\quad = 15 + 6$$
$$= 21 \qquad\qquad = 21$$

Distributive Property:
$$a(b + c) = ab + ac$$

Think about Math

Directions: Use the property listed to simplify the expression.

1. $25(4 \times 7)$; Associative
2. $9(12 - 3)$; Distributive
3. $\left(5 \times \frac{1}{4}\right) \times 20$; Commutative and Associative

---

Perform Operations

Understanding how to simplify expressions using the order of operations is an important skill to have. However, sometimes parentheses can be forgotten, which may or may not change the value of an expression.

For example, the expression $18 + 4 \times 7 + 10 - 3^2 \times 2$ simplifies to 38. If a set of parentheses are added to create the expression $18 + 4 \times 7 + (10 - 3^2) \times 2$, the expression now simplifies to 48.

Rewrite the expression $3 + 4^2 - 5 \times 7 + 11$ adding in one set of parentheses so that the expression equals 25.

Order of Operations

There is a certain order in which everyday tasks must be performed. For example, you cannot put on your shoes before you put on your socks. The same is true with operating with numbers in mathematics. The order in which different operations are performed has a direct impact on the final answer, so certain rules and conventions must be followed.

Understanding the Order of Operations

There may be several operations involved in evaluating an expression. You might reach a different answer depending on the order in which you do those operations.

For the expression $3^2 - 2 \times 3$, the correct value is 3. To make sure there is only one correct answer for any given problem, mathematicians have agreed on an order in which to perform the operations when evaluating expressions with multiple operations. This order is called the **order of operations**.

- First simplify inside parentheses or other grouping symbols.
- Second, evaluate any exponents.
- Next, work in order from left to right to multiply or divide.
- Finally, work in order from left to right to add or subtract.

So, to evaluate the expression $3^2 - 2 \times 3$, we use the following steps:

Step 1 Parentheses (none)		$3^2 - 2 \times 3$
Step 2 Exponents		$= 9 - 2 \times 3$
Step 3 Multiplication/Division		$= 9 - 6$
Step 4 Addition/Subtraction		$= 3$

You can use the letters *PEMDAS* or the phrase *Please Excuse My Dear Aunt Sally* to remember the first letters of the operations in the order that they should be performed: parentheses, exponents, multiplication and division, and addition and subtraction.

Using the Order of Operations

To evaluate an expression containing multiple operations, be sure to follow the order of operations.

$$30 - (6 - 3)^2 + 8 \div 2$$

Step 1 Simplify inside the parentheses. $= 30 - (3)^2 + 8 \div 2$

Step 2 Evaluate exponents. $= 30 - 9 + 8 \div 2$

Step 3 Multiply and divide in order from left to right. $= 30 - 9 + 4$

Step 4 Add and subtract in order from left to right. $= 21 + 4 = 25$

The value of this expression is 25.

Undefined Expressions

Not all numerical expressions can be evaluated to obtain a numerical result. Such expressions are said to be **undefined**.

The most common example of an undefined expression involves division by 0, which is itself undefined.

To understand why, think about the equal-groups representation of division. The expression $8 \div 4$ can be interpreted as separating 8 objects into 4 groups of 2, or into 2 groups of 4. The expression $8 \div 0$ would then mean separating 8 objects into 0 groups, or into groups of 0, which is not possible. Therefore, division by zero is undefined.

When you are evaluating a numerical expression according to the order of operations, be on the lookout for steps that result in division by 0. Such expressions are undefined and cannot be evaluated further.

$$17 + 2^3 \div (16 - 2 \times 8) + 9$$
$$17 + 2^3 \div (16 - 16) + 9$$
$$17 + 2^3 \div 0 + 9$$

undefined

CALCULATOR SKILL

Recognizing when an expression is undefined can help as a check when you are solving problems. One of the most recognizable undefined expressions is any expression that requires division by zero. When you enter an expression into your calculator that requires division by zero, the TI-30XS MultiView™ will show the error DIVIDE BY 0. Enter expression $\frac{2}{((-1)^2 - 1)}$ into your calculator and see what the calculator shows. Can you think of a different mathematical expression that is undefined and will give an error on your calculator?

Think about Math

Directions: Find the value of each expression.

1. $12.5 + 6(15 - 12)^2 - 6.5$
2. $8 \times 5 \div 10 + 50 \div 5 \times (3 - 1)^2$
3. $25 - 30 \div (15 - 3 \times 5)^2$

Vocabulary Review

Directions: Write the missing term in the blank.

addend   factor   greatest common factor
least common multiple   order of operations   undefined

1. The _____ of two numbers is the smallest number for which both numbers are factors of that number.

2. When finding the sum of two numbers, each number is called a(n) _____.

3. The number 6 is the _____ of 24 and 42.

4. A(n) _____ expression is one that has no answer.

5. When using the _____, evaluate parentheses, exponents, multiplication/division, and addition/subtraction in that order, from left to right.

6. The number 12 has six _____: 1, 2, 3, 4, 6, and 12.

Skill Review

Directions: Read each problem and complete the task.

1. Which expression completes the equation?

$5(12 + 23) = 5 \times 12 + $ ____

A. $5 + 23$
B. 5×23
C. $5 + 12$
D. 5×12

2. Which is the value of the expression?

$\frac{1}{8} \times (24 - 22)^3 - 3^2$

A. -8
B. 9
C. 55
D. undefined

3. Which is the greatest common factor of 25 and 45?

A. 25
B. 15
C. 5
D. 1

4. What is the prime factorization of 90?

5. Marquita owns a small business producing wooden toys. She has received an order for 120 toy trains and 95 toy soldiers. Because of the large order, she wants to break it up into equal shipments. How many boxes will she need if each box must contain an equal number of toy trains and an equal number of toy soldiers?

6. Which is the least common multiple of 15 and 20?

A. 300
B. 60
C. 35
D. 1

Skill Practice

Directions: Read each problem and complete the task.

1. Jay is making a painting based on a 10-inch by 15-inch picture. He divides the picture into grid squares. What are the greatest size grid squares he can make?

2. What numbers would make this expression undefined?

   $18 \div (25 - x^2)$

3. To evaluate expressions with several sets of parentheses, find the inner set of parentheses and apply the order of operations within the parentheses before evaluating the entire expression. Which is the value of this expression?

   $10 \times (12 - (3 \times 2)^2 + 6)$
   A. -180
   B. -8
   C. 60
   D. undefined

4. Find the GCF and LCM of $2^4 \times 3^2 \times 7^3 \times 13$ and $2^2 \times 3^3 \times 5 \times 11^2$.

   A. GCF $= 2 \times 3 \times 5 \times 7 \times 11 \times 13$;
      LCM $= 2^6 \times 3^5 \times 5 \times 7^3 \times 11^2 \times 13$
   B. GCF $= 2^2 \times 3^2$;
      LCM $= 2^4 \times 3^3 \times 5 \times 7^3 \times 11^2 \times 13$
   C. GCF $= 2^4 \times 3^2 \times 7^3 \times 13$;
      LCM $= 2^2 \times 3^3 \times 5 \times 11^2$
   D. GCF $= 2 \times 3$
      LCM $= 2 \times 3 \times 5 \times 7 \times 11 \times 13$

5. Which property can be applied to this expression?

   $85 \times (100 - 5)$

6. A store receives shipments each day. Every 4 days it receives a shipment of milk, and every 10 days it receives a shipment with cookies. If they receive a shipment that includes both milk and cookies on April 2, when is the next date that they will receive a shipment that includes both?

7. Ethan is working on his monthly bills. He currently has $1,000 in his savings account. After receiving two paychecks of $1,100 each, he now needs to pay rent and other bills. His rent is $800, cell phone $90, groceries $200, utilities that cost $120 and are paid twice a month, and other expenses which total $350. Which expression shows how much Ethan will have in his savings account after paying bills?

   A. $1,000 + 1,100 - 800 + 90 + 200 + 120 \times 2 + 350$
   B. $1,000 + 1,100 - (800 + 90 + 200 + 120 \times 2 + 350)$
   C. $1,000 + 2 \times 1,100 - (800 + 90 + 200 + 120 \times 2 + 350)$
   D. $1,000 + 2 \times 1,100 - 800 + 90 + 200 + 120 \times 2 - 350$

8. Evaluate this expression, showing each step. Write the property you used or an explanation for each step.

   13×24

   $(10 + 3) \times (20 + 4)$
   rewrite the problem for easier multiplication
   $(10 + 3) \times 20 + (10 + 3) \times 4$

   uses the _____
   $10 \times 20 + 3 \times 20 + 10 \times 4 + 3 \times 4$

   uses the _____
   $200 + 60 + 40 + 12 = 312$ simplified

9. Complete the prime factorization. Then write the prime factorization in exponent form.

   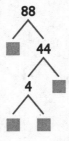

   $88 =$ _____

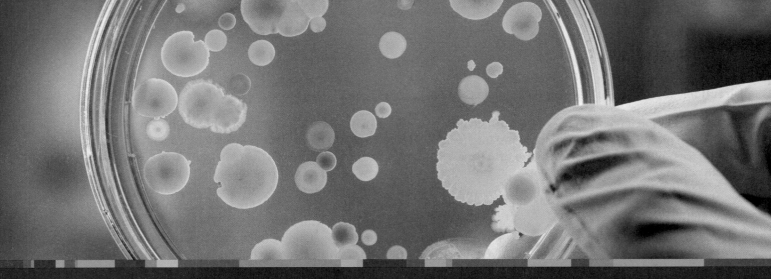

LESSON 1.3 Compute with Exponents

LESSON OBJECTIVES

- Apply rules of exponents to expressions
- Perform operations on numbers written in scientific notation
- Solve real-world problems involving squares and cubes

CORE SKILLS & PRACTICES

- Represent Real-World Problems
- Make Use of Structure

Key Terms

cube
a number raised to the third power

scientific notation
a system of writing a number as the product of a decimal and a power of 10

square
a number raised to the second power

Vocabulary

order of operations
the rules for the order that calculation should be done when evaluating an expression

reciprocals
two numbers or expressions whose product is 1

standard notation
the way in which a number is typically written, using place value

Key Concept

Exponents can be used to represent and solve problems, such as those involving squares and cubes or scientific notation. You can rewrite and simplify expressions involving exponents.

Exponential Notation

If you open a bank account with compound interest, you can use a formula involving an exponent to calculate the amount of money in your account after a certain amount of time.

Defining Powers

Repeated multiplications, like 4×4, can be expressed using powers. In a power, the number that is repeatedly multiplied is called the base. The small raised number is called an exponent. It tells you how many times to use the base as a factor.
You can read the power shown here as "4 to the second power."

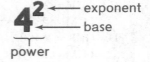

To evaluate a power, simply perform the repeated multiplication. The expressions 4×4 and 4^2 have the same value, 16.

You can evaluate powers involving exponents of 1, 0, or negative numbers. You can also raise decimal and rational numbers to a given power.

$a^1 = a$ — A number to the 1st power uses the base number as a factor only 1 time. It is usually written without the exponent.

$a^0 = 1$ — Any number to the zero power is equal to 1.

$a^{-n} = \frac{1}{a^n}, a \neq 0$ — A nonzero number raised to a negative power is equal to the reciprocal of the number raised to a positive power. The **reciprocal** is any one of two numbers whose product is 1. Find the reciprocal by inverting the number written as a fraction.

Example 1: Examples of Numbers in Exponential Notation
Write each number using exponential notation.
$$32 = 2 \times 2 \times 2 \times 2 \times 2 = 2^5 \qquad \frac{1}{27} = \frac{1}{3 \times 3 \times 3} = \frac{1}{3^3} = 3^{-3}$$

Andreas Reh

Squares and Cubes

The product of a number to the second power is usually called the **square** of the number. This is because if you show the multiplication visually with rows and columns, a square is formed. Similarly, a number raised to the third power is called the **cube** of the number. If you show the multiplication visually, a cube is formed. The expressions shown here can be read as "4 squared" and "4 cubed."

$$4^2 = 4 \times 4 = 16$$

$$4^3 = 4 \times 4 \times 4 = 64$$

Cubes and squares have special names because they frequently appear in real-world problems.

Example 2: Solving Real-World Problems

If an object is dropped, the distance it has fallen after t seconds is given by the expression $16t^2$. Find the number of feet that a dropped object has fallen after 2 and 3 seconds.

Step 1 Substitute $t = 2$ and $t = 3$ into the expression.

After 2 seconds: $16t^2 = 16(2)^2$

After 3 seconds: $16t^2 = 16(3)^2$

Step 2 Rewrite the powers using repeated multiplication.

After 2 seconds: $16(2)^2 = 16 \times 2 \times 2$

After 3 seconds: $16(3)^2 = 16 \times 3 \times 3$

Step 3 Evaluate.

After 2 seconds: $16 \times 2 \times 2 = 64$ ft

After 3 seconds: $16 \times 3 \times 3 = 144$ ft

◣ Think about Math

Directions: Write and evaluate an exponential expression to solve the problem.

1. To the nearest dollar, what is the total cost of installing new carpet in a room that is 15.5 ft by 15.5 ft, if the carpet costs $3.75 per square foot and there is a $50 installation fee?

    Expression: _____ Cost: _____

Represent Real-World Problems
When you are solving a real-world problem, take note of any repeated multiplication that can be represented using exponential shorthand. For example, one cubic foot of granite weighs about 170 pounds. To find the weight in pounds of a cubic yard of granite, use the following diagram to understand how many cubic feet are in one cubic yard.

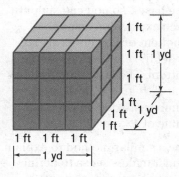

The exponential expression 3^3 gives the number of cubic feet in one cubic yard, so the weight of one cubic yard of granite can be calculated by evaluating the exponential expression 170×3^3.

If one cubic inch of gold weighs about 0.7 pound, what exponential expression represents the weight of one cubic foot of gold?

Make Use of Structure

You can use properties of exponents to understand why any number raised to a power of 0 is 1.

Consider the expressions shown in the table, and examine them in terms of their structure.

Expression	$\frac{4^7}{4^4}$	$\frac{4^7}{4^5}$	$\frac{4^7}{4^6}$	$\frac{4^7}{4^7}$
Power	4^3	4^2	4^1	4^0

The expressions on the top have like bases, so you can simplify each expression using the Quotient of Powers Property to get the expressions on the bottom. The last column shows that $4^0 = 4^7 \div 4^7$, and any number divided by itself has a value of 1.

Use a similar method to explain why it makes sense to define negative exponents using the reciprocal of the positive exponent.

Rules of Exponents

When you go to a science museum you will often see the skeletons of animals from thousands of years ago. When scientists discover these bones they can use a method called "carbon dating" to figure out the age. The mathematics behind carbon dating involves negative exponents, which you can understand by looking at some of the properties of exponents.

Products and Quotients of Powers

You can simplify expressions involving the product of two powers with like bases by rewriting the powers using repeated multiplication. Count the number of times the base is used as a factor and use that as the new exponent. A shortcut for multiplying powers with like bases is to keep the same base and add the exponents of each power for the new exponent. The answers are the same.

$$3^4 \times 3^2 = (3 \times 3 \times 3 \times 3) \times (3 \times 3)$$
$$= 3 \times 3 \times 3 \times 3 \times 3 \times 3$$
$$= 3^6$$

Shortcut:
$$3^4 \times 3^2 = 3^{4+2}$$
$$= 3^6$$

You can also simplify the quotient of two powers with like bases by rewriting the power using repeated multiplication. Simplify by dividing out pairs of factors from the numerator and denominator. Notice that a shortcut for dividing powers with like bases is to keep the same base and subtract the exponents of each power for the new exponent.

$$4^7 \div 4^5 = \frac{4^7}{4^5}$$
$$= \frac{4 \times 4 \times 4 \times 4 \times 4 \times 4 \times 4}{4 \times 4 \times 4 \times 4 \times 4}$$
$$= 4^2$$

Shortcut:
$$4^7 \div 4^5 = 4^{7-5}$$
$$= 4^2$$

These two shortcuts hold for all numbers (except when $a = 0$ for a quotient of powers).

Product of Powers Property
$$a^m \times a^n = a^{m+n}$$

Quotient of Powers Property
$$\frac{a^m}{a^n} = a^{m-n} \quad \text{(for } a \neq 0\text{)}$$

Power of Powers

You can also simplify expressions involving the power of a power. Consider the power inside the parentheses as the base and rewrite using repeated multiplication. Then expand out each of those factors and count to determine the new exponent. A shortcut for finding the power of a power is to keep the base and multiply the exponents for the new exponent.

$$(5^2)^3 = 5^2 \times 5^2 \times 5^2$$
$$= (5 \times 5) \times (5 \times 5) \times (5 \times 5)$$
$$= 5 \times 5 \times 5 \times 5 \times 5 \times 5$$
$$= 5^6$$

Shortcut:
$$(5^2)^3 = 5^{2 \times 3}$$
$$= 5^6$$

This shortcut also holds true for all numbers.

Power of a Power Property
$$(a^m)^n = a^{mn}$$

You can use the Power of a Power Property along with the Product and Quotient of Powers Properties to simplify more complicated expressions involving powers with like bases. Remember to follow the **order of operations**, or the order that the calculation should be done, any time you are evaluating an expression with multiple operations.

Example 3: Using Properties of Exponents

Simplify $\dfrac{(4^2 \times 4^6)^2}{(4^3)^5}$.

Step 1 Use the Product of Powers Property to simplify inside the parentheses of the numerator.

$$\frac{(4^2 \times 4^6)^2}{(4^3)^5} = \frac{(4^{2+6})^2}{(4^3)^5} = \frac{(4^8)^2}{(4^3)^5}$$

Step 2 Use the Power of a Power Property to simplify in the numerator and denominator.

$$\frac{(4^8)^2}{(4^3)^5} = \frac{4^{8\times2}}{4^{3\times5}} = \frac{4^{16}}{4^{15}}$$

Step 3 Use the Quotient of Powers Property to write the expression as a single power.

$$\frac{4^{16}}{4^{15}} = 4^{16-15} = 4^1 = 4$$

Powers of Products and Quotients

You can also simplify expressions that involve the power of a product (or quotient). Use repeated multiplication to expand the expression using the product (or quotient) as the base. Then rewrite the expression using the Commutative and Associative Properties to group like factors, and rewrite again to show each base raised to the same exponent.

$$(5 \times 2)^3 = (5 \times 2) \times (5 \times 2) \times (5 \times 2) \qquad (2 \div 5)^3 = \left(\frac{2}{5}\right)^3$$

$$= (5 \times 5 \times 5) \times (2 \times 2 \times 2) \qquad\qquad = \frac{2}{5} \times \frac{2}{5} \times \frac{2}{5}$$

$$= 5^3 \times 2^3 \qquad\qquad\qquad\qquad\qquad = \frac{2^3}{5^3}$$

You can also use shortcuts to simplify the above expressions. Unlike the previous properties, these properties involve different bases, but the same exponent.

Power of a Product Property
$a^n b^n = (ab)^n$

Power of a Quotient Property
$\dfrac{a^n}{b^n} = \left(\dfrac{a}{b}\right)^n$ (for $b \neq 0$)

Example 4: Using Properties of Exponents

Simplify $\dfrac{(4^2)^3 \times 5^6}{2^6}$.

Step 1 Use the Power of a Power Property to simplify the parentheses.

$$\frac{(4^2)^3 \times 5^6}{2^6} = \frac{4^{2\times3} \times 5^6}{2^6} = \frac{4^6 \times 5^6}{2^6}$$

Step 2 Use the Power of a Product Property to simplify in the numerator.

$$\frac{4^6 \times 5^6}{2^6} = \frac{(4 \times 5)^6}{2^6} = \frac{20^6}{2^6}$$

Step 3 Use the Power of a Quotient Property to write the expression as a single power.

$$\frac{20^6}{2^6} = \left(\frac{20}{2}\right)^6 = 10^6 = 1,000,000$$

Compute with Exponents

Scientific Notation

When you look through a telescope some stars will appear larger than others based on their distance from Earth. These distances are very large numbers of miles and can be written in a shorthand form called scientific notation.

Use Scientific Notation

Scientific notation is a system for writing very large or very small numbers, using exponents. Numbers in scientific notation are written as a product of two factors. The first factor is a decimal number greater than or equal to 1 and less than 10. The second factor is a power of 10.

a decimal number A, where $1 \le A < 10$ {7.5×10^5} a power of 10

The power of 10 factor tells how many places to move the decimal point when changing from scientific notation to **standard notation**, the way numbers are usually written. The sign of the exponent indicates whether the number is greater than 10 or less than 1.

Positive Exponents
$7.5 \times 10^5 = 750,000$

move 5 places to the right

Negative Exponents
$4.2 \times 10^{-3} = 0.0042$

move 3 places to the left

To write a number in scientific notation, move the decimal point until the number is greater than or equal to 1 but less than 10, counting the number of place values moved. Use this decimal number as the first factor. Use the number of place values moved as the exponent in the power of 10.

> **Example 5:** Writing Numbers in Scientific Notation
>
> Write each number in scientific notation.
>
> $60,000,000 \rightarrow 6 \times 10^7$ $0.000075 \rightarrow 7.5 \times 10^{-5}$

Add and Subtract in Scientific Notation

To add or subtract two numbers in scientific notation, the powers of 10 must be the same. If they are not the same, use properties of exponents to rewrite the expression until the exponents are the same. Be sure to write the final answer in scientific notation, where the first factor is greater than or equal to 1 but less than 10.

> **Example 6:** Adding Numbers in Scientific Notation
>
> Add: $(3.4 \times 10^7) + (8.9 \times 10^5)$
>
> **Step 1** Rewrite the numbers so the powers of 10 are the same.
> $$(3.4 \times 10^7) + (8.9 \times 10^5) = (340 \times 10^5) + (8.9 \times 10^5)$$
>
> **Step 2** Use the Distributive Property to factor the power of 10.
> $$(340 \times 10^5) + (8.9 \times 10^5) = (340 + 8.9) \times 10^5$$
>
> **Step 3** Combine inside the parentheses.
> $$(340 + 8.9) \times 10^5 = 348.9 \times 10^5$$
>
> **Step 4** Write in scientific notation. 348.9×10^5 is not in scientific notation because the decimal, 348.9, is greater than 10.
> $$348.9 \times 10^5 = 3.489 \times 10^2 \times 10^5$$
> $$= 3.489 \times 10^{2+5} = 3.489 \times 10^7 \text{ Use the Product of Powers Property.}$$

Follow similar steps when adding or subtracting numbers in scientific notation less than 1. The exponents should always be the same.

Multiply and Divide in Scientific Notation

When you multiply or divide numbers in scientific notation, the powers of 10 do not need to be the same. To multiply two numbers in scientific notation, use the Commutative and Associative Properties of Multiplication to group the decimal factors together and the powers of 10 together. Multiply to combine the decimal factors, and use the Product of Powers Property to combine the powers of 10.

Example 7: Multiplying Numbers in Scientific Notation

Multiply: $(2.1 \times 10^3) \times (6.5 \times 10^5)$

Step 1 Use the Commutative and Associative Properties.

$$(2.1 \times 10^3) \times (6.5 \times 10^5) = (2.1 \times 6.5) \times (10^3 \times 10^5)$$

Step 2 Multiply the coefficients, and use the Product of Powers to multiply the powers of 10.

$$(2.1 \times 6.5) \times (10^3 \times 10^5) = 13.65 \times 10^{3+5} = 13.65 \times 10^8$$

Step 3 Write in scientific notation. 13.65×10^8 is not in scientific notation because the decimal, 13.65, is greater than 10.

$$13.65 \times 10^8 = 1.365 \times 10^1 \times 10^8$$
$$13.65 \times 10^{1+8} = 1.365 \times 10^9 \qquad \text{Use the Product of Powers Property}$$

Follow similar steps to divide two numbers written in scientific notation. Divide the decimal factors, and divide the powers of 10. Use the Quotient of Powers Property, and rewrite until the quotient is properly written in scientific notation as needed.

Example 8: Dividing Numbers in Scientific Notation

Divide: $(8.4 \times 10^6) \div (2.2 \times 10^9)$

Step 1 Write the quotient using a fraction bar.

$$(8.4 \times 10^6) \div (2.2 \times 10^9) = \frac{8.4 \times 10^6}{2.2 \times 10^9} = \frac{8.4}{2.2} \times \frac{10^6}{10^9}$$

Step 2 Divide the coefficients, and use the Quotient of Powers Property to divide the powers of 10.

$$\frac{8.4}{2.2} \times \frac{10^6}{10^9} = 3.82 \times 10^{6-9} = 3.82 \times 10^{-3}$$

Think about Math

Directions: Perform each indicated operation and express the answer in scientific notation.

1. $(2.6 \times 10^3) \times (4.3 \times 10^4)$
2. $(5.1 \times 10^7) + (4.8 \times 10^6)$
3. $(4.4 \times 10^{-3}) - (6.9 \times 10^{-2})$
4. $(8.7 \times 10^8) \div (2.4 \times 10^5)$

21ST CENTURY SKILL

Health Literacy

Red blood cells are responsible for transporting oxygen throughout the body and removing waste. They are the most common of the blood cells, and make up between 40–45% of human blood. They are naturally very small, averaging a length of 7×10^{-6} meters. There are about 2.5×10^{13} red blood cells in the adult human body.

If you laid all of your red blood cells end to end, how long would they be?

Vocabulary Review

Directions: Match each term to its definition.

1. ____ cube

2. ____ order of operations

3. ____ reciprocal

4. ____ scientific notation

5. ____ square

6. ____ standard notation

a. the way in which a number is typically written, using place value

b. a number raised to the third power

c. a system of writing a number as the product of a decimal and a power of 10

d. the rules for the order that calculation should be done when evaluating an expression

e. two numbers or expressions whose product is 1

f. a number raised to the second power

Skill Review

Directions: Read each problem and complete the task.

1. A gardener plants one aster seedling in each square foot of a 12-ft by 12-ft garden. Which expression shows the cost to plant all the asters if each seedling costs $7?

A. 7×12^2

B. 7×12^3

C. 12×7^2

D. 12×7^3

2. Which property can be applied to this expression?

$5^4 \div 2^4$

3. The table shows the distances between several planets and the sun.

Planet	Distance from the Sun (mi)
Venus	6.71×10^7
Earth	9.3×10^7
Mars	1.4×10^8
Jupiter	4.84×10^8

What is the distance between Earth and Mars when they are on opposite sides of the sun?

A. 1.07×10^7

B. 1.07×10^8

C. 2.33×10^7

D. 2.33×10^8

4. Rodrigo invested $2,000 in a fund that returns 8% interest compounded yearly (at the end of the year) and makes no additional deposits or withdrawals. The total value of the fund, including accrued interest, at the end of n years is given by the equation $V = 2{,}000(1.08)^n$. What is the total value of the fund at the end of the third year?

5. What is the value of the expression shown?

$$\frac{(5^2)^3 \times 5^4}{5^{12}}$$

A. 0.04
B. 0.2
C. 5
D. 25

Skill Practice

Directions: Read each problem and complete the task.

1. A party planner will use colored sand to fill 10 large and 5 small cubic vases like the ones shown to use for centerpieces and other decorations.

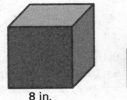

8 in.   5 in.

Which expression shows the number of cubic inches of colored sand she will need to completely fill all the vases?

A. $10 \times 8^2 \times 5^3$
B. $10 \times 8^4 \times 5^3$
C. $10 \times 8^2 \times 5^4$
D. $10 \times 8^3 \times 5^4$

2. Simplify the expression shown, justifying your work using properties of exponents.

$$\frac{5^2 \times 2^6 \times 5^6}{10^6}$$

3. Write an expression using an exponent to show the total cost for carpeting the square room shown if the carpet costs $3.42 per square foot and there is a $100 installation fee.

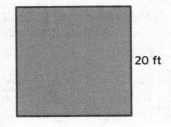

20 ft

4. The table shows the lengths of two bacteria.

Bacterium	Length (cm)
A	2.0×10^{-5}
B	1.2×10^{-4}

How many of the smaller bacteria do you need to place end-to-end to equal the length of the larger bacterium?

A. 4
B. 6
C. 8
D. 10

5. Find the product of the numbers shown by expressing them both in scientific notation.

$250{,}000 \times (7.6 \times 10^{-4})$

6. What value for n makes the expression shown below have a value of 4^3?

$$\frac{4^3 \times (4^2)^n}{4^{10}}$$

A. 1
B. 3
C. 5
D. 7

LESSON 1.4 Compute with Roots

Mark Dierker/McGraw-Hill Education

LESSON OBJECTIVES

- Perform computations with square and cube roots
- Solve real-world problems involving square and cube roots
- Simplify expressions involving roots using the properties of rational exponents

CORE SKILLS & PRACTICES

- Represent Real-World Arithmetic Problems
- Attend to Precision

Key Terms

cube root
a number that, when cubed, equals a given number

rational exponent
an exponent that is a rational number

square root
a number that, when squared, equals a given number

Vocabulary

index
the small number next to a radical sign that indicates the degree of the root

irrational numbers
the set of numbers that cannot be expressed as the ratio of two integers

prime factorization
a number written as the product of its prime factors

Key Concept

Numerical expressions involving roots (often called radicals) can be written using rational exponents and then simplified using the rules of exponents.

Square Roots and Cube Roots

Roots, including square roots and cube roots, often appear in real-world problems.

Defining Roots

The square of a number n can be thought of as the area of a square with side lengths n. Just like subtraction undoes addition, the process of squaring has an inverse operation, called finding the square root. The **square root** of a positive number n is a number which, when squared, equals n. To find the square root of 4, ask, "What number multiplied by itself equals 16?"

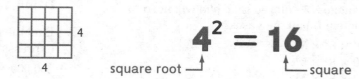

This question actually has two answers; one positive and one negative. For real-world problems, you usually only need to consider the positive square root.

$4^2 = 16 \rightarrow 4$ is a square root of 16. $(-4)^2 = 16 \rightarrow -4$ is also a square root of 16.

Cubing a number also has an inverse. The **cube root** of a given number n is the number which, when cubed, equals n. To find the cube root of 8, ask, "What number multiplied by itself and by itself again equals 8?" Unlike square roots, there is only one cube root of a number.

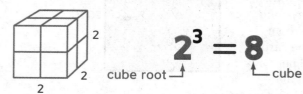

Compute with Roots

You can use similar logic to define the *n*th root of a number. Roots are shown using a radical sign. The small number is called the **index** of the root. If the index is odd, there is only one possible value for the root of a number. If the index is even, you should consider both positive and negative values for the root.

If *n* is odd:
$$a^n = b \rightarrow \sqrt[n]{b} = a$$

If *n* is even:
$$a^n = b \rightarrow \sqrt[n]{b} = a$$
$$(-a)^n = b \rightarrow \sqrt[n]{b} = -a$$

Roots of Perfect Squares and Cubes

Numbers that have whole number square roots are called perfect squares. Perfect squares are easily found by squaring whole numbers.

Example 1: Perfect Squares

$1^2 = 1$	$4^2 = 16$	$7^2 = 49$	$10^2 = 100$	$13^2 = 169$
$2^2 = 4$	$5^2 = 25$	$8^2 = 64$	$11^2 = 121$	$14^2 = 196$
$3^2 = 9$	$6^2 = 36$	$9^2 = 81$	$12^2 = 144$	$15^2 = 225$

Similarly, numbers that have whole-number cube roots are called perfect cubes. You can find perfect cubes by cubing whole numbers.

Example 2: Perfect Cubes

$1^3 = 1$	$4^3 = 64$	$7^3 = 343$	$10^3 = 1{,}000$	$13^3 = 2{,}197$
$2^3 = 8$	$5^3 = 125$	$8^3 = 512$	$11^3 = 1{,}331$	$14^3 = 2{,}744$
$3^3 = 27$	$6^3 = 216$	$9^3 = 729$	$12^3 = 1{,}728$	$15^3 = 3{,}375$

It is good to remember the first several perfect squares and perfect cubes so that you can easily use them to solve real-world problems.

Example 3: Solving Real-World Problems

A sculptor has 1,331 cubic inches of clay to make a single cube to use as part of a large sculpture. Which of the cubes shown could he make?

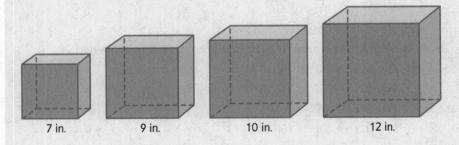

7 in.　　9 in.　　10 in.　　12 in.

Step 1 Determine the index of the root needed to solve the problem.

The volume of a cube is measured in cubic units, and *side length* $= \sqrt[3]{volume}$. Use the cube root.

Step 2 Find the cube root of the total volume.

1,331 is a perfect cube. $11^3 = 1{,}331$, so $\sqrt[3]{1{,}331} = 11$.

Step 3 Interpret the cube root to answer the question.

The largest cube he can make from 1,331 cubic inches is 11 inches on each side. He could make the 7 in., 9 in., or 10 in. cube.

CORE SKILL

Represent Real-World Arithmetic Problems

When you are solving a real-world problem involving roots, you need to be able to determine the *type* of root needed to solve the problem. For problems involving measurements, you can often tell the index of the root that is needed by considering the units of measurement. Cubic units often indicate the need to find a cube root, while square units often imply the need for a square root.

A sculptor uses a thin layer of colored clay to cover a large square tile he is using for a large sculpture. He has enough clay to cover 576 square inches. What are the dimensions of the largest square he could cover with the colored clay?

Your TI-30XS MultiView™ calculator can be used to calculate square roots and cube roots. There is a key specifically for calculating square root.

You must press   to access the square root button. There is not a specific button for calculating cube root, but you can use the feature, which is located above the key. In order to calculate a cube root, such as $\sqrt[3]{125}$, type

  . You should get the answer 5. Use your calculator to find the cube roots of 1 through 10, and verify the approximations in the table.

Approximating Square and Cube Roots

Roots of nonperfect squares and cubes are often **irrational numbers**, the set of numbers that cannot be expressed as the ratio of two integers. The decimal expansion of irrational numbers does not terminate or repeat, but you can approximate them with terminating decimals or whole numbers. You can use what you know about perfect squares and cubes to approximate roots of nonperfect squares and cubes.

Example 4: Approximate a Square Root of a Number

To the nearest whole number, what is $\sqrt{61}$?

Step 1 Identify the perfect squares that the number is between.

$$49 < \mathbf{61} < 64$$

Step 2 Find the square roots of the perfect squares.

$$\sqrt{49} < \sqrt{61} < \sqrt{64}$$
$$7 < \sqrt{61} < 8$$

Step 3 Estimate the square root. Compare to determine the integer to which the root of the number is closer.

Try 7.5 → $7.5^2 = 56.25$

$56.25 < 61 < 64$, so $\sqrt{61}$ is closer to 8 than to 7.

You can use a similar method to approximate cube roots. To obtain more accurate approximations, you can check your decimal estimate by cubing it, and then refine your estimate as needed.

Example 5: Approximate a Cube Root of a Number

To the nearest tenth, what is $\sqrt[3]{187}$?

Step 1 Identify the perfect cubes that the number is between.

$$125 < \mathbf{187} < 216$$

Step 2 Find the cube roots of the perfect cubes.

$$\sqrt[3]{125} < \sqrt[3]{187} < \sqrt[3]{216}$$
$$5 < \sqrt[3]{187} < 6$$

Step 3 Estimate the cube root. Refine your estimate as needed.

Try 5.5 → $5.5^3 = 166.375$

Try 5.6 → $5.6^3 = 175.616$

Try 5.7 → $5.7^3 = 185.193$

Try 5.8 → $5.8^3 = 195.112$

So, $\sqrt[3]{187} \approx 5.7$.

It can be helpful to memorize some of the decimal approximations of common square and cube roots.

Square Roots			
$\sqrt{3} \approx 1.73$		$\sqrt{6} \approx 2.45$	
$\sqrt{4} = 2$		$\sqrt{7} \approx 2.65$	
$\sqrt{5} \approx 2.24$		$\sqrt{8} \approx 2.83$	

Cube Roots			
$\sqrt[3]{3} \approx 1.44$		$\sqrt[3]{6} \approx 1.82$	
$\sqrt[3]{4} \approx 1.59$		$\sqrt[3]{7} \approx 1.91$	
$\sqrt[3]{5} \approx 1.71$		$\sqrt[3]{8} = 2$	

Think about Math

Directions: Choose the best answer to the question.

1. The maximum walking speed in inches per second of an animal with leg length in inches can be approximated by the formula shown.

$$\text{maximum walking speed (in. per sec)} = 19.6\sqrt{\text{leg length (in.)}}$$

Use the formula to approximate the maximum walking speed of a giraffe with a leg length of 72 inches. Approximate any square roots to the nearest whole number.

A. 117.6 inches per second

B. 166.3 inches per second

C. 176.4 inches per second

D. 235.2 inches per second

Radicals and Rational Exponents

As you might expect, the length of time it takes a planet to orbit the sun is related to the distance that planet is from the sun. Johannes Kepler, a seventeenth-century German mathematician and astronomer, derived an equation that indeed relates these two quantities, and it includes what is known as a rational exponent. This relationship is given by the equation $d = t^{\frac{2}{3}}$.

Multiplying Like Radicals

Expressions involving roots are called radical expressions, or sometimes just radicals. There are several properties you can use when operating with radicals that have the same index.

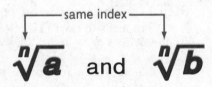

A multiplication property of radicals allows you to write the product of roots (with the same index) as the root of a product. This property comes in handy when you are asked to simplify a square root or a cube root.

Multiplication Property

$\sqrt[n]{a} \times \sqrt[n]{b} = \sqrt[n]{ab}$

> **Example 6: Simplify Roots**
>
> Simplify $\sqrt{90}$.
>
> **Step 1** Rewrite the number under the radical sign as a product. Look for factors of the number that are perfect squares.
>
> $\sqrt{90} = \sqrt{9 \times 10}$
>
> **Step 2** Use the property of multiplying radicals to rewrite the root of the product as the product of roots.
>
> $\sqrt{9 \times 10} = \sqrt{9} \times \sqrt{10}$
>
> **Step 3** Simplify the known root.
>
> $\sqrt{9} \times \sqrt{10} = 3 \times \sqrt{10}$

To attend to precision when computing with radicals, you need to ensure that you use the correct property. You also attend to precision when you identify which of given numbers to use to solve a problem, and when you approximate to a given place value.

An object traveling at a rate of r miles per hour for t hours travels a total of d miles as given by the formula $d = rt$. The table shows the rates and times covered by different runners expressed using square roots.

Runner	r	t
A	$\sqrt{54}$	$\sqrt{5}$
B	$\sqrt{50}$	$\sqrt{6}$
C	$\sqrt{45}$	$\sqrt{8}$

To the nearest tenth of a mile, how far did Runner B travel?

Dividing Like Radicals

You can also divide radicals with the same index using a property of radicals. The quotient of two roots (with the same index) is equal to the root of the quotient.

Division Property

$$\frac{\sqrt[n]{a}}{\sqrt[n]{b}} = \sqrt[n]{\frac{a}{b}} \quad (b \neq 0)$$

You can use the division property along with the multiplication property to help you simplify and approximate radical expressions. It is often helpful to use the **prime factorization** of the number under a radical sign in order to identify perfect squares and cubes that can be removed from underneath the radical. (Remember, the prime factorization shows a number written as a product of its prime factors.)

Example 7: Simplify Radical Expressions

Simplify and approximate $\frac{\sqrt[3]{2,880}}{\sqrt[3]{60}}$.

Step 1 Use the property of dividing radicals to rewrite the quotient of roots as the root of a quotient. Evaluate the quotient.

$$\frac{\sqrt[3]{2,880}}{\sqrt[3]{60}} = \sqrt[3]{\frac{2,880}{60}} = \sqrt[3]{48}$$

Step 2 Write the prime factorization of the number under the radical.

$$\sqrt[3]{48} = \sqrt[3]{2^4 \times 3}$$

Step 3 Use the properties of exponents and the property of multiplying radicals to rewrite the root of the product as the product of roots.

$$\sqrt[3]{2^3} \times \sqrt[3]{6} \approx 2 \times 1.82 \approx 3.64$$

Step 4 Simplify the root of the perfect cube and approximate the root of the nonperfect cube.

Defining Rational Exponents

What if the index of the radicals are not the same? You can still perform operations on these radicals by first writing them in an equivalent form using a rational exponent. A **rational exponent** is an exponent that is a rational number.

$$\sqrt[n]{b} = b^{\frac{1}{n}}$$

Think about why this notation makes sense:

- On the left, the nth root of a number raised to the nth power is the number.
$$\left(\sqrt[n]{b}\right)^n = b$$

- On the right, raising the expression to the nth power using the rules for exponents results in the number raised to the first power, or just the number.
$$\left(b^{\frac{1}{n}}\right)^n = b^{\frac{1}{n} \times n} = b^{\frac{n}{n}} = b^1 = b$$

You can also use non-unit fractions in lowest terms as rational exponents.

$$\sqrt[n]{b^m} = b^{\frac{m}{n}}$$

The same rules of exponents that apply to powers also apply to rational exponents.

- **Product of Powers** To multiply powers with like bases, add the exponents.
  $$a^m \times a^n = a^{m+n}$$

- **Quotient of Powers** To divide powers with like bases, subtract the exponents.
  $$\frac{a^m}{a^n} = a^{m-n} \quad \text{(for } a \neq 0\text{)}$$

- **Power to a Power** To raise a power to a power, multiply the exponents.
  $$(a^m)^n = a^{mn}$$

- **Power of a Product** To multiply two powers with the same exponent, multiply the bases.
  $$a^n b^n = (ab)^n$$

- **Power of a Quotient** To divide two powers with the same exponent, divide the bases.
  $$\frac{a^n}{b^n} = \left(\frac{a}{b}\right)^n \quad \text{(for } b \neq 0\text{)}$$

Simplifying with Rational Exponents

Take a look at this example showing how to use rational exponents to simplify radical expressions. For this radical expression, the index of the radicals are not the same, meaning you cannot just write this quotient of roots as the root of a quotient.

Example 8: Simplify Rational Expressions

Simplify $\frac{\sqrt{64}}{\sqrt[4]{16}}$.

Step 1 Rewrite the radicals using rational exponents. Notice that you cannot yet use the Quotient of Powers Property, as the bases are not the same. However, you should recognize the bases as powers of 2.
$$\frac{\sqrt{64}}{\sqrt[4]{16}} = \frac{64^{\frac{1}{2}}}{16^{\frac{1}{4}}}$$

Step 2 Rewrite each base as a power of 2, and then use the Power to a Power Property.
$$\frac{64^{\frac{1}{2}}}{16^{\frac{1}{4}}} = (2^6)^{\frac{1}{2}}/(2^4)^{\frac{1}{4}} = \frac{2^3}{2^1}$$

Step 3 Use the Quotient of Powers Property and evaluate.
$$\frac{2^3}{2^1} = 2^{3-1} = 2^2 = 4$$

Think about Math

Directions: For each expression involving rational exponents, identify the equivalent radical expression.

1. $6^{\frac{1}{2}} \times 6^{\frac{1}{3}}$

   A. $\sqrt[6]{6}$
   B. $\sqrt[6]{6^5}$
   C. $2 \times \sqrt[3]{6}$
   D. $3 \times \sqrt{6}$

2. $\frac{(8^3)^{\frac{1}{2}}}{64^{\frac{1}{2}}}$

   A. $2\sqrt{2}$
   B. $2\sqrt{8}$
   C. $8\sqrt{2}$
   D. $3\sqrt{8}$

Vocabulary Review

Directions: Write the missing term in the blank.

cube root	index	irrational number
prime factorization	rational exponent	square root

1. In the radical expression $\sqrt[3]{8}$, the _____ is 3.

2. The _____ of a number shows the number written as the product of its prime factors.

3. The radical expression $\sqrt{4}$ can be rewritten using a(n) _____ as $4^{\frac{1}{2}}$.

4. The _____ of 64 is 4.

5. A number that cannot be written as the ratio of two integers is called a(n) _____.

6. A(n) _____ of 16 is −4.

Skill Review

Directions: Read each problem and complete the task.

1. Which property can be applied to this expression?

$\sqrt{\frac{81}{144}}$

2. Four expressions are shown below.

$16^{\frac{3}{2}}, 25^{\frac{3}{2}}, 27^{\frac{2}{3}}, 343^{\frac{2}{3}}$

What is the difference between the expression with the greatest value and the expression with the least value?

A. 219
B. 116
C. 61
D. 40

3. To the nearest hundredth, what is the value of $\sqrt{191}$?

4. A farmer needs to build a fence to enclose a square plot of land with an area of 200 square yards. To the nearest tenth of a yard, how much fencing does the farmer need?

Compute with Roots

5. Kepler's Third Law of Planetary Motion relates the average distance d, in astronomical units (AU), from a planet to the sun to the time t, in years, it takes the planet to orbit the sun. This relationship is given by the equation $d = t^{\frac{2}{3}}$. Which is the best estimate of difference between the average distance of Uranus from the sun and the average distance of Jupiter from the sun? Approximate any roots to the nearest tenth.

Planet	Time to Orbit Sun (years)
Mars	2
Jupiter	12
Saturn	30
Uranus	84
Neptune	165

A. 5 AU
B. 14 AU
C. 19 AU
D. 20 AU

6. Which of the following shows the expression $\sqrt{12} \times \sqrt[4]{18}$ written with a single radical?

A. $3\sqrt{2}$
B. $3\sqrt[4]{2}$
C. $6\sqrt{2}$
D. $6\sqrt[4]{2}$

Skill Practice

Directions: Read each problem and complete the task.

1. To the nearest tenth, what is the side length of a cube with a volume of 439 cubic centimeters?

A. 7.2 cm
B. 7.4 cm
C. 7.6 cm
D. 7.8 cm

2. Evaluate the expression shown, justifying your work using properties of exponents and/or radicals.
$$\frac{\sqrt{9} \times \sqrt[3]{729}}{\sqrt{3} \times \sqrt{27}}$$

3. Which property can be applied to this expression?
$$\sqrt{5} \times \sqrt{40}$$

4. Which of the following expressions is equivalent to $\sqrt{90} \times \sqrt{450}$?

A. $30\sqrt{2}$
B. $90\sqrt{2}$
C. $30\sqrt{5}$
D. $90\sqrt{5}$

5. An artist is painting a large mural comprised of different size squares. She wants to paint one magenta square that is as large as possible, but she only has one can of magenta paint. If the can of paint can cover 220 square feet, what is the side length of the largest square that she can paint if she needs to apply 2 coats of paint? Round to the nearest whole number.

6. Which of the following expressions is not equivalent to $\frac{\sqrt{20} \times \sqrt[4]{4}}{\sqrt{2}}$?

A. $\frac{4\sqrt{5} \times \sqrt[4]{4}}{\sqrt{2}}$

B. $\frac{20^{\frac{1}{2}} \times 4^{\frac{1}{4}}}{2^{\frac{1}{2}}}$

C. $10^{\frac{1}{2}} \times \sqrt[4]{4}$

D. $\frac{2 \times 5^{\frac{1}{2}} \times 4^{\frac{1}{4}}}{2^{\frac{1}{2}}}$

Directions: Choose the best answer to each question.

1. Which is the value of the expression?

$15 + 10(2 + 5)^2 - 12$

A. 48
B. 62
C. 273
D. 493

2. Jen is ordering these expressions, so she would first like to find the expression with the greatest value. Which is the greatest?

A. $\sqrt{9}$

B. $\sqrt[3]{8}$

C. $\sqrt{8}$

D. $\sqrt[3]{10}$

3. Micah is ordering these expressions, so he would first like to find the expression with the least value. Which is the least?

A. $27^{\frac{1}{3}}$

B. $16^{\frac{1}{2}}$

C. $8^{\frac{2}{3}}$

D. $4^{\frac{2}{3}}$

4. Vanessa is making a painting based on a 12-inch by 24-inch picture. She wants to divide the picture into grid squares. What is the greatest size of grid squares she can make?

A. 36 inch by 36 inch
B. 12 inch by 12 inch
C. 6 inch by 6 inch
D. 2 inch by 4 inch

5. Allen works out at the gym every 6 days. Freddy works out at the gym every 4 days. Allen and Freddy see each other at the gym on a Monday. What day of the week will it be when they see each other again?

A. Monday
B. Wednesday
C. Friday
D. Saturday

6. Which is the absolute value of the expression $25 - 4^3$?

A. -39
B. -13
C. 13
D. 39

7. Which expression has the greatest value?

A. $\dfrac{8^5}{8^2}$

B. $8^5 \times 8^2$

C. $\dfrac{8^2}{8^5}$

D. $(8^5)^2$

8. Which shows this expression in simplest form? $\sqrt{12}\ \sqrt{27}$

A. 18
B. $4\sqrt{3} \times 3\sqrt{3}$
C. $2\sqrt{3} \times 3\sqrt{3}$
D. 324

9. Which shows this expression in simplest form?

$\sqrt[3]{\dfrac{16}{27}}$

A. $\dfrac{4}{3}$

B. $\dfrac{\sqrt[3]{16}}{3}$

C. $\dfrac{2\sqrt[3]{2}}{3}$

D. $\dfrac{48}{81}$

Directions: Use the paragraph for Problems 10–11.

Four friends are on a basketball team. During a game, each friend kept track of how many shots they attempted and how many of those attempts they made.

Henry made 0.45 of his shots.

Allison made $\dfrac{4}{15}$ of her shots.

Arthur made $\dfrac{8}{20}$ of his shots.

Trevor missed 58% of his shots.

10. Which friend had the best record for the number of shots made?

A. Henry
B. Allison
C. Arthur
D. Trevor

11. Which friend had the worst record for the number of shots made?

A. Henry
B. Allison
C. Arthur
D. Trevor

12. This expression, $\frac{15 + 85}{4^3 - 8^2}$, is called _____ because when simplifying, there is a 0 in the denominator.

13. The _____ Property is represented by both $12 + 15 = 15 + 12$ as well as $4 \times 5 = 5 \times 4$.

14. Gianna has two savings accounts. One account has a rate of return of 3.75% while the other account has a rate of return of 0.375%. 3.75 is _____ times greater than 0.375.

15. You can read the value of $\sqrt{16}$ as *the* _____ *of sixteen*.

Directions: Use the paragraph for Problems 16–17.

Ellis is a botanist. He found the heights of three different Redwood trees and recorded the information in a table.

Heights of Redwood Trees	
Tree	Height (in centimeters)
1	1.0668×10^4
2	1.1430×10^4
3	9.114×10^3

16. What is the difference between the heights of Tree 1 and Tree 3 in Scientific Notation?
   A. 1.554×10^3 cm
   B. 15.54×10^4 cm
   C. 15.54 cm
   D. 1.554×10^4 cm

17. What is the total of the heights of Tree 2 and Tree 3 in Scientific Notation?
   A. 2.0544×10^4 cm
   B. 20.544×10^3 cm
   C. 1.0257×10^4 cm
   D. 10.257×10^3 cm

18. A _____ number is one that can be written as a ratio of two integers.

19. Tile Company Pro charges $4.15 for each tile that is one square foot. To tile a room that is 12.5 feet wide and 12.5 feet long, the price would be _____ for the tiles for the room.

20. You can read the number 8^3 as eight to the third power, or eight cubed, and can simplify it to _____.

Check Your Understanding

On the following chart, circle the items you missed. The last column shows pages you can review to study the content covered in the question. Review those lessons in which you missed half or more of the questions.

Lesson	Item Number(s)			Review Page(s)
	Procedural	Conceptual	Problem Solving	
1.1 Order Rational Numbers	18	2, 3	10, 11	12–19
1.2 Apply Number Properties	1, 14	12, 13	4, 5, 16, 17	20–27
1.3 Compute with Exponents	6	7, 20	19	28–35
1.4 Compute with Roots	8, 9	15		36–43

Chapter 2

Ratio, Proportion, and Probability

To attract new customers and to get more business, stores will often advertise special sales. They may use newspapers, billboards, commercials, and other media outlets to get the word out about their sale. Discounts may be described as a percentage such as "75% off original prices." Stores may also use fractions such as "half-off." Understanding how to calculate using ratios and percentages will help you be a more informed shopper and will let you know how much you can really save by buying items during sales.

Lesson 2.1
Apply Ratios and Proportions

Ratios compare two different values, such as distance and time. The speed limit sign you see on the highway is an example of a ratio of number of miles per hour. Learn how to simplify ratios and set up proportions to solve real-world problems.

Lesson 2.2
Calculate Real-World Percentages

Percentages describe values as parts of a whole. Thinking about your household budget and finances in terms of percentages can help you manage how you spend your money. Categories with a larger percentage represent your biggest area of expenses and help you budget accordingly. Learn how to solve problems using percentages and convert between ratios and percentages.

Lesson 2.3
Use Counting Techniques

If you wanted to know how many eggs you had left in the fridge, you could quickly count. What happens when you have to count larger numbers of items? How would you count in such scenarios without individually counting each item? Learn how to use counting techniques to quickly calculate the number of items in a set.

Lesson 2.4
Determine Probability

How do you determine the likelihood of something happening? Probability tells you how likely it is that something will happen and is usually described as a ratio or percent. When you listen to the weather forecast and hear the chance of rain reported as a percent, that is a probability. Learn how to calculate probabilities and apply counting techniques to solve probability problems.

Goal Setting

Understanding ratios, proportions, and percentages will help you solve probability problems, which in turn will tell you how likely something is to happen. Think of situations where you encounter probability in your everyday life. Words like *chance* and *likelihood* are good indications that you are dealing with probability. For each situation, list how the probability is indicated—as a percentage, a fraction, or some other way. Based on your list, what is the most common way of indicating probability?

In what situations would a percentage probability make more sense than a fractional probability? How could you use the lessons in this chapter to help you solve problems involving probability?

LESSON 2.1 Apply Ratios and Proportions

■ LESSON OBJECTIVES

- Compute unit rates
- Use scale factors
- Apply ratios and proportions to solve real-world problems

■ CORE SKILLS & PRACTICES

- Compare Unit Rates
- Use Ratio Reasoning

Key Terms

proportion
an equation stating that two ratios are equal

ratio
a comparison of two values

scale factor
a ratio of corresponding parts of similar figures

unit rate
a ratio that compares a quantity to a single unit

Vocabulary

equivalent
equal; having the same value

similar
having the same shape, but not necessarily the same size

Key Concept

A ratio, which is often written as a fraction, is a comparison of the relative sizes of two numbers. Operations on ratios follow the same rules as operations on fractions. When two ratios are equivalent, they are called proportional.

Ratios

How do moviemakers figure out what movies audiences want to watch? How do coffee shops know which coffee roasts to sell? Many businesses conduct surveys, or polls, of people's preferences. They use computer software to translate the results into ratios, such as 3 out of 4 people prefer nonfat milk to full-fat milk in their coffee. This information helps businesses determine what products to develop or sell.

Apply Ratios and Proportions

Write Ratios

A **ratio** is a comparison of the relative value of two numbers. For example, the owner of a popular coffee shop surveyed 140 customers and learned that 70 customers preferred blueberry muffins to all other choices. The ratio of customers who preferred blueberry to all customers surveyed is 70 to 140.

There are three ways to write a ratio:

70 to 140   70:140 $\frac{70}{140}$

Equivalent ratios are ratios that have the same value. The ratio $\frac{70}{140}$ can be simplified to the equivalent ratio $\frac{1}{2}$, which means that one out of every two customers surveyed preferred blueberry muffins.

Unit Rates

A **unit rate** is a ratio that compares an amount or quantity to one unit, such as a unit of weight, time, or distance. In other words, the denominator of a unit rate is always 1.

$2.00 per pound	12 gallons per minute	85 miles per hour	$0.89 per foot
$\frac{2\ dollars}{1\ pound}$	$\frac{12\ gallons}{1\ minute}$	$\frac{85\ miles}{1\ hour}$	$\frac{0.89\ dollars}{1\ foot}$

You can use division to convert any ratio into a unit rate. For example, suppose a dozen doughnuts costs $3.60, or $3.60 per 12 doughnuts. Divide the numerator by the denominator 12, the number of items in a dozen, to calculate the unit rate, or cost per doughnut.

$$\frac{3.60\ dollars}{12\ doughnuts} = 3.60 \div 12 = 0.3$$

One doughnut costs $0.30.

Think about Math

Directions: Answer the following questions.

1. A 2-lb bag of lentils costs $3.70. What is the unit cost per pound?
2. A 10-oz can of corn costs $1.88. What is the unit cost per ounce?
3. A box of 144 pencils costs $5.76. What is the unit cost per pencil?

Proportions

When painting a portrait, an artist often begins by sketching the face. All human faces have the same basic proportions. Each face can be divided into thirds, with hair and eyebrows in the top third; the eyes, nose, and ears in the middle third; and the mouth in the bottom third. Once the basic proportions are sketched, the artist can add the details that make each face unique.

Compute Unit Rates

Unit rates are helpful when determining the best price on an item. For example, when a bakery is buying flour, they can calculate the unit rates for bags of different sizes to determine the best price. A large bag will likely cost more than a small bag, but the unit cost per pound may be less for the large bag, giving the bakery more flour for their money. Suppose a 25-lb bag of flour costs $25.00, a 50-lb bag of flour costs $40, and a 5-lb bag of flour costs $10. Which bag has the least unit cost per pound?

Use Ratio Reasoning

Many problems can be solved by identifying key information in the problem that can be represented as ratios and then using your knowledge of ratios to solve.

An art teacher is buying classroom supplies. Twelve bottles of paint cost $38.28. How much will 36 bottles of paint cost?

$$\frac{12}{\$38.28} = \frac{36}{x}$$

$$\$38.28 \times 36 = 12x$$

$$1378.08 = 12x$$

$$x = \$114.84$$

36 bottles of paint will cost $114.84.

How much will 50 bottles of paint cost?

Testing for a Proportion

A **proportion** is a mathematical statement that two ratios are equivalent: $\frac{a}{b} = \frac{c}{d}$.

Example 1

Determine whether the ratios $\frac{4}{6}$ and $\frac{2}{3}$ form a proportion.

Step 1 Multiply the numerator of the first ratio by the denominator of the second ratio.

$$\frac{4}{6} \stackrel{?}{=} \frac{2}{3} \qquad 4 \times 3 = 12$$

Step 2 Multiply the numerator of the second ratio by the denominator of the first ratio.

$$\frac{4}{6} \stackrel{?}{=} \frac{2}{3} \qquad 2 \times 6 = 12$$

Step 3 If the two products are equal, the ratios form a proportion. Otherwise, the ratios do not form a proportion. These two ratios form a proportion.

Use Proportions to Solve Problems

Understanding the cross-multiplication strategy for determining proportionality makes it possible to solve problems in which you must determine an unknown value.

Example 2

Suppose you can buy 8 paintbrushes for $10.00. How many paintbrushes can you buy for $25.00?

Step 1 Write a proportion representing the situation.

$$\frac{8}{10} = \frac{p}{25}$$

Step 2 Cross-multiply and set the two products equal to each other.

$$8 \times 25 = 10p$$

Step 3 Solve the equation for p.

$$200 = 10p$$

$$\frac{200}{10} = \frac{10p}{10}$$

$$20 = p$$

You can buy 20 paintbrushes for $25.00.

Think about Math

Directions: Use the cross-multiplication strategy to answer the following questions.

1. A small auto company produces 220 vehicles in a 5-day workweek. How many cars will they produce in 15 days?

   A. 1,100
   B. 660
   C. 73
   D. 44

2. A car company pays $112.40 for 8 door handles. How much will they pay for 24 door handles?

   A. $2697.60
   B. $899.20
   C. $337.20
   D. $37.46

Scale

A scale model is a proportional copy of a real object. The model has the same shape as the original object, but is usually a different size. Engineers use scale models to test the performance of a new design without having to create an expensive full-sized prototype. Scale models are also used by architects to show the look of a new skyscraper and how it will fit in with the rest of a city's skyline.

Similarity and Scale Factor

Similar figures have the same shape but may have different sizes. You can use proportions to determine whether two figures or objects are similar.

> **Example 3**
>
> Determine whether $\triangle ABC$ and $\triangle DEF$ are similar. If so, determine the scale factor from $\triangle ABC$ to $\triangle DEF$.
>
>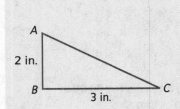
>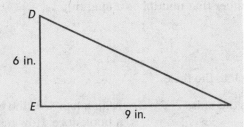
>
> **Step 1** Use the measures of two corresponding sides to write ratios.
>
> $$\frac{AB}{BC} \overset{?}{=} \frac{DE}{EF}$$
>
> $$\frac{2}{3} \overset{?}{=} \frac{6}{9}$$
>
> **Step 2** Perform cross multiplication to determine whether the ratios form a proportion.
>
> $$2(9) \overset{?}{=} 6(30)$$
>
> $$18 = 18$$
>
> The ratios form a proportion, so the triangles are similar.
>
> **Step 3** The difference in size between similar figures is determined by a scale factor, a number which scales, or multiplies, some quantity. The dimensions of $\triangle DEF$ are 3 times the dimensions of $\triangle ABC$, so the scale factor from $\triangle ABC$ to $\triangle DEF$ is 3.

Think about Math

Directions: Answer the following questions.

1. A photography studio enlarged a photograph that was 8 inches wide and 9 inches long to a similar photograph that was 36 inches long. How wide was the enlarged photograph?

2. What scale factor did the photography studio use to enlarge the photograph?

21ST CENTURY SKILL

Civic Literacy

Urban planners often rely on scale models of skyscrapers to examine existing structures and to design new ones. To use the scale of a building or map, write a proportion. Be sure to keep units in mind. For example, to solve the following problem, you need to convert inches to feet.

A model of a skyscraper is built at a scale of 1:800 and the model's height is 14.4 inches. What is the height of the actual skyscraper in feet?

Vocabulary Review

Directions: Write the missing term in the blank.

unit rate	equivalent	scale factor
proportion	ratio	similar

1. A(n) _____ is a comparison of two numbers.

2. Two figures or objects that have the same shape but may have different

sizes are _____ .

3. A(n) _____ compares a quantity to a single unit.

4. Two equivalent ratios form a(n) _____ .

5. Two ratios that are _____ have the same value.

6. A(n) _____ is a number that multiplies a quantity.

Skill Review

Directions: Read each problem and complete the task.

1. A 3-lb bag of rice costs $3.87. What is the unit cost per pound?

    A. $1.19
    B. $1.29
    C. $1.33
    D. $11.61

2. A company is packaging bags of sugar for sale. The unit rate for a pound of sugar is $0.74/lb. What will be the cost of a 5-lb bag of sugar?

    A. $0.15
    B. $1.48
    C. $3.70
    D. $5.74

3. Aliyah has $20. She can buy 3 books for $5.00 at a bookstore. How many books can she buy with her $20?

4. A model airplane has a scale factor of 1:48. The wingspan of the model is 8 inches. What is the wingspan of the actual airplane?

    A. 6 ft
    B. 32 ft
    C. 48 ft
    D. 384 ft

Skill Practice

Directions: Read each problem and complete the task.

1. Explain the difference between a ratio and a proportion. Use examples in your explanation.

2. A school can purchase paper in boxes of 5 reams per box. Each box costs $11.20. How much will the school pay for 20 reams of paper?
   - A. $6.20
   - B. $44.80
   - C. $56.00
   - D. $224.00

3. If 3 pies cost $12, how much will 10 pies cost?
   - A. $30.00
   - B. $40.00
   - C. $48.00
   - D. $60.00

4. Ginnie has two rectangular vegetable gardens that are similar to one another. The smaller garden has a length of 3 feet and a width of 2 feet. The larger garden has a length of 15 feet. What is the width of the larger garden?

5. Samantha is comparing two boxes of cereal. One box contains 12 ounces of cereal and costs $2.88. The other box contains 16 ounces of cereal and costs $3.52. Which box of cereal is the better value? Compare unit rates to explain.

6. An architectural firm is creating a scale model of a new building. The actual building will be 125 feet tall, and the model will have a scale of 1:400 inches. How tall should the model be?
   - A. 0.3125 inches
   - B. 3.13 inches
   - C. 3.75 inches
   - D. 4.5 inches

7. Derek has $24. He wants to buy as many plastic ducks as possible for a carnival. A package of 5 ducks costs $2.50. What is the greatest number of ducks that Derek can buy?
   - A. 29
   - B. 45
   - C. 48
   - D. 60

8. The length of a rectangular swimming pool is 18 feet. The width is half the length. A company is building a similar pool with a width of 4.5 feet. What is the length of the similar pool? What scale factor did the company use to determine the dimensions of the similar pool?

LESSON 2.2 Calculate Real-World Percentages

LESSON OBJECTIVES

- Relate fractions, decimals, and percents
- Compute percent change
- Find a discount
- Calculate simple interest
- Use percent to solve real-world problems

CORE SKILLS & PRACTICES

- Use Tools Strategically
- Use Percent

Key Terms

discount
a decrease or reduction in price

percent
a ratio of a number to 100

simple interest
a charge paid on an original principal

Vocabulary

benchmark
a point of reference from which other measurements or estimates can be made

interest rate
the amount that is earned or charged during a certain amount of time

principal
an amount of money invested or borrowed

Key Concept

A percent is a ratio of a number to 100. In fact, the word *percent* comes from the Latin term *per centum*, meaning "by the hundred," and it is represented by the symbol %. Fractions and decimals are also ratios, and they are related to percents.

Percent of a Number

The U.S. Department of Labor includes a Bureau of Labor Statistics. This organization collects workforce data. One area of focus is the activity of recent high-school graduates. Recently, the bureau analyzed information regarding the post-high school activities of 3.2 million graduates. Among those graduates, 66.2 percent were enrolled in colleges or universities.

The Meaning of Percent

A ratio, often written as a fraction, describes a part of a whole. A **percent** is a ratio of a number to 100.

Graph paper is a useful tool for modeling percents. A block of 10 squares by 10 squares is 100 total squares and represents 100%. You can shade squares to represent any percent, or part of the whole.

$$0.05 = \frac{5}{100}$$
$$= 5\%$$

$$0.18 = \frac{18}{100}$$
$$= 18\%$$

$$0.11 = \frac{11}{100}$$
$$= 11\%$$

Just like a fraction and a percent, a decimal represents a part of a whole. The "whole" in a decimal is 1.0. Therefore, a 10-by-10 block of graph paper squares is a useful model for representing decimals, and for representing the relationships between fractions, decimals, and percents.

Author's Image/PunchStock

Calculate Real-World Percentages

Percents as Decimals

The term "e-commerce" describes any business conducted electronically, especially over the Internet. Statisticians collect e-commerce data to determine how and why people shop online. Businesses use this information to identify and market to potential customers.

Example 1: Use a Decimal to Find Percent of a Number

In a survey of 500 shoppers, 73% reported doing at least half of their shopping online. How many of the surveyed shoppers does this represent?

Step 1 Change the percent to a decimal. $73\% = 0.73$

Step 2 Multiply the decimal by the total $0.73 \times 500 = 365$
number of shoppers.

Of the 500 shoppers surveyed, 365 reported doing at least half of their shopping online.

Percents as Fractions

In the same e-commerce survey, 4 out of 5 shoppers reported shopping online because they can find a larger selection of items.

Example 2: Write a Fraction as a Percent

What percent of the surveyed shoppers reported shopping online because they can find a larger selection of items?

Step 1 Write "4 out of 5" as a fraction. $4 \text{ out of } 5 = \frac{4}{5}$

Step 2 Write an equivalent fraction $\frac{4 \times 20}{5 \times 20} = (80/100) = 80\%$
with a denominator of 100.

80% of the shoppers surveyed reported shopping online because they can find a larger selection of items.

CALCULATOR SKILL

You can change a percent to a decimal on many calculators by using the % button.

For example, to convert 18% to a decimal, press 18 followed by the % button.

On the TI-30XS MultiView™, to find 18 percent of 200, press 18 followed by [2nd] [( ], then multiply by 200.

In mathematics, there are a number of tools you can use to solve a problem. There are physical tools, such as protractors or calculators, and there are mental tools.

Benchmarks are useful mental tools when working with percents. A **benchmark** is a point of reference from which other measurements or estimates can be made. For example, you can estimate 33% of 25 by using a benchmark fraction or decimal in place of 33%. You know that 30% = 0.3, so 33% of 25 is about 0.3 × 25 = 7.5. Because 33% is greater than 30%, 33% of 25 is greater than 7.5. A good estimate is 8.

Use a benchmark to estimate 48% of 170. What benchmark did you use?

Percent as Proportion

You can write and solve a proportion to find a percent of a number.

$$\frac{\text{percent}}{100} = \frac{\text{part}}{\text{whole}}$$

Example 3: Use a Proportion to Find Percent of a Number

Imagine that you and your friends celebrate a special event by eating dinner at your favorite restaurant. The amount of the bill before tax is $54.00. You want to leave your server a 15% tip. How much should you leave?

Step 1 Write a proportion.

$$\text{percent} \rightarrow \frac{15}{100} = \frac{p}{54} \begin{array}{l} \leftarrow \text{part} \\ \leftarrow \text{whole} \end{array}$$

Step 2 Use cross products to solve the proportion for p.

$$15(54) = 100p$$
$$810 = 100p$$
$$\frac{810}{100} = \frac{100p}{100}$$
$$8.1 = p$$

You should leave $8.10 for a tip.

Think about Math

Directions: Find the percent of each number.

1. 18% of 50
2. 63% of 1,000
3. 90% of 150
4. 45% of 500
5. 32% of 250
6. 75% of 900

Percent Change

Many people invest their money in the stock market. Each day, the prices of stocks fluctuate. This change represents a change in percent.

Find Percent Change

Gyms and fitness centers often offer a variety of payment plans, including biweekly plans, monthly plans, and yearly plans.

Example 4: Percent Increase

When a local gym first opened last year, the biweekly fee was $17.99. After a year of operation, the biweekly fee is now $24.95. What is the percent change in cost?

Step 1 Subtract the original fee from the new fee.

$24.95 − $17.99 = $6.96

Step 2 Divide the difference by the original fee.

$6.96 divided by $17.99 is about 0.39.

Step 3 To write the quotient as a percent, multiply by 100.

0.39 × 100 = 39%

The percent change in cost is about 39%. Notice that the percent change is positive. This is because the new fee is greater than the original fee.

Example 5: Percent Decrease

Each year, the gym holds a New Year's Resolution promotion. For members who join during the month of January, the gym lowers its monthly fee from $32.99 to $29.99. What is the percent change in the monthly fee during the month of January?

Step 1 Subtract the original fee from the new fee.

$29.99 − $32.99 = −$3.00

Step 2 Divide the difference by the original fee.

−$3.00 divided by $32.99 is about −0.09.

Step 3 To write the quotient as a percent, multiply by 100.

−0.09 × 100 = −9%

The percent change in cost is about –9%. Notice that the percent change is negative. This is because the new fee is less than the original fee.

Discounts

A **discount** is a decrease, or reduction, in price. Sometimes advertisers describe a discount by stating the new, reduced price. Other advertisers indicate the discount as a percent instead, leaving it to you to calculate how much money you will save.

Example 6: Discounts

The original cost of a bike helmet was $24.99, but the price has now been reduced to $19.99. What is the amount of the discount? What is the discount as a percent?

Step 1 Find the amount of the discount by subtracting the original price from the new, reduced price.

$$\$19.99 - \$24.99 = -\$5.00 \qquad \text{The amount of the discount is } \$5.00.$$

Step 2 Find the percent of the discount by dividing the amount of the discount by the original price and writing the quotient as a percent.

$$\frac{-\$5.00}{\$24.99} \approx -0.20 = -20\%$$

The change in price represents a 20% discount.

 Think about Math

Directions: Answer the following questions.

1. A company's stock started the day priced at $18.99 per share. The company's stock price ended the day at $17.25 per share. What is the percent change in price per share for this particular day?

   A. 9.2%
   B. 0.9%
   C. −0.9%
   D. −9.2%

2. A student studying computer programming printed an online coupon for a software program. At the store, the student saved $6 on the purchase of the software originally priced at $39.99. What was the percent discount?

   A. −84%
   B. −15%
   C. 15%
   D. 84%

Simple Interest

The College Board, the organization that offers the SAT and Advanced Placement, monitors the costs of higher education and trends in student financial aid. Financial aid comes from a variety of sources, including loans. Because of their fixed interest rates, federal loans can be one of the least expensive ways to pay for college.

Use a Formula

Saving and borrowing money both involve interest. If you save money in a bank account, the bank pays you interest for the use of your money. If you borrow money from a bank, you pay the bank interest for the use of its money.

The amount of money that is initially borrowed or saved is called the **principal**. The **interest rate** is a percentage earned or charged during a certain time period. **Simple interest** is the amount of interest charged or earned after the interest rate is applied to the principal.

Simple interest (I) is the product of three values: the principal (P), the interest rate written as a decimal (r), and time (t).

$$I = P \times r \times t$$

Example 7: Simple Interest

A bank offers simple interest loans at an interest rate of 6.5% per year. How much interest will you pay if you borrow $1,500 for 2 years?

Step 1 Write the interest rate as a decimal by dividing the percent by 100.

$6.5\% \div 100 = 0.065$

Step 2 Substitute the values into the simple interest formula and multiply.

$I = P \times r \times t$
$= 1,500 \times 0.065 \times 2$
$= 195$

You will pay $195 in interest.

Think about Math

Directions: Answer the following questions.

1. A car dealership finances simple interest car loans at a fixed rate of 4.2% per year for 3 years. How much interest will a buyer pay if she borrows $13,999.00 for a new car?

   A. $1,763.87
   B. $1,399.90
   C. $587.96
   D. $5,879.58

2. A student borrows $3,000.00 at a fixed 5.7% simple interest rate. The loan period is 2 years. How much interest will the student pay on the loan?

   A. $171.00
   B. $342.00
   C. $300.00
   D. $150.00

21ST CENTURY SKILL

Financial, Economic, Business, and Entrepreneurial Literacy

Three friends invested money in accounts earning simple interest.

- Carlos invested $1,000 for 5 years at a 5% interest rate.
- Molly invested $1,500 for 5 years at a 3% interest rate.
- Jin invested $900 for 10 years at a 3% interest rate.

Carlos says that he will earn the most interest because he has the greatest interest rate. Molly says that she will earn the most interest because she has the greatest principal. Jin says that she will earn the most interest because her investment is for the greatest amount of time. Who is correct?

Vocabulary Review

Directions: Write the missing term in the blank.

benchmark   discount   interest rate
percent   principal   simple interest

1. You can write a decimal as a(n) _____ by multiplying by 100.

2. A(n) _____ is a decrease, or reduction, in price.

3. When investing, the original amount of money invested is the _____.

4. _____ is the amount of interest charged or earned after an interest rate is applied to the principal.

5. The _____ is the percentage that is applied to a principal and paid by borrowers for use of the money they have borrowed.

6. A(n) _____ is a tool you can use to estimate a percent.

Skill Review

Directions: Read each problem and complete the task.

1. Write the decimal, percent, and fraction that this model represents.

_____

2. A bank offers simple interest loans at an interest rate of 6% per year. How much interest will you pay if you borrow $1,200 for 3 years?

A. $21.60
B. $216.00
C. $2,160
D. $21,600

3. Janice and her friend have lunch at a restaurant. The amount of the bill is $25.00, and the friends add an 18% tip. What is the total amount that Janice and her friend spent on lunch?

A. $4.50
B. $9.00
C. $29.50
D. $43.00

4. In 2011, the yearly fee for a fruit-of-the-month club was $59.99. In 2012, the yearly fee was $64.99. What is the percent change in cost?

5. The original cost of a wireless keyboard was $45.50, but the price has now been reduced to $38.68. What is the amount of the discount? What is the discount as a percent?

6. In a survey of 600 shoppers, 55% reported using fabric softener. How many of the surveyed shoppers does this percent represent?

7. In a survey of 200 computer shoppers, 3 out of 4 shoppers reported purchasing a laptop computer. How many of the surveyed shoppers does this represent?

Skill Practice

Directions: Read each problem and complete the task.

1. A day-care center charges $50 per day for each child. The center plans on increasing the cost to $60 per day. What will the percent change in cost be?

2. Elena wants to borrow $2,200 from the bank. They have two simple-interest loans she can choose from. One is a 3-year loan at an interest rate of 6.3%, and the other is a 4-year loan at an interest rate of 5.5%. Elena thinks she will pay less interest on the 4-year loan because the interest rate is lower. Do you agree with Elena? Why or why not?

3. Alex and a friend order meals at a restaurant that each cost $9. There is a 6.5% sales tax. They would like to leave a 20% tip in addition to the taxed amount for the waitress. How much money will the two friends spend in all?

4. A group of 500 people were surveyed about their grocery shopping habits. Of the people surveyed, 6 out of 10 paid with a credit card and 30% paid with cash. How many of the people surveyed paid with some other method?

A. 10 people
B. 50 people
C. 180 people
D. 320 people

5. Emma plans to buy a new computer monitor. She has two coupons that she could use, but the store only allows you to use one coupon per purchase. The first coupon is for a 15% discount. The second coupon is for $25 off any purchase over $125. Do you need any additional information to help Emma decide which coupon to use? If so, what else do you need to know? If not, which coupon should Emma use and why?

6. A fabric store regularly mails coupons to its customers. The store has collected data on coupon use over the last three months.

Month	Number of Shoppers	Percent Who Used Coupon
1	375	32%
2	200	55%
3	310	40%

Which shows the months in order from greatest number of shoppers who used a coupon to least number of shoppers who used a coupon?

A. Month 2, Month 3, Month 1
B. Month 2, Month 1, Month 3
C. Month 3, Month 1, Month 2
D. Month 1, Month 3, Month 2

LESSON 2.3   Use Counting Techniques

LESSON OBJECTIVES

- Apply the Fundamental Counting Principle
- Recognize and calculate factorials
- Determine permutations and combinations

CORE SKILLS & PRACTICES

- Use Counting Techniques
- Model with Mathematics

Key Terms

combination
a selection of objects or values in which order is not important

experiment
an activity or situation in which the results are uncertain

factorial
the product of a series of all descending consecutive positive integers from a given starting point

outcome
a result of an experiment or activity that involves uncertainty

permutation
a selection of objects or values in which order is important

Vocabulary

tree diagram
a branching diagram that shows possible outcomes of an experiment

Key Concept

Certain events can allow for uncertainty. When this occurs, it can be possible to determine the number of possible outcomes by using permutations and combinations.

Factorials

Sometimes you make yourself a list of errands that you must complete and you have to make certain chores a priority. By ordering the list, you are finding one of the many possible ways that the errands can be ordered. Factorials are used to determine how many ways ordering can be done.

The Language of Counting

Learning to use counting techniques is easier when you know the meanings of special terms. An **experiment**, for example, is an activity or situation in which the results are uncertain and an **outcome** is a result of an experiment or activity that involves uncertainty.

Now let's put those terms in the context of a simple experiment. Say you're playing a board game with friends, and it's your turn to roll a die. Rolling the die is an experiment with an uncertain outcome because six possible outcomes exist: 1, 2, 3, 4, 5, or 6.

Say that you roll the die, and the number 4 appears. The number 4 is an outcome, or result of the experiment.

Tree Diagrams

You can use a branching diagram, called a **tree diagram**, that shows possible outcomes of an experiment. Take, for example, the toss of a coin. There are two possible outcomes—heads or tails. Say you want to know all of the possible outcomes of tossing a coin 3 times.

First	Second		Outcomes
		H	HHH
	H		
		T	HHT
H		H	HTH
	T		
		T	HTT
		H	THH
	H		
		T	THT
T		H	TTH
	T		
		T	TTT

The branches of a tree diagram show all of the possible outcomes after each toss. To the right of the diagram is a list of all possible outcomes after 3 tosses. There are 8 possible outcomes.

Another way to look at the outcomes of the tree diagram is by putting them into a chart. Notice there are 8 total results listed on the chart, just like in the tree diagram.

Toss		
1st	2nd	3rd
H	H	H
H	H	T
H	T	H
H	T	T
T	H	H
T	H	T
T	T	H
T	T	T

The Factorial

Suppose you wanted to organize your music collection. How many ways could you list your songs? If you have a large collection of music, this number is also very large. But how would you determine it?

Let's look at the problem on a smaller scale. Suppose you had 4 songs (A, B, C, and D) you wanted to organize. How many options are there for the first song in the order? There are 4 options since there are 4 songs. What about the second song? How many choices are there? Since one song has already been chosen, that leaves 3 songs. The third song? 2 choices, leaving 1 choice for the 4th song.

Song A
Song B
Song C
Song D

1st Song	2nd Song	3rd Song	4th Song
4 options	3 options	2 options	1 option

CORE SKILL

Use Counting Techniques

In more complex situations, you may want to use the Fundamental Counting Principle. This involves multiplying to find the total number of possible outcomes. So, if there are n outcomes for event N and m choices for event M, then there are nm outcomes for the event where both N and M occur. This can be generalized to more than 2 events.

For example, the choices at a frozen yogurt shop are shown below. Customers must select a yogurt type, 1 fruit, and 1 syrup. How many possible mixtures of yogurt, fruit, and syrup can the shop advertise?

Yogurt	Fruit	Syrup
Vanilla	Cherries	Caramel
Chocolate	Blueberries	Marshmallow
Coffee	Pineapples	Hot Fudge
Peach	Mango	Peanut Butter
	Strawberries	

Using the Fundamental Counting Principle, you can determine that there are $4 \times 3 \times 2 \times 1 = 24$ possible outcomes. A product of this kind has a special name. The **factorial** of a number is the product of the series of all descending consecutive integers from the number. It is represented by using the exclamation point (!). Therefore, in our music scenario, $4! = 4 \times 3 \times 2 \times 1 = 24$.

If you had 5 songs to organize, the total possible ways to order the songs would be $5! = 5 \times 4 \times 3 \times 2 \times 1 = 120$.

Think about Math

Directions: Answer the following questions.

1. How many possible sandwiches can be made from 3 types of bread, 5 types of cheese, and 6 types of filling, assuming each sandwich is made with 1 type of bread, 1 type of cheese, and 1 filling type?
   A. 14
   B. 80
   C. 90
   D. 104

2. How many possible ways can you order 5 people in a line?
   A. 15
   B. 120
   C. 625
   D. 3,125

Permutations

If 3 people finish a race in 1st, 2nd, and 3rd, does it matter which order they come in? Of course it does. Knowing the number of ways a set of objects can be ordered is what permutations are used for.

Order Matters

Sometimes, the order in which outcomes are arranged is important. Such a list of outcomes is called a **permutation**.

The formula for finding the number of permutations is easy to understand and can easily be calculated.

$$P(n, k) = \frac{n!}{(n - k)!}$$

P = the number of permutations
n = the number of items
k = the number of items being ordered

Say that there are 10 competitors in a math competition, but only the top 3 receive a prize, as 1st, 2nd, and 3rd place. How many different ways could the competitors finish in 1st, 2nd, and 3rd place?

We can use the Fundamental Counting Principle to determine the possibilities for 1st, 2nd, and 3rd place. How many possibilities exist for each place, starting with 1st?

(10 possibilities for 1st) × (9 possibilities for 2nd) × (8 possibilities for 3rd)
= 10 × 9 × 8 possibilities = 720 possibilities

Another way to solve this problem is to use the permutation formula. In this case, there are 10 competitors and 3 of them are being ordered, so $n = 10$ and $k = 3$:

$$P(10, 3) = \frac{10!}{(10 - 3)!} = \frac{10!}{7!}$$
$$= \frac{10 \times 9 \times 8 \times \cancel{7 \times 6 \times 5 \times 4 \times 3 \times 2 \times 1}}{\cancel{7 \times 6 \times 5 \times 4 \times 3 \times 2 \times 1}} = 10 \times 9 \times 8 = 720$$

Calculating Permutations

Suppose a town council needs to elect a president, vice president, treasurer, and secretary. There are 9 people on the council and no one can hold more than one office. How many ways can the council elect its officers?

Because order matters here (if someone is elected, it matters for which office they were elected), a permutation is what is being calculated.

$$P(9, 4) = \frac{9!}{(9 - 4)!} = \frac{9!}{5!}$$
$$= \frac{9 \times 8 \times 7 \times 6 \times \cancel{5 \times 4 \times 3 \times 2 \times 1}}{\cancel{5 \times 4 \times 3 \times 2 \times 1}} = 9 \times 8 \times 7 \times 6 = 3,024$$

Think about Math

Directions: Answer the following questions.

1. How many ways can 5 different positions be filled by 10 applicants?

    A. 2
    B. 120
    C. 30,240
    D. 3,628,800

2. Which represents the number of ways to order 6 people in a line?

    A. $P(6, 0)$
    B. $P(6, 6)$
    C. $P(0, 6)$
    D. $P(12, 6)$

CORE PRACTICE

Model with Mathematics

Sometimes it is helpful to use a formula to model a mathematical process, such as counting possible permutations.

There are 12 students enrolled in an advanced sculpture class. For their semester project, all students enter one sample of their work in a juried exhibit. The jury of professional artists awards one gold, one silver, and one bronze medal. How many different ways can the awards be distributed among the students?

Computing a Factorial

Some calculators have a way to evaluate factorials. Sometimes the factorial function is under the menu labeled "Probability." For example, on the TI-30XS MultiView™ calculator, the factorial symbol can be found by pressing the ⬭prb key and then the ▢3. But factorials can be calculated even without a specific factorial button. To calculate 5!, simply multiply 5 by every positive integer smaller than it: $5 \times 4 \times 3 \times 2 \times 1$.

When working a problem involving a fraction that has a factorial in both the numerator and the denominator, it will help to write out both factorials and cancel all the common factors on the top and bottom of the fraction. Simplify and then compute on your calculator.

Combinations

When a large group of friends or family goes to an amusement park, they often need to figure out how to break up into smaller groups to go on rides together. It doesn't matter in what order the people are picked, just the final combinations. Likewise in math, combinations calculate the number of possible outcomes to an experiment assuming that the order that it occurs doesn't matter.

Order Doesn't Matter

A **combination** is a selection of objects or values in which order is not important. Combinations are not like permutations in that respect, but you use permutations to find them.

Say that an airport provides pick up and delivery service from different gates at an airport terminal to remote parking lots. A shuttle arrives and picks up passengers every 5 minutes. At the last gate, a shuttle has room left for only 2 passengers, but 4 passengers are waiting to board. How many ways can 2 passengers be chosen from 4 passengers waiting to board? You can use a tree diagram to find the answer.

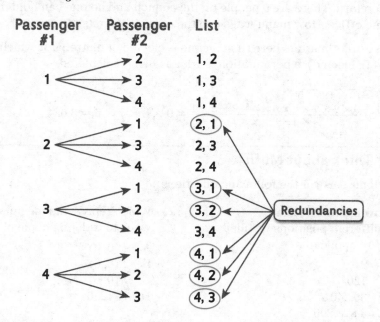

At first count, there are 12 permutations. However, recall that order doesn't matter in this situation. Whether you are chosen 1st or 2nd, you still get to go on the shuttle. So, some permutations in the list are redundant. In other words, they repeat existing combinations. Because each pair shows up twice (1, 2 is the same as 2,1), only half of the list is needed. By eliminating the redundancies, you are left with 6 combinations.

Calculating Combinations

You can use a formula in place of a tree diagram to find combinations. Consider the same problem, in which an airport shuttle has room for 2 passengers, but 4 passengers want to board, and the order of boarding doesn't matter.

$$C(n, k) = \frac{P(n, k)}{k!}$$

C = the number of combinations
P = the number of permutations
n = the number of passengers
k = the number of ways the passengers can be selected

To begin, insert values for n and k.

$$C(4, 2) = \frac{P(4, 2)}{2!}$$

If the order the passengers boarded mattered, it would be enough to find $P(4, 2)$. However, order doesn't matter in combinations. So, divide the value of $P(4, 2)$ by the value $k!$, or the number of ways the boarding passengers can be arranged. This eliminates the redundancies in the list.

$$P(4, 2) = \frac{4!}{2!} = \frac{24}{2} = 12$$

$$C(4, 2) = \frac{P(4, 2)}{2!} = \frac{12}{2}! = \frac{12}{2} = 6$$

There are 6 possible combinations.

Think about Math

Directions: Answer the following questions.

1. On Monday, an online movie service makes 10 new releases available for streaming. Your membership allows you to select 3 of the new releases. How many ways can this be done?

   A. 120
   B. 240
   C. 720
   D. 86,400

2. At an amusement park, there are 8 roller coasters you want to ride. Unfortunately, you only have time to ride 4 of them. How many ways can you choose the 4 coasters you want to ride?

   A. 1,680
   B. 140
   C. 70
   D. 2

Understand the Question

Before you can understand a test question, you must understand the question's special vocabulary. Then you can rewrite the question in your own words.

Marcus was given 4 tickets to a sporting event. He wants to take his friends but must choose who he will take from among 5 of them. Since Marcus will be using a ticket, how many ways can he pick which friends will accompany him to the sporting event?

Before you can understand the question, you need to know the meaning of the word *combinations*. A combination is a collection of objects in which the order of those object doesn't matter. Next, rewrite the question: *How many ways can 3 people be chosen from 5 people, if order doesn't matter?* Draw a tree diagram or use a formula to find the answer.

Vocabulary Review

Directions: Write the missing term in the blank.

combination   experiment   factorial
outcome   permutation   tree diagram

1. A(n) _____ is the product of a series of all descending consecutive positive integers from a given starting point.

2. A(n) _____ is an activity or situation in which the results are uncertain.

3. A(n) _____ is a result of an experiment or activity that involves uncertainty.

4. A selection of objects or values in which order is not important is a(n) _____.

5. You can use a(n) _____ to show the possible outcomes of an experiment.

6. A list of outcomes in a situation where the order of the outcomes is important is

a(n) _____.

Skill Review

Directions: Read each problem and complete the task.

1. A teacher has a box of markers 12 markers that are each different colors. She chooses 1 marker for each of 2 students. How many ways can she choose the 2 markers to give to the 2 students?

A. 66
B. 132
C. 1,188
D. 1,320

2. At a sandwich shop, customers can choose white or whole wheat bread. The shop also offers 5 different vegetables, and 4 different meats. How many possible sandwiches can be made using one vegetable and one meat?

A. 11
B. 20
C. 40
D. 80

3. Six people are scheduled to speak at a conference. Because of a time constraint, only the first 4 people in line can speak. How many ways can the volunteers speak?

A. 720
B. 360
C. 90
D. 15

4. How many possible ways can you order 7 people in a line?

A. 5,040
B. 2,520
C. 720
D. 49

5. How many ways can 5 different positions be filled by 9 applicants?

A. 5
B. 20
C. 126
D. 15,120

6. Which number represents the number of ways to choose 5 people from a group of 15 people for a task force?

A. $P(15, 15)$
B. $P(15, 5)$
C. $C(15, 5)$
D. $C(15, 15)$

Skill Practice

Directions: Read each problem and complete the task.

1. Hillary packs her lunch with three small dishes: one pasta, one fruit, and one vegetable. She can choose from 2 different pasta dishes, 2 different fruit dishes, and 2 different vegetable dishes. Complete this tree diagram, then find how many lunch combinations Hillary can make.

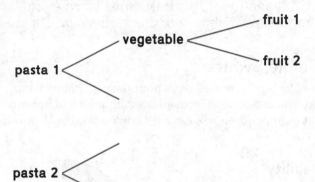

2. There are 12 people who are trying to get 3 spots on a school council. Andrew calculates that there are 1,320 different outcomes. Is Andrew correct? Explain.

3. Anthony is packing clothes for a trip. He packs 4 shirts, 3 pairs of pants, and 2 pairs of shoes. He can wear any shirt with any pair of pants, and with any pair of shoes. How many different combinations of outfits has Anthony packed? Explain your answer.

4. Keisha has 14 songs in a playlist. Keisha wants to know how many different ways she can play the songs in her playlist. What are two different ways to show this?

5. There is another formula for combinations. Explain how it is like the formula in the lesson.

$$C(n, k) = \frac{n!}{k!(n-k)!}$$

LESSON 2.4 Determine Probability

Design Pics/Darren Greenwood

LESSON OBJECTIVES

- Determine the probability of simple events
- Determine the probability of compound events

CORE SKILLS & PRACTICES

- Determine Probabilities

Key Terms

compound event
an event formed by two or more simple events

probability
the study of how likely it is for an event to occur

tree diagram
a branching diagram that shows possible outcomes of an experiment

Vocabulary

complement
an event that shows all the ways that an event cannot happen

dependent event
a second event whose probability depends upon a first event

independent event
a second event whose probability does not depend upon a first event

Key Concept

The probability of a chance event uses a number between 0 and 1 to describe the likeliness that the event will occur. You can use the number of total and favorable outcomes of an event to determine the probabilities of simple or compound events.

Probability of Simple Events

Genetic traits, like eye color, are inherited from your parents; theirs from their parents, and so on. You can analyze possible combinations in order to find the probability that your children will have certain colored eyes or other genetic traits.

The Basics of Probability

Probability is the study of how likely it is for an event to occur. The different results that can occur are called outcomes. This spinner shows two different outcomes: blue and white.

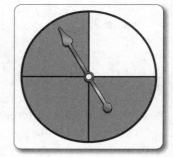

The probability of an event is often expressed with words such as *likely* or *unlikely*. If the outcome is sure to happen, we say it is certain; if there is no chance for the outcome to occur, we say it is impossible.

- Spinning blue is likely.
- Spinning white is unlikely.
- Spinning blue or white is certain.
- Spinning red is impossible.

Probability can also be described as a ratio. The ratio is found by comparing the number of ways that are favorable for an event to occur with all of the outcomes that are possible.

$$\text{probability} = \frac{\text{number of favorable outcomes}}{\text{total number of outcomes}}$$

For this spinner, write the probability of spinning blue as the ratio of the number of blue sections in the spinner (favorable outcomes) to the total number of sections in the spinner (total outcomes). The probability of spinning blue is 3 out of 4.

Determine Probability

Probability as a Number

The number that represents a probability can be expressed using words, as a fraction, as a decimal, or as a percent.

Example 1: Find the Probability

Art, Etan, José, Ryan, and Dwight each write their names on slips of paper and put the papers in a box. One name will be randomly drawn to win a prize. What is the probability Ryan will win?

Step 1 Find the total number drawn.

- There are five names that can be of outcomes.

Step 2 Find the number of favorable outcomes.

- Only 1 name is Ryan's.

Step 3 Write the probability as a fraction, decimal, or percent.

$$\text{probability of Ryan winning} = \frac{\text{favorable outcomes}}{\text{possible outcomes}}$$
$$= \frac{1}{5} = 0.2 = 20\%$$

The probability that Ryan wins is 20%.

You can also write the probability of the **complement** of an event. The complement is an event that shows all the ways that an event cannot happen. To write the probability that Ryan does not win, the number of possible outcomes is still 5, but the number of favorable outcomes—the other names in the box—is 4.

$$\text{probability of Ryan not winning} = \frac{\text{favorable outcomes}}{\text{possible outcomes}}$$
$$= \frac{4}{5} = 0.8 = 80\%$$

So, the probability that Ryan will not win is 80%.

Probability Facts

Here are some other important facts to know about probabilities expressed as numbers.

- If there is no chance that an event can happen, the probability is 0.
- If an event is certain to happen, its probability is 1, or 100%.
- The greater the probability that an event will occur, the closer the number will be to 1 or 100%; the less likely the probability, the closer the number will be to 0.
- The sum of the probability of an event and its complement is 1.
- If two events are equally likely, the probability of each is 50%.

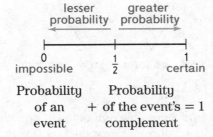

Find Favorable Outcomes

For many probability problems, making a list of possible outcomes is the best first step. Use your list to count total outcomes and favorable outcomes.

Example 2: Find Favorable Outcomes

Jenni, David, and Lauren have tickets to a play and got three seats together in the same row. If each person randomly chooses a ticket, what is the probability that Lauren will sit next to David?

Step 1 Make a list of all possible outcomes.

Step 2 Count the number of possible outcomes.

- There are 6 possible seating arrangements.

Jenni	David	Lauren
Jenni	Lauren	David
David	Jenni	Lauren
David	Lauren	Jenni
Lauren	Jenni	David
Lauren	David	Jenni

Step 3 Count the number of favorable outcomes.

- There are 4 arrangements with Lauren next to David.

Step 4 Write the probability as a fraction.

$$\text{probability of Lauren sitting next to David} = \frac{4}{6} = \frac{2}{3}$$

Tree Diagrams

Another way to count outcomes is to draw a **tree diagram,** which resembles a branching tree. Each path identifies a possible outcome of an experiment.

Example 3: Tree Diagrams

The Clothes Factory has XL sweatshirts on sale. The sweatshirts come only in blue or green. Each sweatshirt comes with either a zipper or a hood. And each may or may not have a store logo printed on it. How many combinations of sweatshirts are available?

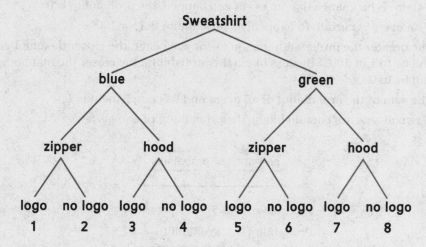

This tree diagram shows 8 possible outcomes. If you wanted to find the probability of a randomly chosen sweatshirt on sale that has a hood, you can use the tree diagram to find there are 4 favorable equally likely outcomes and write the probability $\frac{4}{8}$, or $\frac{1}{2}$.

Predict with Probability

When you flip a coin, roll a number cube, or spin a spinner, you are basing probability on the laws of chance. This is called theoretical probability. Theoretical probability can be used to make predictions.

Example 4: Theoretical Probability

If you roll a number cube 50 times, how many times are you likely to roll the number 4? You can't predict with certainty, but you can say how many 4s are *most likely* to occur.

Step 1 Write the probability of rolling a 4 on one roll.

$$\text{probability for 4 on one roll} = \frac{\text{favorable outcomes}}{\text{possible outcomes}} = \frac{1}{6}$$

Step 2 Multiply by the number of rolls.

Out of 50 rolls, you are likely to roll a 4 about 8 times.

You can also base probability on past performance or on data. This type of probability is called experimental probability.

Example 5: Experimental Probability

The table shows the last 50 sales at Motor City. What is the best prediction for the number of Japanese cars sold out of the next 200 car sales?

Type	Number
Domestic	24
European	6
Japanese	20

Step 1 Use the given data to write the probability that the next car sold is a Japanese car.

$$\frac{\text{favorable outcomes}}{\text{possible outcomes}} = \frac{20}{50} = \frac{2}{5}$$

Step 2 Multiply by the number of cars that will be sold.

$$\frac{2}{5} \times 200 = \frac{400}{5} = 80$$

Out of the next 200 car sales, 80 of the sales are likely to be Japanese cars.

Think about Math

Directions: Answer the following questions.

1. If you flip a pair of coins 48 times, how many times are you most likely to get 2 heads?

2. A basketball player makes 15 free throws out of his last 45 attempts. How many free throws is he most likely to make out of his next 90 attempts?

Determine Probabilities

Regina is playing a game in which she draws marbles from a bag. The bag contains 1 white marble, 2 gray marbles, and 3 green marbles. What is the probability that Regina will draw a gray marble on her first try?

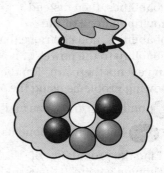

Each marble in the bag represents a possible outcome. For this problem, drawing a gray marble is a favorable outcome. Two of the six marbles in the bag are gray, so there are 2 favorable outcomes out of 6 possible outcomes.

$$\begin{aligned}\text{probability of gray marble} &= \frac{\text{favorable outcomes}}{\text{possible outcomes}} \\ &= \frac{2 \text{ gray marbles}}{6 \text{ total marbles}} \\ &= \frac{1}{3}\end{aligned}$$

Now find each of these probabilities using the bag of marbles shown below: blue, white, gray, not blue, not white, and not gray.

Remember that when calculating the probability of independent events, the probability of each independent event is multiplied. If the independent events have equal probabilities, then instead of multiplying all of the probabilities, you can raise that probability to the power of how many events there are. Try the following problems using the button for fractions and the button for exponents.

1. Find the probability of flipping a coin 20 times and getting tails every time.

2. Find the probability of a basketball player who has a 70% probability of making a free throw making his next seven free throws.

Probability of Compound Events

Probability also has applications in weather forecasting. An "80% chance of rain" may seem like a simple statement, but there is a lot of data and probability of many different events that are at work quantifying the probability with a single number.

Independent Events

So far, you have been working with simple events. Sometimes, you might want to know the probability of an event formed by two or more simple events, like rolling a number cube and spinning a spinner. This is called a **compound event** and is found by multiplying the probability of the first event by the probability of the second.

Example 6: Independent Events

When flipping a penny twice, what is the probability of getting two heads in a row?

Probability of two *heads* in a row $= \frac{1}{2} \times \frac{1}{2}$

probability of a head on first flip ⌐ probability of a head on second flip

Step 1 The probability of getting a head each time you flip the coin is $\frac{1}{2}$.

Step 2 Multiply the probabilities of getting a head on each flip.

The probability of two heads in a row is $\frac{1}{4}$. Events like this double coin flip are **independent events**, meaning the probability of a second event does not depend on the first event.

For each draw, there is a probability of $\frac{6}{9} = \frac{2}{3}$ drawing a white marble.

$$\frac{2}{3} \times \frac{2}{3} = \frac{4}{9}$$

first draw   second draw

Independent events are often found in probability problems involving replacement. An example is when you draw a marble from a bag and then replace it before drawing another. The events are independent. Each marble has the same probability of being drawn on both draws.

Dependent Events

Think about how you might find the probability of randomly choosing a pair of face cards from the four cards shown. For this compound event, the probability of the second event depends on the first event. These events are said to be **dependent events**.

Example 7: Dependent Events

Suppose you randomly take two cards from these four cards: king of hearts, queen of diamonds, five of hearts, eight of diamonds. What is the probability that both cards will be face cards?

Step 1 The probability that the first card is a face card is $\frac{2}{4}$.

Determine Probability

Step 2 For your second card, you need to remove your first card from the possible outcomes. There are now 3 cards remaining, one face card and two number cards. Choosing a face card for your second card occurs with a probability of $\frac{1}{3}$.

Step 3 Multiply the first probability by the second probability.

$\frac{2}{4} \times \frac{1}{3} = \frac{2}{12} = \frac{1}{6}$

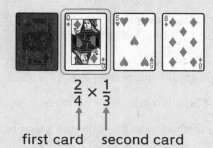

$\frac{2}{4} \times \frac{1}{3}$

first card   second card

Compound probabilities like this one are said to be without replacement, because once the first card is chosen it cannot be chosen again. When you are asked to find probabilities without replacement, you know that the number of possible outcomes for the event decreases, and so the events are dependent.

More with Tree Diagrams

You can also use tree diagrams to help you find probabilities of compound events. For choosing two cards of the four cards in Example 7, let the first level of the diagram show outcomes for the first card, and the second the outcomes for the second card.

Example 8: Tree Diagrams

Step 1 Label each branch of the diagram with its corresponding probability.

FC = face card
NF = not a face card

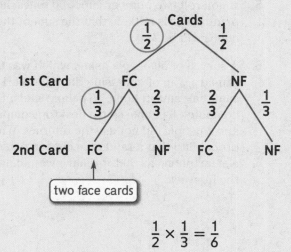

$\frac{1}{2} \times \frac{1}{3} = \frac{1}{6}$

Step 2 Multiply the probabilities along the branches of the favorable outcomes. The result is the same as what you found by analyzing possible outcomes.

So the probability of choosing two face cards is $\frac{1}{6}$.

21ST CENTURY SKILL

Civics Literacy

News media and other organizations often conduct exit polls after people have cast their vote. Voters are asked, for example, which candidates they voted for and why, as well as whether they voted for or against any ballot initiatives. Exit polls can then be used to make predictions about the overall results of the election.

At a recent election, voters were asked to vote for or against two unrelated referendums, Referendum A and Referendum B. Exit poll results showed that 2 out of 3 people voted for Referendum A, and 3 out of 5 people voted for Referendum B. Based on the results of the exit poll, what is the probability of at least one of these Referendums passing?

Vocabulary Review

Directions: Write the missing term in the bank.

compound event	**probability**	**complement**
independent event	**tree diagram**	**dependent event**

1. The probability of two or more simple events happening is a(n) _____ .

2. _____ is the study of how likely it is for an event to occur.

3. A(n) _____ is a way to show possible outcomes that resembles a branching tree.

4. You flip a coin fifteen times. The probability that on the sixteenth flip it will be heads is

   a(n) _____ .

5. The probability that today is not Tuesday is the _____ of the probability that today is Tuesday.

6. An event that depends on the result of a prior event is a(n) _____ .

Skill Review

Directions: Read each problem and complete the task.

1. There are an apple, orange, and pear in a fruit basket. Marcus randomly selects one piece of fruit to eat each day. What is the probability that Marcus will choose an apple before choosing a pear?

2. There is a bag of 4 red marbles, 5 blue marbles, and 1 white marble. Wendy draws a marble, and then places it back in the bag. Then Peter draws a marble. What is the probability that both Wendy and Peter drew a blue marble?

3. Daniel, Tiana, Elise, Mary, and Robert put their names on slips of paper and put the papers in a box. What is the probability that Tiana's name will be picked?

4. There are five rides at a theme park, two of which are roller coasters. Maria wants to ride each ride once. She chooses a roller coaster for her first ride, then randomly chooses the second ride. What is the probability that the second ride is a roller coaster?

5. If you roll two number cubes 30 times, how many times is it likely that the sum of the two cubes will be 7?

6. A survey of 50 people asked which was their primary way of accessing the Internet. 15 people said their smartphone, 25 people said a laptop computer, 8 people said a desktop computer, and 2 people did not use the Internet. What is the probability that a randomly chosen person uses a smartphone as their primary way to access the Internet?

7. Dave packs 8 pairs of pants for a business trip. Three of the pants are khaki, and 5 of the pants are black. Draw a tree diagram to find the probability of Dave randomly choosing khaki pants for his first two days.

8. There is a bag of marbles. Five times a marble has been drawn and placed back in the bag. Of the five draws, 2 were green marbles, 1 was a yellow marble, and 2 were blue marbles. How likely is it that the next drawing will be a blue marble?

Skill Practice

Directions: Read each problem and complete the task. Use the tree diagram to answer questions 1–3.

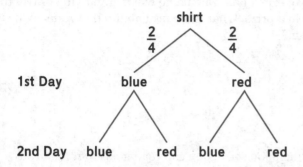

1. Hannah drew this tree diagram to help her find probabilities of wearing a red or blue shirt over 4 days. She wears a shirt for one day and then chooses from the remaining shirts the next day. Is this probability situation independent or dependent?

2. Complete the tree diagram. Extend the branches to show each day Hannah wears the shirts.

3. What shirt does Hannah wear on the first day if the probability of wearing a blue shirt on the second day will be $\frac{1}{3}$?

4. Draw a spinner with 12 sections so that these are the probabilities of spinning each color.

Red: $\frac{1}{3}$ Blue: $\frac{1}{12}$ Green: $\frac{1}{4}$ Yellow: $\frac{1}{6}$ Purple: $\frac{1}{6}$

5. Walter owns a parking lot with 25 spaces. Only 1 car uses a parking space each day. Design an experiment to predict the number of red cars that will park in his lot on Friday.

6. A red and a blue number cube are used for one turn of a game. If the outcome of rolling the red number cube is 4, you then get to roll the blue number cube. If 24 people are playing the game, how many people are likely to roll an even number on the blue number cube?

7. The table shows the transportation method used by the employees in an office. What is the probability that the next two employees that join the office staff will take the bus?

Transportation	Number
car	18
bus	6
walk	6

A. 0.04
B. 0.12
C. 0.33
D. 0.36

8. Julian randomly chose a penny from the three shown. Then Matt chose one from the two that remained. What is the probability that Matt chose tails?

A. $\frac{1}{6}$

B. $\frac{1}{3}$

C. $\frac{1}{2}$

D. $\frac{2}{3}$

Directions: Choose the best answer to each question

1. Amber bought 10 pounds of potatoes for $4.50. What is the price per pound of the potatoes?

   A. $45.00 per pound
   B. $14.50 per pound
   C. $4.50 per pound
   D. $0.45 per pound

2. In a survey of 300 shoppers, 40% reported bringing a bottle of water with them. How many of the surveyed shoppers does this represent?

   A. 75
   B. 120
   C. 260
   D. 340

3. The probability that it will rain tomorrow and the probability that it will not rain tomorrow is 1, or _____.

4. A spinner has four equal parts. One part is red, one part is blue, one part is yellow, and one part is green. If you spin the spinner 64 times, you are likely to spin a red part _____ times.

5. Anna and her friend had a dinner that cost $45 and they would like to leave an 18% tip. Which equation can you use to find the amount of the tip?

   A. $\frac{18}{45} = \frac{x}{100}$
   B. $18 \times 100 = 45x$
   C. $\frac{18}{100} = \frac{x}{45}$
   D. $\frac{100}{45} = \frac{x}{18}$

6. Ellis spends $\frac{17}{20}$ of his allowance on toys. Marie spends 85% of her allowance on toys. $\frac{17}{20}$ and 85% are _____ because both numbers can be written as the same ratio.

7. Jen borrows $4,000 from her bank. She has a simple interest loan at a 5% interest rate and will pay it back over 4 years. How much interest will she pay on this loan?

   A. $8,000
   B. $4,009
   C. $800
   D. $50

8. Lisa is a video editor. She has 4 videos that she needs to edit. How many different orders are there for her to edit the videos?

   A. 48
   B. 24
   C. 10
   D. 6

9. Liam, Mia, Aiden, Lily, Avery, and Logan each write their name on a slip of paper and place the slips of paper in a box. One name will be randomly drawn to win a prize. What is the probability that Aiden will win?

   A. 6%
   B. $\frac{1}{5}$
   C. $\frac{1}{6}$
   D. 60%

10. When you see 8! you can read it as *eight factorial* and it means to multiply _____.

11. A bag has 3 blue marbles and 5 red marbles. After a marble is drawn, it is placed back in the bag. What is the probability of drawing a blue marble twice?

    A. $\frac{9}{64}$
    B. $\frac{9}{25}$
    C. $\frac{3}{8}$
    D. $\frac{3}{5}$

12. This model shows the fraction $\frac{26}{100}$, the percentage _____, and the decimal 0.26.

13. Mrs. Albert buys 10 boxes of crayons for $30. How many boxes of crayons can she buy for $75, if the ratio stays the same?

    A. 20 boxes
    B. 25 boxes
    C. 60 boxes
    D. 75 boxes

14. Mr. Thomas is sketching a scale model of a park. The length of the park is 500 yards and the width of the park is 200 yards. He drew the length as 25 inches and the width as _____ inches.

15. A bag has 5 yellow marbles and 7 green marbles. After a marble is drawn, it is set aside. What is the probability of drawing a green marble twice?

   A. $\frac{35}{144}$
   B. $\frac{7}{22}$
   C. $\frac{49}{144}$
   D. $\frac{7}{12}$

16. A gym charges $40 a month for membership in 2013. In 2014 the price increases to $45 a month. What was the percent change in the price?

   A. −12.5%
   B. 11.1%
   C. 12.5%
   D. 88.9%

17. A club is electing a president, vice president, and treasurer. There are 7 people running for positions and no one can hold more than one position. How many ways can the positions be filled?

   A. 1.75
   B. 210
   C. 840
   D. 5,040

18. This tree diagram shows all possible outcomes of rolling a _____ .

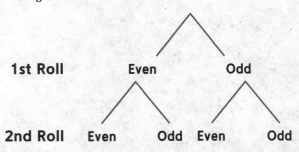

19. A company has 4 summer intern positions. There are 12 applicants. How many ways can the positions be filled?

   A. 19,958,400
   B. 11,880
   C. 495
   D. 384

20. The two triangles are _____ because the measures of the sides are proportional.

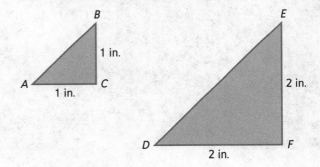

Check Your Understanding

On the following chart, circle the items you missed. The last column shows pages you can review to study the content covered in the question. Review those lessons in which you missed half or more of the questions.

Lesson	Item Number(s)			Review Page(s)
	Procedural	Conceptual	Problem Solving	
2.1 Apply Ratios and Proportions	5	6, 12, 20	1, 13	48–53
2.2 Calculate Real-World Percentages	14		2, 7, 16	54–61
2.3 Use Counting Techniques	19	10	8, 17	62–69
2.4 Determine Probability	4, 15, 18	3	9, 11	70–77

Chapter 3

Linear Equations and Inequalities

An unknown variable is critical in solving some real-world problems because the unknown is what you are trying to figure out. You can use variables to stand for anything you don't know when you use an equation to represent and solve a real-world problem. A variable stands in for the answer until you can use the order of operations and number properties to determine the answer. Once you understand variables, you can represent and solve more complex algebraic problems.

Fuse/Getty Images

Lesson 3.1
Evaluate Linear Expressions

How can you use math to describe a situation? How can you figure out whether a situation can be modeled by addition, subtraction, or another operation? Learn how to identify key words in a real-world problem and translate it to a linear expression that represents the situation in the problem.

Lesson 3.2
Solve Linear Equations

When you travel to a new place, you can find your way home by following the directions in reverse. You are essentially "undoing" each turn. When you solve a linear equation you want to "undo" every operation to solve for the variable. Learn how to solve linear equations.

Lesson 3.3
Solve Linear Inequalities

What happens when a problem has more than one solution? Linear inequalities have a range of values as their solution. Because they have more than one answer, their solutions are often graphed on number lines. Learn how to solve and graph linear inequalities.

Lesson 3.4
Use Expressions, Equations, and Inequalities to Solve Real-World Problems

Where do you start when you need to solve a real-world problem? How can you write a real-world problem as an equation or inequality that you can solve? Learn how to translate and solve real-world problems using linear equations and inequalities.

Goal Setting

Think about the last time you had to decide between two options. It might have been choosing between cell-phone plans or comparing two different housing options. What information did you know? What information was unknown? What did you need in order to make the decision? How could you use algebra to represent the two options mathematically and help you make your decision?

How else can you use algebra to help you find solutions to problems you encounter at work or at home? The world is full of unknowns. Algebra can help you make sense of what you know and what you don't know but need to find out.

LESSON OBJECTIVES

- Use algebraic symbols to represent unknown quantities
- Perform operations on linear expressions
- Evaluate linear expressions

CORE SKILLS & PRACTICES

- Perform Operations
- Evaluate Expressions

Key Terms

variable
a letter that is used to represent an unknown value

constant
an expression that stays the same

coefficient
a number that is multiplied by a variable

algebraic expression
a mathematical statement containing letters and numbers organized as terms but with no equal sign

Vocabulary

evaluate
to substitute values for variables

distribute
to use multiplication over addition or subtraction

Key Concept

There are a lot of unknowns around us. In math we do not always know the total we are solving for or the values we are calculating. These unknowns are expressions. Evaluating linear expressions means substituting values (numbers) for variables (letters).

Algebraic Expressions

If you earn $10.50 an hour, how long will it take you to earn $600? What about $800? Questions like this come up all the time. As situations change we need to calculate differently.

Expressions and Variables

An **algebraic expression** is a mathematical statement containing letters and numbers organized as terms. Addition and subtraction separate terms. Terms can be numbers, letters, or letters that are multiplied or divided by numbers. The letters you see in algebraic expressions are called **variables**. The variable represents different values, so the value of an expression can change depending on the value assigned to the variable.

Algebraic Expressions:
1 Term:	$2q$	q is the variable.
2 Terms:	$5n + 4$	n is the variable.
3 Terms:	$6w + 21r - 10$	w and r are both variables.

Parts of an Algebraic Expression

In an algebraic expression, numbers that stand alone as a term are called **constants**. They are constants because their value stays the same.

Algebraic Expressions: $5n + 4$ 4 is "alone." It is a constant.

Numbers that are multiplied by variables are called **coefficients**. The coefficient of a variable is usually written in front of the variable. If a variable does not have a number in front of it, the coefficient is 1.

Algebraic Expressions:
$6w + 21r - 10$ The coefficient of w is 6. The coefficient of r is 21.

Algebraic expressions must contain at least one variable and one number. They do not contain an equal sign.

Translating Between Phrases and Expressions

Translating words to algebraic expressions is an important skill when solving real-world problems. You can begin to practice this skill by translating simple phrases. To do so, use the meanings of the operations.

Mathematical Operation	Key Phrases
Addition +	Sum, Increase, Add, All together, Total
Subtraction −	Subtract, Decrease, Difference, Minus, Fewer
Multiplication × or •	Times, Multiply, Product
Division ÷ or /	Divide, Per, Quotient

The following examples show how phrases are translated to algebraic expressions and how algebraic expressions are translated to words.

Example 1: Translating Phrases to Algebraic Expressions

Phrases	Expressions
5 less than twice a number	$2n - 5$
A number increased by 10	$x + 10$
The product of two and a number	$2r$
Six times the sum of a number and five	$6(w + 5)$

Example 2: Translating Algebraic Expressions to Words

Expressions	Phrases
$50 + w$	"the sum of 50 and w" "50 increased by w"
$20 - x$	"the difference of 20 and x" "x less than 20"
$6n$	"the product of a number and 6" "6 times n"
$\dfrac{10}{r}$	"the quotient of 10 and r" "10 divided by r"

Think about Math

Directions: Fill in the blank with the correct expression. Use n to represent an unknown.

1. The phrase "5 less than a number" can be written as _____.

2. The phrase "the sum of −7 and 2 times a number" can be written as _____.

3. The phrase "the product of 10 and the sum of 2 and a number" can be written as _____.

Use Reasoning

In business, algebraic expressions are used to determine many things, including weekly pay, costs, discounts, and benefits. If you are paid on a commission basis, the amount you earn depends on the amount of total sales you produce. Some employers offer salaries that pay a set amount in addition to a commission. In this case, total sales represent a value that varies, and the set amount is a constant. When calculating total pay, the commission percentage is multiplied by total sales, then the set amount is added.

For example, Hilary, a clothing retail saleswoman, earns $250 per week plus a 10% commission on her sales. Write an expression, using s to represent sales, that can be used to calculate the amount of money Hilary earns each week.

Linear Expressions

Linear expressions are used to represent many different situations. For example, cell phone plans are represented by linear expressions because there is a constant monthly charge and additional charges for taxes or overage charges.

Identifying Linear Expressions

A linear expression is one type of algebraic expression. In linear expressions, no term can have two or more variables, nor can they have square roots or exponents.

These are examples of linear expressions:

$x + 7$	$3x + 7$	$3x + 7y$

These are <u>not</u> linear expressions:

x^2	No exponents on variables
$3xy + 5$	Can't multiply two variables
$\frac{x}{y} + 4$	Can't divide two variables
$2\sqrt{x}$	No square root sign on variables

In linear expressions you can only add or subtract like terms. Like terms have the same variables but may have different constants. You can simplify linear expressions by combining like terms.

Example 3: Combining Like Terms

Simplify the expression: $7w - 2 + 3w + 5$

Step 1 Rearrange the expression so like terms are next to each other.

$7w - 2 + 3w + 5$
$7w + 3w - 2 + 5$

Step 2 Combine the whole numbers.

$7w + 3w - 2 + 5$
$7w + 3w + 3$

Step 3 Combine the like terms with variables by combining their coefficients.

$7w + 3w + 3$
$10w + 3$

Adding Linear Expressions

When adding two linear expressions, like terms need to be combined in one expression.

> **Example 4: Adding Linear Expressions**
>
> Add: $(5x - 3) + 3(-x + 2)$
>
> **Step 1** First, we **distribute** the coefficients to remove the parentheses. To distribute means to multiply over addition or subtraction. In this case, we will multiply 1 by $(5x - 3)$, and 3 by $(-x + 2)$.
>
> $(5x - 3) + 3(-x + 2)$
> $1(5x - 3) + 3(-x + 2)$
> $1(5x) + 1(-3) + 3(-x) + 3(2)$
> $5x + (-3) + (-3x) + 6$
>
> **Step 2** Rearrange the expression so like terms are near each other.
>
> $5x + (-3) + (-3x) + 6$
> $5x + (-3x) + (-3) + 6$
>
> **Step 3** Combine the coefficients for like terms and simplify the expression.
>
> $5x + (-3x) + (-3) + 6$
> The answer is $2x + 3$.

Subtracting Linear Expressions

Subtracting linear expressions is similar to adding them, except that you will have to multiply a negative, or distribute the minus sign, before combining like terms.

> **Example 5: Subtracting Linear Expressions**
>
> Subtract: $(7r - 1) - (2r + 6)$.
>
> **Step 1** Distribute.
>
> $(7r - 1) - 1(2r + 6)$
> $7r - 1 - 1(2r) - 1(6)$
> $7r - 1 - 2r - 6$
>
> **Step 2** Rearrange the expression so like terms are near each other.
>
> $7r - 1 - 2r - 6$
> $7r - 2r - 1 - 6$
>
> **Step 3** Combine like terms and simplify the expression.
>
> $7r - 2r - 1 - 6$
> $5r - 7$

CORE SKILL

Perform Operations

You will often be asked to perform operations on expressions. Every operation has a different method to follow. Those methods could involve multiplying expressions through parentheses using the Distributive Property, simplifying expressions by combining like terms, or evaluating the expression by substituting a value into the variable. The first step always is to identify what operations you are performing and then follow the steps for that particular operation, paying attention to the order of operations while you simplify. Using your knowledge of the operations as well as simplifying expressions, add the linear expressions: $2(3n - 1)$ and $5(3n + 2)$.

Directions: Select the most appropriate answers.

1. Multiply the rational coefficient
    by the linear expression:
    $-12(3b - 4)$

    A. $36b + 4$
    B. $-36b - 4$
    C. $-36b - 48$
    D. $-36b + 48$

2. Multiply the rational coefficient
    by the linear expression:
    $-6(2x + 10)$

    A. $-12x - 60$
    B. $-12x + 60$
    C. $-12x + 10$
    D. $-8x - 16$

Evaluating Linear Expressions

Ever wonder how many calories you burn while exercising? The number will actually depend on the type of exercise and the amount of time you spend exercising. When you find out this information, you can write and evaluate a linear expression to find out how many calories you are actually burning.

Evaluate Linear Expressions

When you **evaluate** linear expressions, you are substituting a number for a variable in the expression then simplifying the expression.

Example 6: Evaluate a Linear Expression

Evaluate: $9x + 10$ for $x = 5$

Step 1 First, determine what number will replace the variable. In this instance, 5 will be replacing x.

$9x + 10$ for $x = 5$
$9(5) + 10$

Step 2 Next, multiply 9 by 5.

$9(5) + 10$
$45 + 10$

Step 3 Finally, add the terms together. The value of the expression when $x = 5$ is 55.

$45 + 10 = 55$

Problem Solving Practice

Compare the following taxi cab charges. Based on the information given, which company is best for someone who needs to travel 30 miles?

Example 7: Evaluate a Linear Expression

Taxi Company #1: $0.25 per mile plus $3.00 service fee
Taxi Company #2: $0.30 per mile, no service fee

Step 1 Write an expression that represents Taxi Company #1.
Let m represent the number of miles.

$0.25m + 3.00$

Step 2 Write an expression that represents Taxi Company #2.
Let m represent the number of miles.

$0.30m$

Step 3 Evaluate each expression when $m = 30$.

Taxi Company #1:
$0.25m + 3.00 = 0.25(30) + 3.00 = 7.50 + 3.00 = 10.50$

Taxi Company #2:
$0.30m = 0.30(30) = 9.00$

In the case of someone who travels 30 miles, Taxi Company #2 is a better option. This plan saves the customer $1.50 compared to Taxi Company #1.

Think about Math

Directions: Select all appropriate answers.

1. When evaluated at $p = 3$, which of the following have a value greater than 9?

    A. $3p$
    B. $2p + 4$
    C. $-2p + 7$
    D. $5p - 4$

2. Which expression has a value of 15 when evaluated at $q = 7$?

    A. $2(q + 1) + 5q - 8$
    B. $3(q + 1) - 17 + q$
    C. $4(q - 2) - q + 2$
    D. $q + 12 - (q - 4)$

CORE SKILL

Evaluate Expressions

Some linear expressions have more than one variable. Evaluating these expressions still follows the same process as for evaluating expressions with only one variable. The only difference occurs at the beginning; you must first substitute both values for their respective variables and then simplify. Be sure to substitute accordingly. If you substitute an incorrect value for a variable or a value for the wrong variable, the resulting expression will be incorrect. What is the value of the expression, $3g + 6r$, when $g = -2$ and $r = 8$?

CALCULATOR SKILL

Entering complex calculations into a calculator can seem rather hard, especially when parentheses must be entered to guarantee the correct expression is calculated. While using the TI-30XS MultiView™, press the (and) buttons to place an expression in parentheses. Simplify the expression $2.25(2 + 1.758 - 6.4)$.

Vocabulary Review

Directions: Match the terms with their description.

variable   **constant**   **distribute**
coefficient   **evaluate**   **algebraic expression**

1. the number that is multiplied by a variable _____

2. to use multiplication over addition or subtraction _____

3. an expression that stays the same _____

4. to substitute values for variables _____

5. a letter that is used to represent an unknown _____

6. a mathematical statement containing letters and numbers organized

as terms _____

Skill Review

Directions: Write expressions that describe the following phrases.

1. Seven less than twice a number

2. The product of -3 and a number

Directions: Follow the prompt for each of the linear expressions.

3. Add: $(3n - 4) + (2n - 5)$

4. Subtract: $(9w - 3) - (6w - 3)$

5. Multiply: $-4(5p - 1)$

6. Evaluate: $4n + 3$, when $n = 6$

Directions: Answer the following questions.

7. When evaluated at $r = 3$, which of the following expressions result in an even number?

   A. $2(r - 1)$
   B. $r + 2 - 3r$
   C. $3(r + 1) + 1$
   D. $(r - 2) + 7r - 1$

8. Which statement represents the expression $3n + 7$?

   A. The product of a number and the quantity 3 plus 7.
   B. The sum of a number and 7 times 3.
   C. Seven plus the product of 3 and a number.
   D. Three times the sum of a number and 7.

Evaluate Linear Expressions

Skill Practice

Directions: Read each problem and complete the task.

1. Susan had to write an algebraic expression for the following phrase: "the difference between two times a number and negative 5," using n to represent the unknown. She then had to evaluate the expression when $n = -5$. Her final answer was -5. Was Susan correct?

2. A volunteer group is planning a celebration for its second anniversary. The local hall charges $200 for rent of the space and $15 per person for food and beverages. Write an expression that represents this situation. Let p represent people.

3. Perform the operations.
$$(3x - 4) + (-2x - 1) - 5(x - 3)$$

4. A couple is saving money to purchase a new car. They do not want to finance any amount. The couple has already saved $12,000, and they are saving an additional $500 each month. The expression that represents this situation is: $500m + 12{,}000$, where m is the number of months. If they save for one year, how much will they be able to spend total?

5. A local gym has a special promotion offering new members three free months if they sign a one-year contract. After the first three months, the gym membership will cost $20 per month. A new gym has just opened and is offering one-month free membership and $18 per month thereafter. Compare the cost of both gyms after one year of membership. Which is more expensive?

6. When evaluated at $p = 6$ and $q = -3$, which of the following expressions have a value greater than -6?
   A. $3p + 2q$
   B. $2p - 3q$
   C. $-p + q$
   D. $p - 4q$

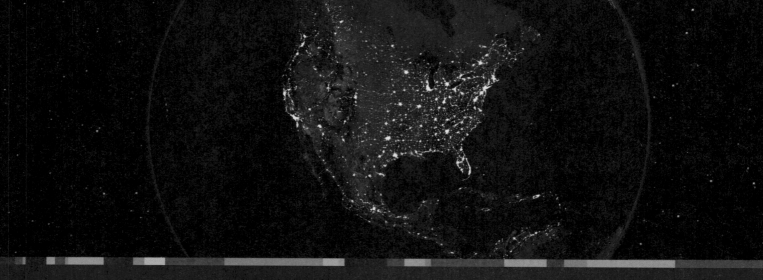

LESSON OBJECTIVES

- Write and solve one-step equations
- Solve multi-step equations

CORE SKILLS & PRACTICES

- Solve Simple Equations by Inspection
- Solve Linear Equations

Key Terms

equation
a mathematical statement that two expressions are equal

solution of an equation
a value for the variable that makes the equation true

Vocabulary

expression
a mathematical statement that contains numbers, operations, and/or variables but no equal sign

inverse operations
operations that undo each other

reciprocal
the number that has a product of 1 when multiplied by the original number

variable
an unknown quantity

Key Concept

You can solve an equation by performing inverse operations on both sides of the equation. The solution can be checked using substitution.

One-Step Equations

Ask anyone who has ever taken algebra what he or she remembers most from the class. Most people will probably mention solving equations. Solving equations is a fundamental concept that is used in almost all branches of mathematics as well as in various other fields, such as science and economics.

Equations and Solutions

A **variable** is a letter or symbol that represents an unspecified number. An **expression** is a mathematical phrase that contains numbers, operations, and/or variables. An **equation** is a mathematical statement that two expressions are equal. It is important to recognize the difference between an expression and an equation. An equation always has an equal sign.

Equations	Expressions
$16 - 9 = 7$	$2q$
$5 + x = 25$	$5n + 4$
$2w = 10x$	$6w + 21r - 10$

An equation that contains a variable may be true or false depending on the replacement value for the variable. A **solution of an equation** with one variable is a value that, when substituted for the variable, makes the equation true. You solve an equation when you find the solution for the variable.

Equation	Solution
$r + 20 = 30$	The solution is $r = 10$ because $10 + 20 = 30$.
$6 - t = 2$	The solution is $t = 4$ because $6 - 4 = 2$.

Solve Linear Equations

Writing Equations

It is important to be able to write equations from verbal descriptions. Real-world problems are stated using verbal descriptions, and being able to translate these situations into equations creates an opportunity to find mathematical solutions. The correct solution can be found only if the equation is written correctly.

Translating from Words to Equations

Look for key words that indicate the operation being performed. Also look for words and phrases that mean "equals," such as *is* and *is equal to*. Represent an unknown number by using any letter as a variable.

Words	Equations
A number increased by 32 is equal to 40.	$n + 32 = 40$
Four times a number is 36.	$4x = 36$
Seven less than what number equals 15?	$k - 7 = 15$
The product of a number and 3 is −15.	$3w = -15$
6 divided by a number is equal to 2.	$\frac{6}{a} = 2$

Equations can be written by breaking up the words and translating each part.

"A number is multiplied by three, and then nine is subtracted to give a result of twelve."

A number → n
is multiplied by three → $3n$
and then nine is subtracted → $3n - 9$
to give a result of → $3n - 9 =$
twelve → $3n - 9 = 12$

Solving One-Step Equations

An equation is a perfect balance between what is on the left side and what is on the right side of the equal sign. If you make any changes on the left side, you must make the same changes on the right side. Think of an equation as a balanced scale.

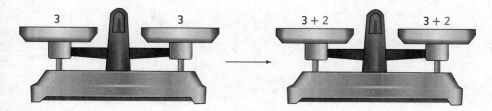

Inverse operations are operations that undo each other. For example, the inverse operation of adding 5 is subtracting 5, and the inverse operation of dividing by 10 is multiplying by 10. To solve an equation, perform inverse operations to isolate the variable. This means that the variable is by itself on one side of the equation.

Solve Simple Equations by Inspection

To solve an equation by inspection means to find the solution mentally, without using pencil and paper or a calculator. Simple equations like $5x = 20$ can be solved by inspection because only one operation is involved and because the numbers are familiar.

The solution of this equation can be found by remembering the multiples of 5. What multiple of 5 is equal to 20?

$$5 \times 1 = 5$$
$$5 \times 2 = 10$$
$$5 \times 3 = 15$$
$$5 \times 4 = 20$$

The solution of $5x = 20$ is $x = 4$.

Solve each of the following equations by inspection.

- $2n = 6$

- $10 = 5r$

- $q + 6 = 12$

- $\frac{6}{g} = 2$

Example 1: Equations Involving Addition or Subtraction

Solve each equation.

$$y + 12 = 18 \qquad n - 7 = 13$$

Step 1 Identify the operation performed on the variable.

12 is added to y.

7 is subtracted from n.

Step 2 Perform the inverse operation on both sides of the equation.

Subtract 12 from both sides.

Add 7 to both sides.

$$\begin{array}{r} y + 12 = 18 \\ -12 = -12 \\ \hline y \quad = \quad 6 \end{array} \qquad \begin{array}{r} n - 7 = 13 \\ +7 = +7 \\ \hline n = 20 \end{array}$$

Step 3 Substitute the answer back into the original equation and check that the equation is true.

$$\begin{array}{l} y + 12 = 18 \\ 6 + 12 = 18 \\ \quad 18 = 18 \checkmark \end{array} \qquad \begin{array}{l} n - 7 = 13 \\ 20 - 7 = 13 \\ \quad 13 = 13 \checkmark \end{array}$$

Example 2: Equations Involving Multiplication or Division

Solve each equation.

$$4r = 20 \qquad \qquad \frac{x}{2} = 9$$

Step 1 Identify the operation performed on the variable.

r is multiplied by 4.

x is divided by 2.

Step 2 Perform the inverse operation on both sides of the equation.

Divide both sides by 4.

$$4r = 20$$
$$\frac{4r}{4} = \frac{20}{4}$$
$$r = 5$$

Multiply both sides by 2.

$$\frac{x}{2} = -9$$
$$2 \times \frac{x}{2} = 2 \times (-9)$$
$$x = -18$$

Step 3 Substitute the answer back into the original equation and check that the equation is true.

$$4r = 20$$
$$4(5) = 20$$
$$20 = 20 \checkmark$$

$$\frac{x}{2} = -9$$
$$\frac{-18}{2} = -9$$
$$-9 = -9 \checkmark$$

Think about Math

Directions: Write an equation to represent each statement. Use the variable n to represent unknown numbers. Then solve your equation.

1. Ten less than a number is 62.

2. A number increased by 20 is equal to 30.

3. A number is multiplied by 3 and the result is 12.

4. The sum of −7 and a number is 2.

5. A number divided by 4 is equal to −5.

Multi-Step Equations

Equations are useful tools not only in mathematics but also in the sciences. Scientists balance chemical equations following the concept of performing the same actions on both sides.

Two-Step Equations

When solving equations that contain more than one operation, the goal is still to find the value of the variable that makes the equation true. Work backwards using inverse operations to undo each operation until the variable is isolated.

Solving Equations Step by Step

- Simplify expressions on both sides of the equations.
- Use addition and/or subtraction to gather all of the variable terms on the left side of the equation.
- Use inverse operations to undo addition and subtraction on the variable term.
- Use inverse operations to undo multiplication and division on the variable term.

Example 3: Two-Step Equations

Solve $5n - 4 = 6$.

Step 1 Add 4 to both sides of the equation.

$$\begin{array}{r} 5n - 4 = 6 \\ +4 \quad +4 \\ \hline 5n = 10 \end{array}$$

Step 2 Divide both sides of the equation by 5.

$$\frac{5n}{5} = \frac{10}{5}$$
$$n = 2$$

Step 3 Substitute the answer back into the original equation and check that the equation is true.

$$5n - 4 = 6$$
$$5(2) - 4 = 6$$
$$10 - 4 = 6$$
$$6 = 6 \checkmark$$

Using Reciprocals

The inverse operation of multiplying by a fraction is multiplying by the reciprocal of that fraction. A reciprocal is a number that has a product of 1 when multiplied by the original number. The reciprocal of a fraction can be found by interchanging the numerator and the denominator. For example, the reciprocal of $\frac{3}{5}$ is $\frac{5}{3}$ because $\frac{3}{5} \times \frac{5}{3} = \frac{3 \times 5}{5 \times 3} = \frac{15}{15} = 1$.

CALCULATOR SKILL

You can use a calculator to check your solutions to equations. A calculator may be especially useful for equations that require several steps to solve. For example, suppose you solved $3(2d + 7) + 5(d - 8) = 80$ and found the solution $d = 9$. To check whether $d = 9$ is correct, substitute 9 for d on the left side of the equation and evaluate. Press the following sequence of buttons on the TI-30XS MultiView™ calculator,

$$3 \times (2 \times 9 + 7) + 5 \times (9 - 8)$$

then press equals/enter. The display should read 80, which means that $d = 9$ is the correct solution.

Use a calculator to check whether $x = 20$ is the solution of $3(x + 4) - 3(x - 1) = 114$.

Example 4: Multiplication by the Reciprocal

Solve $16 = \frac{1}{2}x + 10$.

Step 1 Subtract 10 from both sides of the equation.

$$16 = \frac{1}{2}x + 10$$
$$\underline{-10 \qquad -10}$$
$$6 = \frac{1}{2}x$$

Step 2 The variable is multiplied by a fraction. To undo this operation, multiply both sides of the equation by the reciprocal of the fraction. The reciprocal of $\frac{1}{2}$ is $\frac{2}{1}$, or 2.

$$2 \times 6 = 2 \times \frac{1}{2}x$$
$$12 = x$$

Step 3 Substitute the answer back into the original equation and check that the equation is true.

$$16 = \frac{1}{2}x + 10$$
$$16 = \frac{1}{2}(12) + 10$$
$$16 = 6 + 10$$
$$16 = 16 \checkmark$$

Simplifying Before Solving

Some equations may require you to simplify one or both sides of the equation before performing inverse operations. For example, you may need to use the Distributive Property.

Example 5: Distributive Property

Solve $5(r - 3) = 21$.

Step 1 Distribute 5 to each term in parentheses.

$$5(r - 3) = 21$$
$$5(r) - 5(3) = 21$$
$$5r - 15 = 21$$

Step 2 Add 15 to both sides of the equation.

$$5r - 15 = 21$$
$$\underline{+15 \quad +15}$$
$$5r = 36$$

Step 3 Divide both sides of the equation by 5.

$$\frac{5r}{5} = \frac{36}{5}$$
$$r = 7.2$$

Step 4 Substitute the answer back into the original equation and check that the equation is true.

$$5(r - 3) + 21$$
$$5(7.2 - 3) = 21$$
$$5(4.2) = 21$$
$$21 = 21 \checkmark$$

Example 6: Combining Like Terms

Solve $2x + 5 - 4x = 15$.

Step 1 Identify and combine like terms on the left side of the equation.

$$2x + 5 - 4x = 15$$
$$-2x + 5 = 15$$

Step 2 Subtract 5 from both sides of the equation.

$$\begin{array}{r} -2x + 5 = 15 \\ -5 \quad -5 \\ \hline -2x = 10 \end{array}$$

Step 3 Divide both sides of the equation by -2.

$$\frac{-2x}{-2} = \frac{10}{-2}$$
$$x = -5$$

Step 4 Substitute the answer back into the original equation and check that the equation is true.

$$2x + 5 - 4x = 15$$
$$2(-5) + 5 - 4(-5) = 15$$
$$-10 + 5 - (-20) = 15$$
$$-5 + 20 = 15$$
$$15 = 15 \checkmark$$

Think about Math

Directions: Solve each equation.

1. $\frac{1}{3}x + 6 = 2$
2. $4(r + 2) = -16$
3. $5y - 2y + 1 = 13$

CORE SKILL

Solve Linear Equations

Some equations, like $5x + 4 = 3x - 2$, have variables on both sides of the equal sign. Just as with all other equations, the goal when solving these equations is to isolate the variable. In order to do this, you will need to first collect all of the variable terms on one side of the equation:

$$\begin{array}{ll} 5x + 4 = 3x - 2 & \text{Subtract } 3x \\ -3x \quad\quad -3x & \text{from both} \\ \hline & \text{sides.} \\ 2x + 4 = \quad -2 & \text{Combine like} \\ & \text{terms on the} \\ & \text{left side:} \\ & 5x - 3x = 2x. \end{array}$$

Once you have collected the variable terms on one side of the equation, continue to solve the equation using inverse operations until the variable is isolated. Check your answer by substituting back into the original equation.

Finish the solution of the equation above. Then solve $10w - 10 = 5w + 15$.

Vocabulary Review

Directions: Write the missing term in the blank.

| equation | variable | solution of an equation |
| reciprocal | expression | inverse operation |

1. A mathematical phrase is a(n) _____ .

2. $\frac{2}{3}$ is the _____ of $\frac{3}{2}$.

3. The _____ of division is multiplication.

4. A(n) _____ contains an equal sign.

5. A letter that is used to represent an unknown number is called a(n) _____ .

6. A _____ that contains one variable is a value for the variable that makes the equation true.

Skill Review

Directions: Read each problem and complete the task.

1. Tell whether each item below is an expression or an equation.
- $-3x + 94$
- $12 = 9 + 3$
- $-2 - (-10)$
- 1.55
- $2x - 3y = 16$
- 8 more than y
- The sum of a and -13 is b.
- $-5n$

Directions: Write an equation to represent each verbal description.

2. Four less than twice a number is equal to 7.

3. Two more than 6 times a number is equal to 6.

Directions: Solve each equation.

4. $x + 3 = 7$

5. $-4x = 32$

6. $\frac{3}{4}x + 3 = 30$

7. $7x + 14 = 35$

8. Martin likes to run. He has been training to run a race next month. He is able to run 5 miles in 35 minutes. Assuming he can run as many miles as he wants at the same pace, write an equation that models the number of minutes, y, it takes Martin to run x miles. How long will it take him to run 8 miles?

9. Louis called a plumber to fix his broken sink. In addition to a $50 fee for the visit, the plumber charges $22 per hour to fix Louis's sink. Write an equation that models this situation and determine how many hours the plumber took if Louis's total bill was $116.

10. If $7x = -28$, what is the value of $x - 8$?
   A. -29
   B. -12
   C. 27
   D. 188

11. If $-3(n + 2) = 6$, what is the value of $12n$?
   A. -48
   B. -16
   C. 0
   D. 9

Skill Practice

Directions: Read each problem and complete the task.

1. The sum of a number and 4 is multiplied by -2 and the result is -6. What is the number?
   A. -7
   B. -1
   C. 1
   D. 2

2. When 10 is added to 3 times a number, the result is 100. Find the number.
   A. 8
   B. 30
   C. 36
   D. 270

3. Solve the equation $(3x - 1) + (-2x - 1) = 2$.

4. Silvia needs $2,100 for a vacation next summer. She plans to save $350 per month. The equation $350m = 2,100$ represents this situation, where m is the number of months Silvia saves. Solve the equation to determine the number of months it will take Silvia to save enough for her vacation.

5. Andrew had a gift card worth $10 to his favorite clothing store. He bought one shirt, and his total cost after using the gift card was $18.05, which included $2.55 in sales tax. The equation $s - 10 + 2.55 = 18.05$ represents this situation, where s is the original cost of the shirt. Solve the equation to find the original cost of the shirt.

6. Jermaine solved the equation $2r - 4 = -7$ as shown below. Identify Jermaine's error. What is the correct solution?

$$2r - 4 = -7$$
$$2r = -3$$
$$r = -6$$

7. In the equation $x - c = 100$, c is a positive number. Is the solution of the equation greater than 100 or less than 100? Explain your reasoning.

LESSON 3.3 Solve Linear Inequalities

■ LESSON OBJECTIVES

- Solve linear inequalities
- Represent solutions of linear inequalities on a number line

■ CORE SKILLS & PRACTICES

- Represent Real-World Problems
- Solve Inequalities

Key Terms

inequality
a mathematical statement showing that two quantities are not equal

inequality signs
symbols used to show the relationship between the expressions in an inequality ($<$, $>$, $\leq$, or $\geq$)

solution of an inequality
the numbers that, when substituted for the variable in an inequality, make the inequality statement true

Vocabulary

equation
a mathematical statement showing that two quantities are equal

inverse operations
operations that reverse the effect of other operations

variable
a symbol used to represent an unknown value

Key Concept

Solving linear inequalities is very similar to solving linear equations, except the solution to a linear inequality will include a range of values, called the solution set. The solution set can be graphed on a number line.

Inequalities

Roller coasters require riders to be a minimum height. Getting a B on the final exam requires a minimum percentage of questions answered correctly. Auditoriums have a maximum capacity. These situations can be described with inequalities because they require numerical values to fall within a certain range.

Inequalities and Signs

An **inequality** is a statement that two expressions are not equal. In an inequality one expression can be compared to another expression as greater, greater than or equal, less than, or less than or equal. **Inequality signs** are symbols used to show the relationship between the expressions in the inequality. You read an inequality from left to right as indicated in the table.

$x < y$	$x \leq y$	$x > y$	$x \geq y$
x is less than y	x is less than or equal to y	x is greater than y	x is greater than or equal to y

Solutions of Inequalities

The **solutions of an inequality,** also called the solution set, are the numbers that, when substituted for the variable in an inequality, make the inequality true. Recall that a **variable** is a symbol used to represent an unknown value.

Checking Solutions

We will review how to find the solution of an inequality in the next sections. For now, consider this inequality and the particular values of the variable. You can test if the values make the inequality true.

Consider the inequality $x + 1 > 2$ and its solution, $x > 1$ (x is greater than 1).

Substituting any number greater than 1 will make the inequality true.

$x = 2$	$x = 1.5$
$2 + 1 > 2$	$1.5 + 1 > 2$
$3 > 2$ ✓	$2.5 > 2$ ✓

Substituting any number that is less than 1 will make the inequality false.

$x = 0$	$x = -1$
$0 + 1 > 2$	$-1 + 1 > 2$
$1 > 2$ ✗	$0 > 2$ ✗

Graphing Solutions

The solution of an inequality that contains one variable can be graphed on a number line. For example, here n is a variable and c is a value on the number line. Notice the circles, or end points, are filled in when the value is part of the solution set.

Graphing Inequalities	
$n < c$	$n > c$
The value of n is less than c. Shaded arrow points left.	The value of n is greater than c. Shaded arrow points right.
$n \leq c$	$n \geq c$
The value of n is less than or equal to c. Shaded arrow points left and the circle is filled in.	The value of n is greater than or equal to c. Shaded arrow points right and the circle is filled in.

Writing Inequalities

It is important to be able to write inequality statements from verbal descriptions. This is easier to do if you recognize that certain phrases indicate inequalities.

$<$	$\leq$	$>$	$\geq$
• less than	• less than or equal	• greater than	• greater than or equal
• fewer than	• no more than	• more than	• no less than
• smaller than	• at most	• larger than	• at least

When writing inequalities, look for phrases that indicate which symbol to use. Also, look for key words that indicate any operation between quantities.

"A number increased by 4 is greater than 10."
a number increased by 4 ⟶ $n + 4$
is greater than 10 ⟶ $n + 4 > 10$

"Two times a number is no more than that number plus 12."
two times a number ⟶ $2q$
is no more than ⟶ $2q \leq$
that number plus 12 ⟶ $2q \leq q + 12$

"Seven less than a number is at least 50."
seven less than a number ⟶ $r - 7$
is at least 50 ⟶ $r - 7 \geq 50$

Solve Linear Inequalities

Represent Real-World Problems

Inequalities can be used to represent many situations. Consider the following scenario. A department store credit card monthly payment must be at least 15% of the account balance. A shopper has charged a total of $600 to her credit card. Use a number line to show payment amounts that are at least 15% of the credit card balance.

To use a number line that represents the solution to a real-world problem, consider the situation and think of how the number line will look. Begin by calculating the payment amount.

• credit card balance is $600
• the payment must be at least 15% of $600
($600)(0.15) = $90

The solution will show a circle filled in at 90 with an arrow going to the right. This means the payment must be at least $90.

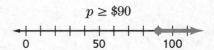
$p \geq \$90$

Now you try. Tasha has a monthly maximum spending budget of 20% of her take-home pay. Last month her take-home pay totaled $2,600. Graph the solution set for her spending budget.

Think about Math

Directions: Answer the following questions.

1. Write an inequality that represents this situation: "The thermostat is always set at a temperature that is below 75 degrees."

2. Is the number 10 in the solution set for $x - 20 \leq -10$?

One-Step Inequalities

Major highways often have safety weigh stations for large trucks on the road. To pass inspection, the weight of the truck plus the weight of the cargo cannot exceed the maximum allowed weight. Because the sum of the weights must not exceed a certain number, then using the truck weight and cargo weights as variables, a linear inequality can be written.

Solve Using Addition and Subtraction

An **equation** is a mathematical statement showing that two quantities are equal. To solve an equation, **inverse operations** are applied to reverse the effect of other operations. When solving inequalities, inverse operations are performed in the same manner as when solving equations. Also, what you do to one side, you will have to do to the other. For example, if $a < b$ and c is any number, then $a + c < b + c$ and $a - c < b - c$. This means that you can add or subtract any number to both sides of the inequality and not change the inequality.

For example, you know that $3 < 7$, so the following are also true.

$$3 + 8 < 7 + 8 \qquad \text{or} \qquad 3 - 8 < 7 - 8$$
$$11 < 15 \qquad\qquad\qquad -5 < -1$$

Example 1: Inequalities Involving Addition or Subtraction

Solve these inequalities using inverse operations.

$$y + 4 \geq 5 \qquad\qquad | \qquad p - 8 < -4$$

Step 1 Identify the operation being used on the variable.

$y + 4 \geq 5$	$p - 8 < -4$
4 is added to y	8 is subtracted from p

Step 2 Perform the inverse operation on both sides of the inequality.

Subtract 4 from both sides	Add 8 to both sides
$y + 4 \geq 5$	$p - 8 < -4$
$\underline{-4 \quad -4}$	$\underline{+8 \quad +8}$
$y \geq 1$	$p < 4$

Check: Test values that are in the solution set. Any number in the solution set will make the inequality true.

Substitute values greater than or equal to 1.	Substitute values less than 4.
$y + 4 \geq 5$	$p - 8 < -4$
$1 + 4 \geq 5$	$3 - 8 < -4$
$5 \geq 5 \checkmark$	$-5 < -4 \checkmark$
$2 + 4 \geq 5$	$0 - 8 < -4$
$6 \geq 5 \checkmark$	$-8 < -4 \checkmark$

Solve Using Multiplication and Division

Inequalities that contain multiplication or division are also solved using inverse operations. Again, what you do to one side, you will have to do to the other. For example, if $a < b$ and $c > 0$, then $ac < bc$ and $\frac{a}{c} < \frac{b}{c}$. This means that multiplication or division by a positive number does not change the inequality.

For example, you know that $2 < 9$ so the following are true.

$$(2)(5) < (9)(5) \qquad \text{or} \qquad \frac{2}{5} < \frac{9}{5}$$
$$10 < 45 \qquad\qquad\qquad\qquad 0.4 < 1.8$$

Example 2: Inequalities Involving Multiplication or Division

Step 1 Identify the operation being performed on the variable.

$9q \le 72$ $\frac{b}{4} > 12$

q is multiplied by 9. b is divided by 4.

Step 2 Perform the inverse operation on both sides of the inequality.

Divide both sides by 9.   Multiply both sides by 4.

$\frac{9q}{9} \le \frac{72}{9}$ $(4)\frac{b}{4} > 12(4)$

$q \le 8$ $b > 48$

Check: Test values that are in the solution set. Any number in the solution set will make the inequality true.

Substitute values less than or equal to 8.	Substitute values greater than 48.
$9q \le 72$	$\frac{b}{4} > 12$
$9(8) \le 72$	$\frac{50}{4} > 12$
$72 \le 72$ ✓	$12.5 > 12$ ✓
$9(0) \le 72$	$\frac{100}{4} > 12$
$0 \le 72$ ✓	$25 > 12$ ✓

Solve Using Multiplication and Division with Negatives

When the inverse operation includes multiplying or dividing by a negative number, the inequality symbol must be reversed for the inequality to be true. For example, if c is negative, the direction of the inequality must be reversed when we multiply or divide both sides of the inequality by c. If $a < b$ and $c < 0$, then $ac > bc$ and $\frac{a}{c} > \frac{b}{c}$. The following illustrates this: $6 < 13$ and $-2 < 0$, so $-12 = -2(6) > -2(13) = -26$. Also, $-3 = \frac{6}{-2} > \frac{13}{-2} = -6.5$.

Example 3: Inequalities Involving Multiplication or Division and Negatives

Step 1 Identify the operation being performed on the variable.

$-10q \le 150$ $\frac{b}{-2} > 7$

q is multiplied by -10. b is divided by -2.

Step 2 Perform the inverse operation on both sides of the inequality and reverse the direction of the inequality.

Divide both sides by -10.   Multiply both sides by -2.

$-10q \le 150$ $\frac{b}{-2} > 7$

$\frac{-10q}{-10} \le \frac{150}{-10}$ $-2\left(\frac{b}{-2}\right) > -2(7)$

$q \ge -15$ $b < -14$

Inequality is reversed!   Inequality is reversed!

Solve Inequalities

A student determined that the number 2 was a solution to the inequality $-2r \le -2$. Was the student correct? In order to answer the question, you must use inverse operations to solve the inequality.

Begin by dividing by -2. Note that since you are dividing by a negative number you will need to reverse the direction of the inequality sign.

$$\frac{-2r}{-2} \le \frac{-2}{-2}$$
$$r \ge 1$$

The answer is $r \ge 1$. The student was correct in saying that 2 was a solution to the inequality because 2 is greater than or equal to 1.

Now you try. Determine if -10 is a solution to the inequality $-10n > 5$.

CALCULATOR SKILL

Many scientific calculators do not have the inequality symbols. Instead, you have to change the inequality into an equation, replacing the inequality sign with an equal sign. You can then use your calculator to find the solution to the equation. Then, use your reasoning to decide which direction the inequality must face based on the operations used to solve the equation.

Use the TI-30XS MultiView™ calculator to solve the inequality $-1.5x + 6.725 \geq 4.25$. First, rewrite as $-1.5x + 6.725 = 4.25$. When you find the solution and need to write it as an inequality, note that you have divided by the negative number -1.5. What is the solution to the inequality?

Check: Test values in the solution set. Any number in the solution set will make the inequality true.

Substitute values greater than or equal to -15.	Substitute values less than -14.
$-10(-15) \leq 150$ $150 \leq 150$ ✓	$\frac{-20}{-2} > 7$ $10 > 7$ ✓
$-10(-10) \leq 150$ $100 \leq 150$ ✓	$\frac{-15}{-2} > 7$ $7.5 > 7$ ✓

Multi-Step Inequalities

Inequalities can model situations that involve personal budgets and money. Using them can help you manage your finances and bring you closer to reaching financial goals, such as buying a car. These types of inequalities might extend beyond the one-step process.

Two-Step Inequalities

When solving inequalities that contain more than one operation, you must isolate the variable on one side of the inequality to find the solution set. Isolate variables using inverse operations.

Example 4: Two-Step Inequality

Solve the inequality, $2x + 4 > 6$.

Step 1 Subtract 4 from both sides of the inequality.

$$\begin{array}{r} 2x + 4 > 6 \\ -4 \quad -4 \end{array}$$

Step 2 Divide both sides of the inequality by 2.

$$\frac{2x}{2} > \frac{2}{2}$$
$$x > 1$$

Check: Test values that are in the solution set. Any number in the solution set will make the inequality true. Since the answer is x is greater than 1, check values greater than 1.

$2x + 4 > 6$ $2(1.1) + 4 > 6$ $2.2 + 4 > 6$ $6.2 > 6$ ✓	$2x + 4 > 6$ $2(4) + 4 > 6$ $8 + 4 > 6$ $12 > 6$ ✓

Inequalities in which the variable has a negative coefficient will require multiplication or division by a negative number in order to solve. This means the direction of the inequality symbol will have to be reversed.

Example 5: Inequalities with a Negative Fractional Coefficient

Solve the inequality, $10 - \frac{1}{2}n \leq 4$.

Step 1 Subtract 10 from both sides of the inequality.

$$\begin{array}{r} 10 - \frac{1}{2}n \leq 4 \\ -10 \quad -10 \end{array}$$

Step 2 Multiply both sides by the reciprocal, -2, and turn the symbol.

$$-\frac{1}{2}n \leq -6$$
$$-2\left(-\frac{1}{2}\right)n \geq (-6)(-2)$$
$$n \geq 12$$

Solve Linear Inequalities

Check: Substitute values that fall in the solutions range. Again, any number in the solution set will make the inequality true. Since the answer is n is greater than or equal to 12, check values greater than or equal to 12.

$$10 - \frac{1}{2}(12) \le 4$$
$$10 - 6 \le 4$$
$$4 \le 4 \checkmark$$

$$10 - \frac{1}{2}(20) \le 4$$
$$10 - 10 \le 4$$
$$0 \le 4 \checkmark$$

Simplify Before Solving

Some inequalities need to be simplified on each side before solving.

Example 6: Distribution and Variables on Both Sides

Solve the inequality $5(g + 1) \le 3(g + 2)$.

Step 1 Distribute the 5 on the left and the 3 on the right.

$$5(g + 1) \le 3(g + 2)$$
$$5g + 5 \le 3g + 6$$

Step 2 Subtract 3g from both sides.

$$\frac{-3g \qquad -3g}{2g + 5 \le 6}$$

Step 3 Subtract 5 from both sides.

$$\frac{-5 \quad -5}{}$$

Step 4 Divide both sides by 2.

$$\frac{2g}{2} \le \frac{1}{2}$$
$$g \le \frac{1}{2}$$

Check: Test values that are in the solution set. Any number in the solution set will make the inequality true. In this case, we will check values less than or equal to $\frac{1}{2}$.

$$5(g + 1) \le 3(g + 2)$$
$$5(0 + 1) \le 3(0 + 2)$$
$$5(1) \le 3(2)$$
$$5 \le 6 \checkmark$$

$$5(g + 1) \le 3(g + 2)$$
$$5(0.5 + 1) \le 3(0.5 + 2)$$
$$5(1.5) \le 3(2.5)$$
$$7.5 \le 7.5 \checkmark$$

Think about Math

Directions: Solve the following inequalities.

1. $2b + 6 < -7$

2. $-\frac{2}{3}q - 3 \le 6$

3. $8(q + 1) > 2(q - 2)$

4. $-3x < 3(5x + 2)$

21ST CENTURY SKILL

Financial Literacy

When buying a home, most people obtain loans called mortgages to help pay for the home. Lenders usually require the homebuyer to have a 20% down payment that goes toward the purchase of the property. Other factors that a lender will consider before giving a loan include credit history, job history, and total debt. Consider the following situation:

A young couple wishes to purchase a home. Both have good credit, a good job history, low debt, and are focused on saving money for a 20% down payment. They are budgeting to buy a home that costs $80,000. They are able to save $400 a month and already have $8,000 saved. How many months will it be until they have enough money saved for the down payment?

To solve, begin by calculating the down payment needed:

$(0.20)(80,000) = 16,000$
20% of $80,000 = $16,000

Use the inequality to answer the question:

m represents months

$$400m + 8,000 \ge 16,000$$

How many months will it take until the couple has enough money for the down payment?

Vocabulary Review

Directions: Write the missing term in the blank.

| inequality | solution of an inequality | inequality sign |
| equation | inverse operations | variable |

1. A mathematical statement showing that two quantities are equal is a(n) _____.

2. A(n) _____ is a symbol used to write an inequality.

3. _____ are operations that reverse the effect of other operations.

4. A(n) _____ is a number that, when substituted for the variable in an inequality, makes the inequality statement true.

5. A letter that represents an unknown quantity is a(n) _____.

6. A(n) _____ is a mathematical statement showing that two quantities are not equal.

Skill Review

Directions: Read each problem and complete the task.

1. Translate the following: "A number increased by 4 is less than 6."

2. Translate the following: "Three times a number is at least that number plus 1."

3. Solve. Then graph the solution: $-x + 3 < 9$

4. Solve. Then graph the solution: $2(x - 1) \geq 6x$

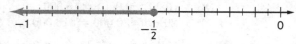

5. Solve. Then graph the solution: $-3q + 1 < -2(q - 2)$

6. Is 0 a solution of $-x \geq 0$?

7. Is 10 a solution of $-2(x - 4) \leq -10$?

8. Is -2 a solution of $\frac{1}{2}x + 6 \geq 4$

Solve Linear Inequalities

Skill Practice

Directions: Read each problem and complete the task.

1. A major credit card company requires monthly payments equal to or greater than 8% of the total balance. A consumer has a credit card balance of $320. Write an inequality that represents acceptable payment amounts.

2. Solve the inequality:
$-3(n + 1) - 2(n + 4) > 6n$

3. Is 0.4 a solution to the inequality?
$-\frac{1}{2}(x + 4) + 2 \leq 4x - 2$

4. Martin's summer allowance of $400, and his spending per week of $25 can be represented by the expression, $400 - 25w$. Martin needs to have at least $175 at the end of the summer. Using the expression, write an inequality that can be solved to determine how many weeks Martin can spend money and still have $175 at the end of the summer. Then solve the inequality.

5. Create a situation involving saving money that can be represented by the inequality:
$2,000 + 30x \geq 9,000$.

6. Michael solved the inequality $5(r + 2) < -10r$ as shown below. Identify Michael's error. What is the correct solution?

$$5(r + 2) < -10r$$
$$5r + 10 < -10r$$
$$15r < -10$$
$$r > -\frac{2}{3}$$

7. To pass a state nursing exam, students must answer at least 70% of all the questions correctly. This year, the exam has 150 questions. Write an inequality that can be solved to determine all the possible numbers of incorrect answers a student can get and still pass the this year's exam. Then solve the inequality and state the answer in a complete sentence.

8. Write the correct symbols to complete each property of inequalities.

   a. If $a < b$ and c is a positive number, then ac ____ bc and $\frac{a}{c}$ ____ $\frac{b}{c}$.

   b. If $a < b$ and c is a negative number, then ac ____ bc and $\frac{a}{c}$ ____ $\frac{b}{c}$.

9. Solve the inequality $5x - 4 > -16x + 3$. Then describe how the graph of the solution should look on a number line.

10. Emily is conducting an experiment. She starts with a solution that has a temperature of 44°F. She lowers the temperature by 6°F each hour. The temperature of the solution cannot go below 20°F. Write an inequality that can be solved to determine the maximum number of hours Emily can lower the temperature. Then solve the inequality and state the answer in a complete sentence.

11. Why do you need to reverse the direction of an inequality symbol when you multiply or divide both sides by a negative number?

12. Jane solved the inequality $bx < 5b$ as shown below. Explain the error that Jane made.

$$bx < 5b$$
$$\frac{bx}{3b} < \frac{5b}{b}$$
$$x < 5$$

LESSON OBJECTIVES

- Write algebraic expressions to represent real-world situations
- Solve real-world problems involving linear equations
- Write linear equations to represent real-world problems
- Solve real-world problems using inequalities

CORE SKILLS & PRACTICES

- Evaluate Expressions
- Solve Real-World Problems

Key Terms

algebraic expression
an expression that contains at least one variable

Vocabulary

equation
a mathematical statement showing that two quantities are equal

inverse operations
operations that undo each other

inequality
a mathematical statement showing that two quantities are not equal

Key Concept

Real-world problems can be translated into algebraic expressions, equations, and inequalities. Mathematical methods can then be used to find real-world solutions.

Expressions and Equations

Expressions and equations are used to model real-world problems all the time. For example, when painting a room, it is important to know how much paint to buy. If you purchase too much, you will waste money. If you don't purchase enough, you will waste time returning to the store for more. You can use expressions and equations to calculate the amount of paint you need.

Real-World Expressions

A real-world situation can be translated into an **algebraic expression**, an expression that contains at least one variable.

Example 1: Rental Car Charges

A car rental agency charges $29.99 plus $0.39 per mile to rent a compact car. A customer rented a compact car and drove 220 miles. What will be the total charge?

Step 1 Identify the variable quantity and assign a variable.

The number of miles driven will vary.

Let m = number of miles.

Step 2 Use the variable and the information given in the problem to write an expression for the total charge.

$29.99 + 0.39m$

$29.99 plus $0.39 per mile

$$29.99 \quad + \quad 0.39 \quad \times \quad m$$

Step 3 To find the total charge for 220 miles, substitute 220 for m and simplify.

$29.99 + 0.39(220)$
$= 29.99 + 85.8$
$= 115.79$

The total charge for 220 miles is $115.79.

Example 2: Weekly Savings

Susan has saved $40 to purchase a new mountain bike. She plans to save an additional $15 each week. How much money will Susan have saved after 6 weeks?

Step 1 Identify the variable quantity and assign a variable.

Let w = number of weeks.

The number of weeks will vary.

Step 2 Use the variable and the information given in the problem to write an expression that represents the total amount saved.

$40 + 15w$

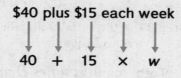

$$40 + 15 \times w$$

Step 3 To find the amount saved after 6 weeks, substitute 6 for w and simplify.

$$40 + 15(6)$$
$$= 40 + 90$$
$$= 130$$

After 6 weeks, Susan will have saved $130.

Real-World Equations

An equation is a mathematical statement that two expressions are equal to each other. For example, $2 + 3 = 4 + 1$ and $3x + 3 = 12$ are equations. Many real-world problems can be represented and solved using equations.

Example 3: Buying a Truck

Roger wants a new truck that costs $15,999 plus an additional $1,600 for taxes, title, and registration fees. Roger has saved $6,000 and plans to borrow the rest of the money he needs. How much money does Roger need to borrow?

Step 1 Identify the unknown quantity and assign a variable. The amount that Roger must borrow is unknown.

Let b = amount Roger must borrow.

Step 2 Use the variable and the information given in the problem to write an equation.

Amount saved plus amount borrowed equals cost of truck plus fees

$$6,000 \quad + \quad b \quad = \quad 15,999 \quad + 1,600$$

Step 3 Use inverse operations to solve the equation. Inverse operations are operations that undo each other. In this case, subtract 6,000 from both sides of the equation.

Subtract 6,000 from both sides:
$$\begin{array}{r} 6,000 + b = 17,599 \\ -6,000 \quad\quad -6,000 \\ \hline b = 11,599 \end{array}$$

Roger must borrow $11,599.

CORE SKILL

Evaluate Expressions

Gilbert works at a restaurant and gets paid $50 per week plus $9 per hour. The expression $50w + 9h$ represents Gilbert's pay for working h hours in w weeks. This week he is scheduled to work 35 hours. Evaluate the expression to determine Gilbert's pay this week.

When you evaluate an expression, you are finding a value. Substitute given numbers for the variables in the expression and then use the order of operations to simplify.

Expression:	$50w + 9h$
Substitute 1 for w and 35 for h:	$50(1) + 9(35)$
Multiplication first:	$50 + 315$
Addition:	365

Gilbert's pay this week will be $365.

Now you try. Jackie works at a convenience store. She earns $11 per hour, but each week $5 is deducted from her paycheck for uniform cleaning services. Write an expression that represents Jackie's pay for working h hours in w weeks. How much will Jackie earn if she works 40 hours during one week? How much will she earn if she works 40 hours during 2 weeks?

Financial, Economic, Business, and Entrepreneurial Literacy

When you borrow money from a bank, you must pay back the amount borrowed plus an additional percentage. When you invest money, you will be paid back the amount invested plus an additional percentage. In both cases, the additional percentage is called interest. There are many methods of computing interest, and different methods result in different amounts of interest paid or earned. Recall that one way to compute interest is to use the simple interest formula.

$$I = Prt$$

I = amount of interest earned or charged

P = initial amount borrowed or invested (called the principal)

r = annual interest rate (as a decimal)

t = time that money is borrowed or invested (in years)

Keely paid 6.5% annual simple interest on a loan for 5 years. She paid a total of $308.75 in interest. Using the simple interest formula, write and solve an equation to find the amount of money that Keely borrowed.

Example 4: Prepaid Cell Phone

Cathy has a prepaid cell phone. Last month she deposited $55.00 into her cell phone account and was able to talk for 700 minutes before running out of credit. What is the charge per minute on Cathy's cell phone? Round to the nearest cent and assume no other fees apply.

Step 1 Identify the unknown quantity and assign a variable. The charge per minute is unknown.

$$\text{Let } m = \text{charge per minute.}$$

Step 2 Use the variable and the information given in the problem to write an equation.

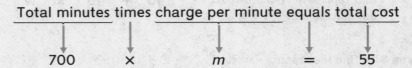

Total minutes times charge per minute equals total cost

| 700 | × | m | = | 55 |

Step 3 Use inverse operations to solve the equation. Round to the nearest cent.

$$700m = 55$$

Divide both sides by 700: $\dfrac{700m}{700} = \dfrac{55}{700}$

$$m \approx 0.08$$

The charge per minute is about $0.08.

Equations with Multiple Operations

For some real-world problems, you will need to write and solve an equation with more than one operation.

Example 5: Gym Membership

One year ago, Gloria joined a gym. She paid a $50 enrollment fee as well as a monthly membership fee. Her total gym expenses for the year were $410. What was Gloria's monthly membership fee?

Step 1 Identify the unknown quantity and assign a variable. The monthly membership fee is unknown.

$$\text{Let } m = \text{monthly membership fee.}$$

Step 2 Use the variable and the information given in the problem to write an equation.

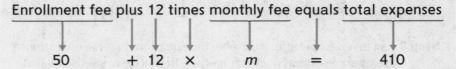

Enrollment fee plus 12 times monthly fee equals total expenses

| 50 | + 12 | × | m | = | 410 |

Step 3 Use inverse operations to solve the equation.

$$50 + 12m = 410$$

Subtract 50 from both sides:

$$\begin{array}{r} 50 + 12m = 410 \\ -50 \qquad\quad -50 \\ \hline 12m = 360 \end{array}$$

Divide both sides by 12: $\dfrac{12m}{12} = \dfrac{360}{12}$

$$m = 30$$

The monthly fee was $30.

Think about Math

Directions: Solve the following problems.

1. Roselda runs x miles three times per week. Which expression represents the number of miles that Roselda runs in 4 weeks?

   A. $3x + 4$
   B. $4x + 3$
   C. $4(3x)$
   D. $4(3) + x$

2. At a local bakery, a cupcake costs $1.25, and a cake costs $14.50. Lori bought two cakes and some cupcakes for a party. She paid $54.00 in total. How many cupcakes did Lori buy?

   A. 20
   B. 30
   C. 45
   D. 65

Inequalities

A credit card limit is the maximum amount that can be charged to a credit card. Once the limit is reached, your credit card will be denied for purchases. Inequalities can be used to determine how much you can charge to your card without exceeding the limit.

Real-World Inequalities

An **inequality** shows the relationship between two expressions that are not equal. In an inequality, one expression will be greater than, less than, greater than or equal to, or less than or equal to the other expression. The inequality symbols $>$, $<$, $\geq$, and $\leq$ are used to represent these relationships.

Inequalities

$<$	$\leq$	$>$	$\geq$
• less than • fewer than • smaller than • below	• less than or equal to • no more than • at most • maximum	• greater than • more than • larger than • above	• greater than or equal to • no less than • at least • minimum

Example 6: Real-World Inequalities

Write an inequality to represent each situation.

a. Drivers may drive at a speed s no greater than the posted speed limit of 70 miles per hour.

$$s \leq 70$$

   Speed   Not greater than

b. To enlist in the United States military, a person's age a must be at least 17 years (with parental approval).

$$a \geq 17$$

   Age   At least

Solve Real-World Problems

To solve a real-world problem, translate the given information into numbers and mathematical symbols. When writing and solving an inequality, compare one quantity to another using inequality signs and assign variables for unknown values.

On most major highways in the United States, the weight limit for an 18-wheeler truck is 80,000 pounds. This includes the weight of the truck, the trailer, and the cargo. Trucks must stop at weigh stations located along the highway and, if a truck exceeds the weight limit, the driver can be fined and prevented from continuing his or her trip.

An 18-wheeler truck weighs 17,000 pounds and is pulling a trailer that weighs 11,000 pounds. Write and solve an inequality to find the allowable cargo weights c that this truck and trailer can carry.

Example 7: Free Delivery

A local furniture store offers free delivery of items if the total purchase amount is at least $500 before taxes. Kara went to the store and found a couch for $399. Kara is also looking for an end table to place next to the couch. How much does the end table need to cost in order for Kara to receive free delivery?

Step 1 Identify the unknown quantity and assign a variable. The cost of the end table is unknown.

$$\text{Let } t = \text{cost of table}$$

Step 2 Use the variable and the information given in the problem to write an inequality.

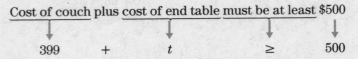

Cost of couch	plus	cost of end table	must be at least	$500
399	+	t	$\geq$	500

Step 3 Use inverse operations to solve the inequality.

$$399 + t \geq 500$$

Subtract 399 from both sides:
$$\underline{-399 \qquad -399}$$
$$t \geq 101$$

The cost of the end table must be at least $101.

Inequalities with Multiple Operations

For some real-world problems, you will need to write and solve an inequality with more than one operation.

Example 8: Plumbing Repairs

The Wongs have budgeted $570 for some plumbing repairs. A plumber charges a $75 service fee plus $45 per hour. For how many hours can the Wongs afford to hire the plumber and stay within their budget?

Step 1 Identify the unknown quantity and assign a variable. The number of hours the Wongs can afford is unknown.

$$\text{Let } h = \text{the number of hours.}$$

Step 2 Use the variable and the information given in the problem to write an inequality.

Service fee	plus	$45	times	number of hours	cannot be more than	$570
75	+	45	$\times$	h	$\leq$	570

Step 3 Use inverse operations to solve the inequality.

$$75 + 45h \leq 570$$

Subtract 75 from both sides:
$$\underline{-75 \qquad\qquad -75}$$
$$45h \leq 495$$

Divide both sides by 45:
$$\frac{45h}{45} \leq \frac{495}{45}$$
$$h \leq 11$$

The Wongs can afford to hire the plumber for no more than 11 hours.

Inequalities with Negative Numbers

Remember that when you multiply or divide both sides of an inequality by a negative number, you must reverse the inequality symbol.

Example 9: Spending Money

John is at band camp for 8 weeks. His parents have given him $600 for spending money. John wants to have at least $320 left at the end of camp so he can buy a trumpet. How much money can John spend each week and still be able to buy the trumpet at the end of camp?

Step 1 Identify the unknown quantity and assign a variable. The amount of money that John can spend each week is unknown.

Let w = amount of money John can spend per week.

Step 2 Use the variable and the information given in the problem to write an inequality.

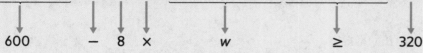

Starting amount minus 8 times amount per week must be at least $320

$$600 \quad - \quad 8 \quad \times \quad w \quad \geq \quad 320$$

$$600 - 8w \geq 320$$

Step 3 Use inverse operations to solve the inequality.

$$600 - 8w \geq 320$$

Subtract 600 from both sides:
$$\underline{-600 \qquad\qquad -600}$$
$$-8w \geq -280$$

Divide both sides by -8.
Reverse the inequality symbol:
$$\frac{-8w}{-8} \geq \frac{-280}{-8}$$
$$w \leq 35$$

John can spend no more than $35 each week.

Think about Math

Directions: Solve the following problems.

1. R.J. has $40 to spend at a carnival. The admission price is $7, and he plans to spend $15 on souvenirs. He also plans to spend $10 to eat dinner at the carnival. Each carnival ride cost $3. Which inequality can be used to determine the number of rides r that R.J. will be able to enjoy?

    A. $7 + 15 + 10 + 3 \geq 40r$
    B. $7 + 15 + 10 + 3r \geq 40$
    C. $7 + 15 + 10 + 3r \leq 40$
    D. $7 + 15 + 10 + 3 < 40r$

2. A cleaning service company charges $15 per hour plus a traveling fee of $20 per month. Gina has $100 per month budgeted for cleaning services. The company rounds the number of hours up to the next whole hour when calculating charges. What is the greatest number of hours that Gina can afford to hire the cleaning company each month?

    A. 4
    B. 5
    C. 6
    D. 8

Use Expressions, Equations, and Inequalities to Solve Real-World Problems

Vocabulary Review

Directions: Write the missing term in the blank.

algebraic expression   **inverse operations**
equation   **inequality**

1. To solve an equation, you must isolate the variable using _____.

2. An example of an _____ is $5p - 1$.

3. If a mathematical statement shows that two expressions are equal to each other, the

statement is an _____ .

4. If a mathematical statement shows one expression to be greater than another, the

statement is an _____ .

Skill Review

Directions: Read each problem and complete the task.

1. A book club charges $10 per book plus a $5 shipping and handling fee per order.
   a.   Write an expression that represents the total cost of b books.
   b.   What is the total cost of 8 books?
   c.   Sally placed an order and paid $35. How many books did she order?

2. In order to save money, Yvette clips coupons each week. She finds a 20% off coupon
for her favorite detergent. Let d represent the detergent's usual price. Which expression
represents the amount Yvette will pay if she uses the coupon?
   A. $0.20d$
   B. $0.20d - d$
   C. $d - 0.20d$
   D. $d + 0.20d$

3. Shane has $125 in his savings account and is saving $25 each week from his paycheck.
How long will it take Shane to save $400?

Directions: For Questions 4–6, write an inequality to represent each situation.

4. The minimum height *h* for a carnival ride is 3 feet.

5. The maximum amount *s* that may be charged on a credit card is $2,000.

6. Andrea, a real estate agent, wants to sell *h* houses this year. She would like to sell more houses this year than she sold last year. Last year, Andrea sold 16 houses.

7. Abdul borrowed $10,000 at a simple interest rate of 5% per year and paid a total of $2,500 in interest. Use the simple interest formula $I = prt$ to determine how long Abdul borrowed the money.

Skill Practice

Directions: Read each problem and complete the task.

1. Gladys is paid $4 per hour plus tips at a local restaurant. On average, Gladys earns 15% of her total sales in tips. Last Friday Gladys had total sales of $1,200 during a 6-hour work shift. How much money did Gladys earn last Friday?

    A. $204
    B. $260
    C. $720
    D. $1,224

2. Lacey begins each week with an $84 allowance for meals. So far this week, she has spent $12 per day and she has $48 left. How many days have passed since the beginning of this week?

    A. 3
    B. 4
    C. 5
    D. 7

3. Kaley has been away at college for two months and will remain there until the end of the semester. Her parents gave her $1,200 in spending money for the entire 5-month semester, but Kaley wants to save at least $100 for her summer break. She has already spent $250 each month that she has been at college. Write and solve an inequality to determine how much money Kaley can spend per month for the rest of the semester and still have at least $100 remaining at the end.

4. Cassie received a paycheck for $620 for the time that she worked last week. This amount equals her total earnings *e* minus 20% in deductions for taxes and benefits. Which equation represents this situation?

    A. $0.20e = 620$
    B. $e - 0.20e = 620$
    C. $(0.20)(620) = e$
    D. $e + 0.20e = 620$

5. Cal invested $10,000 in an account earning simple interest. Three years later, the total amount in his account was $11,200. Cal wrote and solved the equation $11,200 = 10,000 \times 3 \times r$ and said that the interest rate was $37\frac{1}{3}\%$. Describe Cal's error. What is the correct interest rate as a percent?

6. Emilio has a board that is 52 inches long. He needs to cut the board into two pieces so that one piece is 8 inches longer than the other. Write and solve an equation to determine the length of the shorter board.

Directions: Choose the best answer to each question.

1. Allen uses an expression to represent how much a bag of apples cost with a coupon. The expression $5x - 8$ can be translated using a phrase such as

   _____.

2. Abby uses an equation to represent how much she owed after a discount using this equation: $\frac{x}{10} = -12$. Which is the value of x?

   A. 120
   B. 22
   C. −2
   D. −120

3. Olivia saves $15 each week and has $20 in an account. She is saving for a bicycle that costs $110. How many weeks will it take for her to have enough money for the bicycle?

   A. 145
   B. 75
   C. 8
   D. 6

4. To participate in a music class, a student needs to be older than 8 years old. This graph represents the inequality _____.

5. This expression represents how many toys a company will make in a number of hours: $25x - 6$. So, _____ toys will be made in 8 hours.

6. A custom jewelry manufacturer uses this inequality to help them decide how many new orders to make each month: $45 - \frac{3}{4}x \leq 18$. Which represents the values for x?

   A. $x \leq 36$
   B. $x \leq 84$
   C. $x > 84$
   D. $x \geq 36$

7. A photographer charges $75 for a family portrait session. It cost $300 for her camera. Which expression represents the amount of money she will make?

   A. $300 + 75x$
   B. $75x - 300$
   C. $(75 + 300)x$
   D. $300x + 75$

8. Ryan uses the expression $15x + 50$ to represent how much he will earn in a month. Zoe uses the expression $20(x - 2)$ to represent how much she will earn in a month. In both expressions x represents the number of hours worked. How much more does Zoe earn in a month than Ryan?

   A. $5x - 90$
   B. $5x - 10$
   C. $35x + 10$
   D. $35x - 90$

9. The equation $\frac{85}{x} = 17$ can be translated using a phrase such as _____.

10. Conner uses the equation $45 = \frac{3}{4}x + 15$ to represent how many songs he needs to sell to make $45. What is the value of x?

A. 80

B. 60

C. 45

D. 40

11. A cab ride costs $0.20 per mile and has a $3.00 fee. Conner has $15.00 to spend on the cab ride. Conner uses the equation _____ to find how many miles he can ride and stay within his budget.

12. Use the inequality $8(4g + 2) > 24g$. Which represents the values for g?

A. $g < -2$

B. $g > -2$

C. $g \le 2$

D. $g \ge 2$

13. Brooklyn uses the expression $30x - 100$ to represent how much money she'll make in a week. Dylan uses the expression $18x + 50$ to represent how much money he'll make in a week. For both expressions, x represents the number of hours worked. How much do Brooklyn and Dylan earn in a week?

A. $12x - 150$

B. $48x - 50$

C. $12x + 150$

D. $48x + 50$

14. Use the equation $6(x - 8) = -24$. What is the value of x?

A. -4

B. -12

C. 4

D. 22

15. Use the inequality $15 + x \ge 32$. Which represents the values for x?

A. $x \ge 17$

B. $x \le 17$

C. $x \le 47$

D. $x \ge 47$

16. To pass a test, Amber needs to have a score of 70 points or better. Each question is worth 10 points. She uses the inequality _____ to find the possible number of questions she can get right to pass the test. She found that she needs to answer 7 or more questions correctly.

Check Your Understanding

On the following chart, circle the items you missed. The last column shows pages you can review to study the content covered in the question. Review those lessons in which you missed half or more of the questions.

Lesson	Item Number(s)			Review Page(s)
	Procedural	Conceptual	Problem Solving	
3.1 Evaluate Linear Expressions	5	1, 7		82–89
3.2 Solve Linear Equations	2, 14	9	3	90–97
3.3 Solve Linear Inequalities	6, 12	4	16	98–105
3.4 Use Expressions, Equations, and Inequalities to Solve Real-World Problems	10	11	8, 13	106–113

Chapter 4

Polynomials and Rational Expressions

In baseball, pitchers need to know exactly where they are throwing the ball to strike out the opposing team. This skill develops over time with practice and an understanding of their own particular pitch. Mathematically, you can model the movement of dropped or thrown objects using polynomials. While a linear expression can be used to model many simple situations, a polynomial is critical for modeling more complex situations. You are already familiar with variables and linear expressions, but when they are combined by multiplication and division you create polynomials and quadratic and rational expressions.

Getty Images

Lesson 4.1
Evaluate Polynomials

How do you calculate and work with numbers raised to exponents? How can you apply what you observed with real numbers to variables raised to exponents? Learn how to identify and classify different types of polynomials.

Lesson 4.2
Factor Polynomials

How do you approach a difficult problem at work? Often you might break the problem down into smaller chunks that are simpler to solve. In math, understanding how to factor polynomials will help you solve more complex problems. Learn tricks and methods for factoring polynomials.

Lesson 4.3
Solve Quadratic Equations

Linear equations are solved by applying inverse operations to get the variable alone. In a quadratic equation, new solving methods are needed because the variable is squared. Learn how to solve quadratic equations by using factoring and formulas.

Lesson 4.4
Evaluate Rational Expressions

When you learned about fractions you were told never to have 0 in the denominator. What happens when a variable expression is in the denominator? How can you avoid having that expression be 0? Learn how to simplify and restrict the values for rational expressions.

Goal Setting

Think about being presented with math problems to solve. When did you have the easiest time? What made you feel confident about solving the problem? Did you know a shortcut or process to solve the problem? What are some examples of math tools, tips, or shortcuts that you have learned in the past? How did you use them to solve problems?

How might breaking down a quadratic equation help you solve it? What previous knowledge might you need to solve quadratic equations?

LESSON 4.1 Evaluate Polynomials

LESSON OBJECTIVES

- Identify different polynomials
- Evaluate polynomials
- Add, subtract, multiply, and divide polynomials

CORE SKILLS & PRACTICES

- Use Math Tools Appropriately
- Evaluate Expressions

Key Terms

polynomial
an algebraic expression consisting of one or more terms in which each term is a number or a product of numbers and variables with whole-number exponents

degree
the value of the greatest exponent in a polynomial

standard form
the form of a polynomial that shows the terms listed from left to right with the powers of the variables from greatest to least

Vocabulary

opposite polynomial
the polynomial with all of its signs changed to their opposites

substitute
to replace a variable in an expression with a numerical value

Key Concept

Polynomials are special types of variable expressions with one or more terms. Each term has a variable raised to a whole number exponent or is a constant.

Identifying Polynomials

Small business owners can use linear expressions to model some simple costs and everyday situations. However, in order to model their profit and other more complex situations, they need to use squares, cubes, and other higher-order variables. These situations can be modeled with polynomials.

Types of Polynomials

Polynomials are algebraic expressions with one or more terms. Each term is a number or a product of numbers and variables with whole-number exponents. The **degree** of a polynomial is the value of the greatest exponent in the polynomial. A linear expression is a polynomial of degree 1. A term that contains only a number is a constant term and has a degree of 0.

$4x^2$	$2z - 6$	$3c^4 + 2c^2 - 1$	$2a^3 + 7a + a$
degree 2	degree 1	degree 4	degree 3

Three types of polynomials are named by the number of terms in the polynomial expression. The names of these polynomials are chosen because of their prefixes. The prefix *poly-* means many.

A *monomial* has one term.	$3x$
A *binomial* has two terms.	$4x + 2$
A *trinomial* has three terms	$2x^2 + x - 5$

A polynomial is written in **standard form** when the terms are listed from left to right with the powers of the variables from greatest to least.

- The polynomial $4x + 5 + 3x^2$ is not written in standard form.
- The polynomial $3x^2 + 4x + 5$ is written in standard form.

Simplifying Polynomials

In addition to writing polynomials in standard form, simplifying polynomials can also make them easier to work with. Polynomials are simplified when all like terms have been combined. Like terms have the same variable raised to the same power. They can be combined by combining their coefficients.

Example 1: Combining Like Terms

Simplify and write in standard form: $11c^2 - 3c + 4c^2 - 2c^3 + c$

Step 1 Use the commutative and associative properties to group like terms.

$$11c^2 - 3c + 4c^2 - 2c^3 + c = 11c^2 + (-3c) + 4c^2 + (-2c^3) + c$$
$$= (11c^2 + 4c^2) + (-3c + c) + (-2c^3)$$

Step 2 Combine like terms by adding the coefficients.

$$(11c^2 + 4c^2) + (-3c + c) + (-2c^3) = (11 + 4)c^2 + (-3 + 1)c + (-2c^3)$$
$$= 15c^2 + (-2c) + (-2c^3)$$

Step 3 Write the polynomial in standard form.

$$15c^2 + (-2c) + (-2c^3) = -2c^3 + 15c^2 - 2c$$

Think about Math

Directions: Choose the best answer to each question.

1. Which of the following binomials is not written in standard form?

   A. $8m^3 + 4$
   B. $7 - n$
   C. $x^2 + x$
   D. $3y^4 - 2y$

2. What is the degree of the polynomial $4x - 2x^2 + 1 + 5x^3$?

   A. 0
   B. 1
   C. 2
   D. 3

Use Math Tools Appropriately

"Standard forms" exist for a reason. It is usually easiest to work with numbers and other mathematical objects when they are expressed in standard forms. Polynomials are no exception; they should be simplified and written in standard form when you need to add, subtract, or multiply them. Remember: *Simplify* means to combine like terms, or terms that share the same variable raised to the same power. *Standard form* refers to the terms written with decreasing exponents from left to right.

Simplify the polynomial by combining like terms and express the result in standard form.

$5 + 3m - 2m^3 + 6m - 5m^2 - 3m^3 + 7m$

Being able to evaluate expressions is a useful skill in solving real-world problems. For example, suppose a baseball player hits a ball with an initial speed of 100 feet per second. The height of the ball t seconds after the ball is hit can be represented by the polynomial $-32t^2 + 100t + 6$. To find the height of the ball 2 seconds after it was hit, you can substitute $t = 2$ into the expression and evaluate.

$$-32t^2 + 100t + 6$$
$$= -32(2)^2 + 100(2) + 6$$
$$= -32(4) + 100(2) + 6$$
$$= -128 + 200 + 6$$
$$= 78$$

What is the height of the ball 3 seconds after it has been hit?

Evaluating Polynomials

Polynomials can be used to represent different real-world phenomena, such as the height of a falling object at a specific time. To find the height of an object at a specific time, you simply need to evaluate the polynomial for the specific value of the variable.

Substitution for Variables

To evaluate a polynomial for a given value of the variable, you **substitute** the value into each place the variable appears in the polynomial. Then, evaluate the expression according to the order of operations.

Example 2: Evaluate a Polynomial

Find the value of $x^2 - 3x + 1$ when $x = -2$.

Step 1 Substitute the value -2 in the polynomial for each x.
$$x^2 - 3x + 1 = (-2)^2 - 3(-2) + 1$$

Step 2 Simplify using the order of operations.

$(-2)^2 - 3(-2) + 1 = 4 - 3(-2) + 1$	Exponents
$= 4 + 6 + 1$	Multiplication
$= 11$	Addition

Operations with Polynomials

You can calculate the distance an object has traveled by multiplying its speed by the time it has traveled. If the speed of an object and the time it travels are each given by polynomials, then we can determine its distance by multiplying the two polynomials.

Adding Polynomials

Adding polynomials is similar to simplifying polynomials. When adding polynomials, use properties to group and simplify like terms.

Example 3: Add Polynomials

Find the sum of the polynomials: $(4x^2 - 3x + 1) + (5 + 2x^2)$

Step 1 Use the commutative and associative properties to group like terms.

$$(4x^2 - 3x + 1) + (5 + 2x^2) = 4x^2 - 3x + 1 + 2x^2 + 5$$
$$= 4x^2 + (-3x) + 1 + 2x^2 + 5$$
$$= (4x^2 + 2x^2) + (-3x) + (1 + 5)$$

Step 2 Combine like terms and write the sum in standard form.

$$(4x^2 + 2x^2) + (-3x) + (1 + 5) = (4 + 2)x^2 + (-3x) + (1 + 5)$$
$$= 6x^2 - 3x + 6$$

You could also show the addition vertically, using coefficients of 0 for any missing terms in order to keep the like terms aligned.

$$\begin{array}{r} 4x^2 - 3x + 1 \\ + \; 2x^2 + 0x + 5 \\ \hline 6x^2 - 3x + 6 \end{array}$$

The Opposite of a Polynomial

To subtract one polynomial from another, you need to add the **opposite polynomial**. The opposite polynomial simply reverses the sign of each term of the polynomial.

- **Polynomial:** $6t^3 + 3t^2 - 4t - 2$
- **Opposite Polynomial:** $-6t^3 - 3t^2 + 4t + 2$

Example 4: Subtract Polynomials

Find the difference of the polynomials: $(3x^2 + 6x - 5) - (2x^2 - 4x + 2)$

Step 1 Rewrite the subtraction as addition of the opposite polynomial.

$$(3x^2 + 6x - 5) - (2x^2 - 4x + 2) = (3x^2 + 6x - 5) + (-2x^2 + 4x - 2)$$

Step 2 Use the commutative and associative properties to group like terms.

$$(3x^2 + 6x - 5) - (2x^2 - 4x + 2) = 3x^2 + 6x + (-5) + (-2x^2) + 4x + (-2)$$
$$= (3x^2 + (-2x^2)) + (6x + 4x) + (-5 + (-2))$$

Step 3 Combine like terms and write in standard form.

$$(3x^2 + (-2x^2)) + (6x + 4x) + (-5 + (-2))$$
$$= (3 + (-2))x^2 + (6 + 4)x + (-5 + (-2))$$
$$= 1x^2 + 10x + (-7)$$
$$= x^2 + 10x - 7$$

Multiplying Polynomials

Multiplying two polynomials is similar to multiplying numerical expressions. To multiply polynomials, use the Distributive Property to multiply each pair of terms from the polynomials.

Example 5: Multiply Polynomials

Find the product of the polynomials: $(5x - 7)(2x^2 + 6x - 3)$

Step 1 Use the Distributive Property to multiply each term of the first polynomial by the second polynomial.

$$(5x - 7)(2x^2 + 6x - 3) = 5x(2x^2 + 6x - 3) - 7(2x^2 + 6x - 3)$$

Step 2 Use the Distributive Property again to multiply each monomial by the second polynomial. Use caution when distributing any negative terms.

$$5x(2x^2 + 6x - 3) - 7(2x^2 + 6x - 3)$$
$$= 5x(2x^2) + 5x(6x) + 5x(-3) - 7(2x^2) - 7(6x) - 7(-3)$$
$$= 10x^3 + 30x^2 + (-15x) + (-14x^2) + (-42x) + 21$$

Step 3 Use the Commutative and Associative properties to combine like terms.

$$10x^3 + 30x^2 + (-15x) + (-14x^2) + (-42x) + 21$$
$$= 10x^3 + 30x^2 + (-14x^2) + (-15x) + (-42x) + 21$$

Step 4 Combine like terms and write in standard form.

$$10x^3 + 30x^2 + (-14x^2) + (-15x) + (-42x) + 21$$
$$= 10x^3 + (30 - 14)x^2 + (-15 - 42)x + 21$$
$$= 10x^3 + 16x^2 + (-57x) + 21$$
$$= 10x^3 + 16x^2 - 57x + 21$$

Vocabulary Review

Directions: Match each term to its definition.

1. _____ degree

2. _____ opposite polynomial

3. _____ polynomial

4. _____ standard form

5. _____ substitute

a. to replace a variable in an expression with a numerical value

b. the value of the greatest exponent in a polynomial

c. the form of a polynomial that shows the terms listed from left to right with the powers of the variables from greatest to least

d. the polynomial with all of its signs changed to the opposite sign

e. an algebraic expression consisting of one or more terms in which each term is a number or a product of numbers and variables raised to whole-number exponents

Skill Review

Directions: Read each problem and complete the task.

1. Which gives the area of the rectangle as a polynomial in standard form?

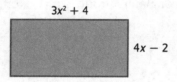

$3x^2 + 4$

$4x - 2$

A. $12x^2 - 6x - 8$
B. $12x^2 - 16x + 8$
C. $12x^3 + 6x^2 + 16x^2 - 8$
D. $12x^3 - 6x^2 + 16x - 8$

2. Explain how to determine the degree of a polynomial that is written in standard form. Give an example in your explanation.

3. Determine whether the statement below is true or false. Explain your reasoning.
The product of 2 monomials is a binomial.

4. What is the value of the polynomial expression below when $y = 4$?

$-3y^3 + 2y^2 - 5y + 7$

A. -173
B. -133
C. 147
D. 187

5. What is the difference of these two polynomials?

$(6x^5 + 7x^3 - 4x^2 + 9) - (2x^4 + 7x^5 + x^2)$

A. $13x^5 - 2x^4 + 7x^3 - 3x^2 - 9$
B. $-x^5 - 2x^4 + 7x^3 - 5x^2 + 9$
C. $8x^5 + 3x^2 - 9$
D. $4x^5 - 5x^2 + 9$

Skill Practice

Directions: Read each problem and complete the task.

1. Which of the following expressions has the greatest value when $x = -1$?

A. $x^3 - 4x^2 + 5$
B. $-x^3 + 3x + 3$
C. $2x^3 - 3x^2 + x$
D. $-2x^3 + x^2 - 4x + 1$

2. What is the degree of the product of a polynomial of degree 2 and a polynomial of degree 3?

A. 2
B. 3
C. 5
D. 6

3. Is it possible to subtract two polynomials of degree 4 and get a polynomial of degree 2? Give an example to support your explanation.

4. What is the degree of the product of two linear polynomials?

5. The table shows a company's costs for labor and the costs of materials to produce x items. What polynomial expression represents the company's total cost to produce x items?

Costs for x Items	
Labor	$3x^2 + 300x + 10$
Materials	$x^2 + 80x + 100$

A. $2x^2 + 220x - 90$
B. $3x^2 + 400x + 90$
C. $4x^2 + 380x + 110$
D. $4x^2 + 220x - 110$

6. What is the degree of the sum of these two polynomials?

$(5x^3 - 2x^2 + 1) + (2x^2 - 6x^4 - 5)$

A. 2
B. 3
C. 4
D. 5

7. Is the following statement sometimes, always, or never true? Give examples to support your answer.

The sum of two monomials is a binomial.

8. Which gives the area of the trapezoid as a polynomial in standard form? (Remember, the area of a trapezoid with bases b_1 and b_2 and height h is given by the expression $\frac{1}{2}(b_1 + b_2)h$.)

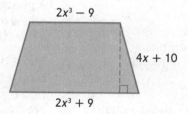

A. $8x^4 + 20x^3$
B. $4x^4 + 10x^2$
C. $4x^3 + 10x^2$
D. $8x^2 + 20x$

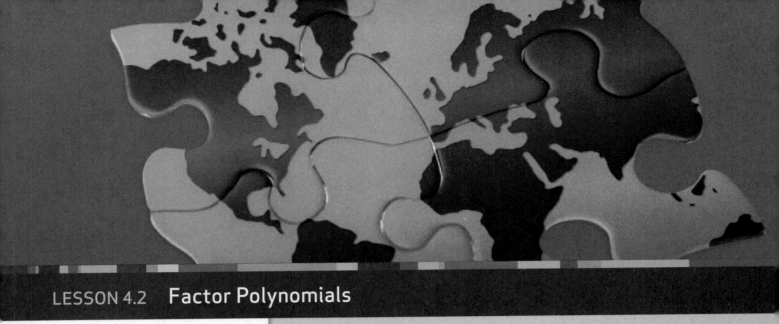

LESSON OBJECTIVES

- Read, write, and evaluate expressions with variables
- Identify the parts of an expression
- Factor polynomials
- Factor quadratic expressions

CORE SKILLS & PRACTICES

- Build Lines of Reasoning
- Make Use of Structure

Key Terms

coefficient
the number that appears before a variable that multiplies the variable in an algebraic expression

degree
the term in a polynomial with the greatest exponential power

polynomial
an expression with one or more monomial(s), or term(s)

Vocabulary

factor
to divide a monomial by another monomial with no remainder

leading coefficient
the coefficient accompanying the first term in a polynomial that has been written in standard form.

monomial
An expression with one term, such as 10, $2x$, and $3xy$

Key Concept

People practicing a variety of professions and hobbies write, simplify, and evaluate polynomial expressions. Polynomial expressions can be classified by their number of terms or by the greatest exponential power.

Factoring Out Monomials

Just as a puzzle is made up of individual pieces that make the puzzle whole, a polynomial is also made up of pieces—its factors. **Monomials**, or polynomials with one term, are certain types of polynomials that can be factored out of larger polynomials to help make them easier to work with.

Polynomial Language

A **polynomial** is a set of terms (or expressions) that include one or more variables raised to a whole-numbered power. Any term may also include a number called a **coefficient**. In the term $5x^2$, for example, the coefficient is the number 5.

The terms in a polynomial written in standard form are ordered from the term with the greatest power to the term with the least power. Because the first term in a polynomial written in standard form has the greatest power, its coefficient is called the **leading coefficient**.

$$\overset{\text{leading coefficient}}{} \qquad \overset{\text{constant}}{}$$

$$3x^3 + 2x^2 + x - 9$$

Classifying Polynomials

One way to classify polynomials is by the number of terms in the expression.

Classification	Number of Terms	Example
monomial	1	$5x^3$
binomial	2	$5x^3 + 8x^2$
trinomial	3	$5x^3 + 8x^2 + 12$

You can also classify polynomials by degree. The **degree** of a polynomial is the greatest power of the variable. When polynomials are written in standard form, the degree appears in the first term.

Classification	Degree	Example
constant	0	$3x^0 = 3$ ($x^0 = 1$ for $x \neq 0$)
linear	1	$2x + 9$
quadratic	2	$5x^2 - 2x + 7$

Factoring Using the Greatest Common Factors

The greatest common factor, or GCF of two or more polynomials is the polynomial with the greatest degree that divides evenly into both polynomials. When factoring a polynomial, treat each term as a separate polynomial and find the GCF of the terms.

Example 1: Find the GCF of a Polynomial

Find the GCF of the polynomial $4x^2y^3 - 2xy^2$.

Step 1 Find the GCF of the coefficients and the GCF of each variable. The GCF is the greatest power shared by every variable that appears in every term.

- The GCF of the coefficients 4 and 2 is 2.
- The greatest power that x^2 and x share is 1, so the GCF of the power of x is x.
- The greatest power that y^3 and y^2 share is 2, so the GCF of the power of y is y^2.

Step 2 Multiply these GCFs to determine the GCF of the polynomial.

The GCF is $2 \times x \times y^2 = 2xy^2$.

Once you have found the GCF of a polynomial, you can use it to **factor** the polynomial, rewriting it as a product of its smaller parts.

Example 2: Factor a Polynomial

Factor the polynomial $4x^2y^3 - 2xy^2$.

Step 1 Find the GCF of the two terms.

The GCF is $2 \times x \times y^2 = 2xy^2$.

Step 2 Divide both terms by the GCF.

$$4x^2y^3 - 2xy^2 = \frac{4x^2y^3}{2xy^2} - \frac{2xy^2}{2xy^2}$$

Step 3 Rewrite the problem by expressing the exponents in the denominator as negative exponents and combining all exponents. (Remember, a number raised to the 0 power is equal to 1.)

$$\frac{4x^2y^3}{2xy^2} - \frac{2xy^2}{2xy^2} = \frac{4}{2}x^{2-1}y^{3-2} - \frac{2}{2}x^{1-1}y^{2-2} = 2xy - 1$$

Step 4 Multiply the GCF of the polynomial, or $2xy^2$, by $(2xy - 1)$ to rewrite the polynomial.

$$4x^2y^3 - 2xy^2 = 2xy^2(2xy - 1)$$

CORE SKILL

Build Lines of Reasoning

When you divide exponents that share the same base, you apply the following rule:

$$\frac{x^p}{x^q} = x^{p-q} \text{ for } x \neq 0$$

To understand this rule, you can show the process of division in nonstandard form, showing all multiples of the variables. For example, you could simplify the expression from Example 2 $\frac{4x^2y^3}{2xy^2} - \frac{2xy^2}{2xy^2}$ as follows.

$$\frac{4x^2y^3}{2xy^2} - \frac{2xy^2}{2xy^2} = \frac{4xxyyy}{2xyy} - \frac{2xyy}{2xyy}$$

$$= \frac{\overset{2}{4} \, x \, x \, y \, y \, y}{2 \, x \quad y \, y} - \frac{\overset{1}{2} \, x \, y \, y}{2 \, x \, y \, y}$$

$$= 2xy - 1$$

For the expression $\frac{8x^3 4y^2}{4xy} + \frac{4xy}{4xy}$, verify that the result is the same using the rule for dividing two powers as well as expanding each power and dividing out common factors.

Make Use of Structure

Recognizing that the coefficients of $x^2 + cx + d$ are either a sum or product of parts of the factors helps to check your work easily. To factor the quadratic expression $x^2 - 3x + 2$, you must find two numbers whose product is 2 and whose sum is -3. First, find two numbers whose product is 2 (1 and 2, or -1 and -2) Then from that list, determine which numbers sum to -3 (-1 and -2).

So, you can factor the quadratic expression $x^2 - 3x + 2$ as $(x + (-1))(x + (-2))$, or $(x - 1)(x - 2)$.

Factor the quadratic expression $x^2 + x - 12$.

Think about Math

Directions: Rewrite each expression by factoring the GCF of the terms.

1. $14x^3 + 4x^9$

2. $2x^7y - 3x^2y^3$

3. $4x^3y^2 + 2x^4y^4 - 6x^2y^3$

Factoring Quadratic Expressions

Quadratic expressions help model most scenarios that we know, from the equations of the orbits of planets to the paths of ballistic objects.

Quadratic Expressions

The names "second degree polynomials," "quadratic trinomials," and "quadratic expressions" all mean the same thing. They all have a variable raised to the second power, as in $x^2 + cx + d$, where c and d are real numbers. The following are examples of quadratic expressions.

$$x^2 + 8x - 4 \qquad 2x^2 + 3x + 5 \qquad 3x^2 + 5x - 2 \qquad 10x^2 - 12x - 8$$

When you factor the quadratic expression $x^2 + cx + d$ for integers c and d, you rewrite the expression as the product of two binomials $(x + a)(x + b)$. If an expression can be factored, then you will be able to find two numbers a and b whose sum equals c, and whose product equals d.

Example 3: Factor a Quadratic Expression

Factor the quadratic $x^2 + 6x + 8$.

1	Is the constant 8 positive or negative?	positive
2	What numbers can you multiply to get a positive product?	two positive or two negative numbers
3	Is the coefficient 6 positive or negative?	positive
4	Review the answer to Question 2. You're looking for two positive or two negative numbers. Can you add two negative numbers to get a positive number?	no
5	So, what two positive numbers can you add to make the coefficient 6?	1 + 5; 2 + 4; 3 + 3
6	Of the numbers that add to make 6, which two can you multiply to make the product 8?	2 × 4
7	What numbers replace the question marks in the terms $(x + ?)(x + ?)$?	2 and 4

So, you can factor $x^2 + 6x + 8$ as $(x + 2)(x + 4)$.

Check your work by expanding your factored expression.

$$\begin{aligned} (x + 2)(x + 4) &= x(x + 4) + 2(x + 4) \\ &= x^2 + 4x + 2x + 8 \\ &= x^2 + 6x + 8 \end{aligned}$$

Leading Coefficients Not Equal to 1

Sometimes, it is possible to factor across a quadratic expression with a leading coefficient not equal to 1. In that case, apply the GCF first. Then factor the expression.

Example 4: Factor the GCF First

Factor the quadratic $2x^2 + 4x - 6$.

Step 1 Factor the GCF of the terms, 2.

$$2x^2 + 4x - 6 = 2(x^2 + 2x - 3)$$

Step 2 The quadratic inside the parentheses has a leading coefficient of 1. Look for the factors of -3.

Factors of -3: 1 and -3, -1 and 3

Step 3 Identify the factors of -3 that sum to 2.

From the previous list, only -1 and 3 have a sum of 2.

Step 4 Since the leading coefficient is now 1, we can use our basic binomial factoring to complete the factoring.

$$2(x^2 + 2x - 3) = 2(x + 3)(x - 1)$$

Sometimes the leading coefficient in a quadratic expression is not equal to 1. You can write such expressions as $ax^2 + bx + c$. Because the leading coefficient is not equal to 1, you must deal with 3 coefficients, a, b, and c.

Example 5: Factor $ax^2 + bx + c$

Factor the quadratic $4x^2 + 14x - 8$ as the product of two binomials.

Step 1 Unlike simpler quadratic expressions, the product of the two numbers in this example must equal ac. In this case, ac = -32.

Factors of -32: 1 and -32, -1 and 32, 2 and -16, -2 and 16, 4 and -8, -4 and 8

Step 2 The sum of the two factors you select must equal b, or in this case, 14.

From the previous list, only -2 and 16 have a sum of 14.

Step 3 Now you are ready to factor by grouping. Rewrite the quadratic by writing the middle term as the sum of the factors. Group the terms in pairs, and factor the GCF out of each pair.

$$4x^2 + 14x - 8 = 4x^2 - 2x + 16x - 8$$
$$= (4x^2 - 2x) + (16x - 8)$$
$$= 2x(2x - 1) + 8(2x - 1)$$

Step 4 Now, each term has the binomial $(2x - 1)$ in it as a factor. So factor out $(2x - 1)$ as the GCF.

$$2x(2x - 1) + 8(2x - 1) = (2x - 1)(2x + 8)$$

Vocabulary Review

Directions: Match each term to its definition.

1. _____ coefficient

2. _____ degree

3. _____ factor

4. _____ leading coefficient

5. _____ monomial

6. _____ polynomial

a. The exponent in the term of a polynomial with the greatest exponential power

b. A polynomial with one term

c. The number that appears before a variable that multiplies the variable in an algebraic expression

d. An expression with one or more monomials

e. The coefficient accompanying the first term in a polynomial that has been written in standard form

f. To divide a monomial by another monomial exactly, meaning with no remainder

Skill Review

Directions: Read each problem and complete the task.

1. What is the leading coefficient of the polynomial shown below?

   $4x^2 - 2x^3 - 5$

   A. -5
   B. -2
   C. 3
   D. 4

2. If $x^2 + bx + c = (x + n)(x + m)$, what is the value of $n + m$?

   A. b
   B. c
   C. bc
   D. $b + c$

3. Write the following polynomial expression as the product of linear terms.
   Show your work.

   $3x^2 - 8x + 4$

4. Which factor pair can you use to find r and s in the equation $(x + r)(x + s) = x^2 - 8x + 15$?

   A. -2 and 4
   B. -4 and 2
   C. 3 and 5
   D. -3 and -5

5. Marquise wrote the quadratic $4x^2 + 10x - 6$ in the factored form $(ax + b)(x + c)$, where a, b, and c are nonzero integers. Which of the following correctly gives the values of a, b, and c in order from least to greatest?

   A. a, b, c
   B. b, a, c
   C. b, c, a
   D. c, a, b

Skill Practice

Directions: Read each problem and complete the task.

1. If $x^2 + bx + c = (x + n)(x + m)$, what is the value of nm?

   A. b
   B. c
   C. bc
   D. $b + c$

2. Rebecca said that she was able to factor the expression $x^2 + x + 1$ as the product of 3 linear factors. Do you agree or disagree with her statement? Explain your reasoning.

3. Which of the following shows the correct factorization of $4x^2 + 8x - 60$?

   A. $2(x - 3)(x - 5)$
   B. $2(x + 3)(x - 5)$
   C. $4(x - 3)(x + 5)$
   D. $4(x + 3)(x + 5)$

4. Write the following polynomial expression as the product of linear terms. Show your work.
   $4x^3 + 2x^2y - 2xy^2$

5. Given that $-5x - 21 + 6x^2 = (ax + b)(cx + d)$, what is the value of bd?

   A. -21
   B. -5
   C. 6
   D. 11

6. Which of the following is not a factor of the polynomial $6x^2 + 15x + 6$?

   A. 3
   B. $(x - 1)$
   C. $(x + 2)$
   D. $(2x + 1)$

7. Determine whether the statement below is true or false. Give an example to demonstrate your reasoning.

   Factoring a quadratic expression always results in the product of two linear binomials.

8. For the quadratic expression $ax^2 + bx + c$ where a, b, and c are all nonzero integers, the product of the leading coefficient and the constant term is 36. Which of the following could not be a value of b?

   A. -13
   B. 15
   C. 20
   D. 35

LESSON 4.3   Solve Quadratic Equations

■ LESSON OBJECTIVES

- Solve a quadratic equation by inspection, by factoring, by completing the square, and by using the quadratic formula

■ CORE SKILLS & PRACTICES

- Reason Abstractly
- Solve Real-World Problems

Key Terms

quadratic formula
a formula that can be used to solve any quadratic equation by substituting the coefficients of the equation

discriminant
the part of the quadratic formula that is under the square root

Vocabulary

completing the square
a technique of manipulating quadratic equations so that they can be solved by taking the square root of both sides

solving by inspection
determining the solution(s) of an equation simply by looking at the equation

perfect square trinomial
a quadratic expression that can be written as a perfect square of a linear expression

Key Concept

Quadratic equations can be solved in several ways. Simple quadratic equations can be solved by inspection. More complex ones can be solved by factoring, completing the square, or using the quadratic formula.

Solving a Quadratic Equation by Factoring

Quadratic equations can be used to describe the motion of objects. Solving a quadratic equation can tell you when an object will land on the ground after being thrown, or even how high it will be off the ground during its travel. Factoring is one way to solve these types of quadratic equations.

Solving a Factored Quadratic Equation

You can find the roots of a quadratic equation by setting the equation equal to zero and applying the zero-product principle: If $a \times b = 0$, then $a = 0$ or $b = 0$ or both a and $b = 0$. In other words, if you know that the product of two factors is 0, then one or both of the factors must be 0.

Example 1: Use the Zero-Product Principle

Solve the equation $(x - 1)(x + 5) = 0$.

Step 1 Set the factors equal to 0. $x - 1 = 0$ $x + 5 = 0$

Step 2 Solve for x. $x = 1$ $x = -5$

The solutions are 1 and -5.

Factoring to Solve a Quadratic Equation

If you are given a quadratic equation that contains a trinomial that is equal to 0, you first need to find the factors. Then you can set the factors equal to 0 and find the solutions to the equation.

Example 2: Solve by Factoring

Solve the quadratic equation $x^2 + 14x + 48 = 0$.

Step 1 Factor the left side of the equation.

$$(x + 6)(x + 8) = 0$$

Step 2 Set each factor equal to 0 and solve for x.

$$x + 6 = 0 \qquad x + 8 = 0$$
$$x = -6 \qquad x = -8$$

The solutions are -6 and -8.

Think about Math

Directions: Solve each quadratic equation.

1. $(3n + 6)(n - 2) = 0$
2. $x^2 + 4x - 21 = 0$

Completing the Square

When you are cooking, you might follow recipes and measure ingredients carefully, or you may know the recipe so well that you can add ingredients by feel. Similarly, some quadratic equations require special tools and methods to solve, and others you can see the solution quickly without any additional help.

Solving by Inspection

Not all quadratic equations can be factored. Some quadratic equations can be solved by taking the square root of both sides. For simple equations, you may be able to do this mentally. This is called **solving by inspection**.

Example 3: Solve by Inspection

Solve the equation $x^2 = 49$.

In this example, 49 is a perfect square. Taking the square root of both sides is easy to do mentally. Remember that a positive number has two square roots, one positive and one negative. The solutions are 7 and -7.

$$x^2 = 49$$
$$\sqrt{x^2} = \sqrt{49}$$
$$x = \pm 7$$

Example 4: Take the Square Root of Both Sides

Solve the equation $x^2 = 77$.

You can take the square root of both sides even though the right side is not a perfect square. In this example, the solutions, $\sqrt{77}$ and $-\sqrt{77}$, are not integers.

$$x^2 = 77$$
$$\sqrt{x^2} = \sqrt{77}$$
$$x = \pm\sqrt{77}$$

Gather Information

When taking an exam, it is always important to understand not only what is being asked but also how to convert verbal information to an algebraic equation.

The area of a rectangular patio is 144 ft^2 and the length of the patio is 10 feet longer than the width. You can write and solve an equation to find the length and width of the patio. Let x represent the width of the patio. Then the length of the patio is $x + 10$. Remember that the formula for the area of a rectangle is $A = \ell w$.

$$x(x + 10) = 144$$
$$x^2 + 10x = 144$$
$$x^2 + 10x - 144 = 0$$
$$(x + 18)(x - 8) = 0$$
$$x + 18 = 0 \qquad x - 8 = 0$$
$$x = -18 \qquad x = 8$$

There are two solutions to this quadratic equation, one positive and one negative. Because x represents the width of a rectangle, only positive values make sense. Therefore, the width is 8 feet and the length is $8 + 10 = 18$ feet.

Suppose you're designing a kitchen island. You want the width to be 2 feet shorter than the length, with a total area of 15 ft^2. What should the length and width be?

A trinomial that can be factored as a square is a **perfect square trinomial**. If the quadratic expression in a quadratic equation is a perfect square trinomial, you can take the square root of both sides.

Example 5: Perfect Square Trinomial

Solve the equation $x^2 + 4x + 4 = 9$.

Step 1 Factor the left side. $\qquad (x + 2)(x + 2) = 9$

Step 2 Write the left side as a square. $\qquad (x + 2)^2 = 9$

Step 3 Take the square root of both sides. $\qquad \sqrt{(x + 2)^2} = \sqrt{9}$
$$x + 2 = \pm 3$$

Step 4 Solve for x. $\qquad x + 2 = 3 \text{ or } x + 2 = -3$
$$x = 1 \text{ or } \qquad x = -5$$

The solutions are 1 and -5.

Solving by Completing the Square

How do you take the square root of a quadratic trinomial that cannot be written as a square? You can use a technique called **completing the square** to make the quadratic trinomial into a perfect square trinomial.

For example, consider the equation $x^2 + 12x + 11 = 0$. Notice that the left side of the equation cannot be written as a square, so we cannot solve by taking the square root of both sides. However, we can manipulate the left side so that it can be written as a square.

The goal is to write the left side in the form $x^2 + 2bx + b^2$, so that we can write it as a square, $(x + b)^2$. We must find the value of b.

Example 6: Completing the Square

Solve the equation $x^2 + 12x + 11 = 0$.

Step 1 Isolate the terms with variables on one side. In this case, that means subtracting 11 from both sides.

$$x^2 + 12x + 11 = 0$$
$$x^2 + 12x = -11$$

Step 2 Find the values of b and b^2. The x-term is $12x$, so $2bx = 12x$.

$$2bx = 12x$$
$$b = 6$$
$$b^2 = 36$$

Step 3 Use the value of b^2 to write the left side of the equation as $x^2 + 12x + 36$. To keep the equation balanced, add 36 to the right side of the equation.

$$x^2 + 12x + 36 = -11 + 36$$
$$x^2 + 12x + 36 = 25$$

Step 4 Write the left side of the equation as a square.

$$(x + 6)^2 = 25$$

Step 5 Take the square root of both sides.

$$\sqrt{(x + 6)^2} = \sqrt{25}$$
$$x + 6 = \pm 5$$

Step 6 Solve for x.

$$x + 6 = 5 \text{ or } x + 6 = -5$$
$$x = -1 \text{ or } \quad x = -11$$

The solutions are -1 and -11.

Think about Math

Directions: Solve each quadratic equation.

1. $(x + 5)^2 = 49$
2. $x^2 + 6x = 0$
3. $x^2 + 4x = 12$

Reason Abstractly

Consider the equation $x^2 = -16$. What happens when we take the square root of both sides?

$$x^2 = -16$$
$$x = \pm\sqrt{-16}$$
$$x = ???$$

There is no real number that can be squared to produce a negative number, so a negative number does not have a real square root. If at any point in solving an equation you must take the square root of a negative number, the equation has no real solutions.

Which equation or equations below have no real solutions?

Equation 1: $x^2 + 100 = 0$

Equation 2: $x^2 + 4x + 10 = 4$

Equation 3: $-x^2 = -81$

The Quadratic Formula

Solving a Quadratic Equation with the Quadratic Formula

The **quadratic formula** is a formula that allows you to solve any quadratic equation. All you need to do is identify each of the coefficients in the equation and substitute them in the formula to find the solution or solutions.

The Quadratic Formula

For the quadratic equation $ax^2 + bx + c = 0$, the solution is $x = \frac{-b \pm \sqrt{b^2 - 4ac}}{2a}$.

Note that a quadratic equation must be in the form $ax^2 + bx + c = 0$ before you can use the quadratic formula.

Example 7: The Quadratic Formula

Solve the equation $x^2 - 5x - 14 = 0$.

Step 1 Identify the values of a, b, and c. $\qquad a = 1, b = -5, c = -14$

Step 2 Substitute the values of a, b, and c into the quadratic formula and simplify.

$$x = \frac{-(-5) \pm \sqrt{(-5)^2 - 4(1)(-14)}}{2(1)}$$
$$= \frac{5 \pm \sqrt{25 - (-56)}}{2}$$
$$= \frac{5 \pm \sqrt{81}}{2}$$
$$= \frac{5 \pm 9}{2}$$
$$x = 7 \text{ or } x = -2$$

The solutions are 7 and -2.

Knowing When a Quadratic Equation has No Real Solutions

You can tell whether a quadratic equation has no real solutions without solving it. You just need the part of the quadratic formula that is under the square root. This expression, $b^2 - 4ac$, is called the **discriminant**.

$$x = \frac{-b \pm \sqrt{b^2 - 4ac}}{2a}$$

- If the discriminant is positive, then the equation has two real solutions.
- If the discriminant is 0, then the equation has one real solution.
- If the discriminant is negative, then the equation has no real solutions.

Example 8: Use the Discriminant

Without solving, tell how many real solutions each equation has.

a. $x^2 + 10x + 25 = 0$ b. $2x^2 - 10x + 9 = 0$ c. $x^2 - 3x + 7 = 0$

Step 1 Identify the values of a, b, and c.

a. $a = 1, b = 10, c = 25$ b. $a = 2, b = -10, c = 9$ c. $a = 1, b = -3, c = 7$

Step 2 Find the discriminant, $b^2 - 4ac$.

a.
$$b^2 - 4ac = (10)^2 - 4(1)(25)$$
$$= 100 - 100$$
$$= 0$$

b.
$$b^2 - 4ac = (-10)^2 - 4(2)(9)$$
$$= 100 - 72$$
$$= 28$$

c.
$$b^2 - 4ac = (-3)^2 - 4(1)(7)$$
$$= 9 - 28$$
$$= -19$$

Step 3 Use the discriminant to determine the number of real solutions.

a. The discriminant is 0, so the equation has one real solution.

b. The discriminant is positive, so the equation has two real solutions.

c. The discriminant is negative, so the equation has no real solutions.

◣ Think about Math

Directions: Use the quadratic equation $x^2 + 2x - 8 = 0$ to answer the following questions.

1. What are the values of a, b, and c?
2. What is the discriminant?
3. How many real solutions does the equation have?
4. Give the real solutions, if they exist.

Solve Real-World Problems

Sometimes when you are solving a real-world problem, one or more of the solutions may not make sense in the real-world situation. Always check that your solutions are reasonable in the context of the problem.

A rock is thrown upward at a speed of 38 ft/sec from a height of 5 ft. Its height in h feet after t seconds is given by the equation $-16t^2 + 38t + 5$. When will the rock hit the ground?

When the rock hits the ground, its height will be 0. Solve the equation $0 = -16t^2 + 38t + 5$.
$a = -16, b = 38, c = 5$

$$t = \frac{-(38) \pm \sqrt{(38)^2 - 4(-16)(5)}}{2(-16)}$$

$$= \frac{-38 \pm \sqrt{1764}}{-32}$$

$$= \frac{-38 \pm 42}{-32}$$

$t = \frac{5}{2} = 2.5$ or $t = -\frac{1}{8} = -0.125$

There are two possible solutions. Remember that t represents time. Time cannot be negative, so only the positive solution makes sense. The rock will hit the ground after 2.5 seconds.

Now you try. An object is shot into the air. Its height is given by the equation $h = -5t^2 + 30t$, where h represents height and t represents elapsed time in seconds. How long will it take the object to reach the ground?

Vocabulary Review

Directions: Write the missing term in the blank.

completing the square	discriminant	perfect-square trinomial
quadratic formula	solving by inspection	

1. _____ is a way to manipulate a quadratic equation so that one side is

   a _____, which is an expression that can be written as a square.

2. Solving simple equations mentally is called _____.

3. The _____ is a formula that can be used to find the solutions of any quadratic equation.

   The part of this formula that is under the square root is called the _____ and it can be used to determine the number of real solutions of the equation.

Skill Review

Directions: Read each problem and complete the task.

1. Solve the equation $(x + 7)(x - 3) = 0$.

2. Solve the equation $x^2 - 15x + 36 = 0$ by factoring.

3. Solve the equation $x^2 + x - 72 = 0$ by factoring.

4. Solve the equation $(x + 4)^2 = 100$ by taking the square root of both sides.

5. A company is installing a swimming pool in a customer's backyard. In order for the pool to fit in the yard, the area must be 195 square feet and the width of the pool must be 2 feet shorter than the length. What should the width of the pool be?

   A. 11 feet
   B. 13 feet
   C. 15 feet
   D. 17 feet

6. Which of the following equations has no real solutions?

   A. $x^2 = 49$
   B. $(x - 7)^2 = 256$
   C. $(x + 11)^2 = -121$
   D. $-4x^2 = -100$

7. What must be added to the expression below to make it a perfect square trinomial?

   $x^2 + 24x +$ _____

   A. 12
   B. 48
   C. 144
   D. 576

8. How many real solutions does the equation $x^2 + 6x - 12 = 0$ have?

9. Solve the equation $-3x^2 - 5x + 2 = 4$ using the quadratic formula.

Skill Practice

Directions: Read each problem and complete the task.

1. Solve the equation $2x^2 + 10x - 3 = x^2 + 15x - 9$.

2. The first several steps to solve the equation $3(x^2 - 4) - 6 = -3(3x + 2)$ are shown below. Complete the solution process to solve the equation.

$$3(x^2 - 4) - 6 = -3(3x + 2)$$
$$3x^2 - 12 - 6 = -9x - 6$$
$$3x^2 - 18 = -9x - 6$$
$$3x^2 + 9x - 18 = -6$$

3. Solve the equation $(x + 3)^2 + 5 = -2x - 2$.

4. A rocket is launched into the air at a velocity of 256 feet per second from ground level. The height h of the rocket in feet after t seconds is given by the equation $h = -16t^2 + 256t$. How long will it take the rocket to reach the ground?

   A. 0 seconds
   B. 4 seconds
   C. 16 seconds
   D. 40 seconds

5. You can use the methods in this lesson to solve formulas for a variable. The formula for the area of a circle is $A = \pi r^2$, where r is the radius of the circle.

   a. Solve this formula for r.
   b. Ana said that there are two solutions for r, one positive and one negative. Is Ana correct? If so, explain why. If not, describe Ana's error.
   c. Find the radius of a circular tabletop whose area is 12 ft². Round your answer to the nearest whole number.

6. The distance d in feet that a dropped object falls in t seconds is given by the equation $d = 16t^2$. If an object is dropped from a height of 900 feet, how long will it take to reach the ground?

   A. 7.5 seconds
   B. 8.7 seconds
   C. 56 seconds
   D. 144 seconds

7. Raul has 100 feet of fencing that he wants to use to make a rectangular pen with an area of 525 ft². He needs to determine the necessary length and width of the rectangle. The equation $w(50 - w) = 525$ models this situation. What do the two solutions to this equation represent? What are the length and width of the rectangle that Raul should make?

8. Yuri operates a food truck that sells sandwiches. After reviewing her financial information, she has determined that her daily profit p in dollars is modeled by the function $p = 70s - s^2 - 1225$, where s is the number of sandwiches sold. How many sandwiches does Yuri have to sell each day to break even? (Hint: Yuri will break even when her profit is $0.)

9. For what value of c does the equation $x^2 + 6x + c = 0$ have one real solution? Explain how you found your answer.

10. For the quadratic equation $3x^2 - 2x + 1 = 4x - 3$, Zach said that $a = 3$, $b = -2$, and $c = 1$. What error did Zach make? Explain how to find the correct values of a, b, and c, and identify these values. Then solve the equation.

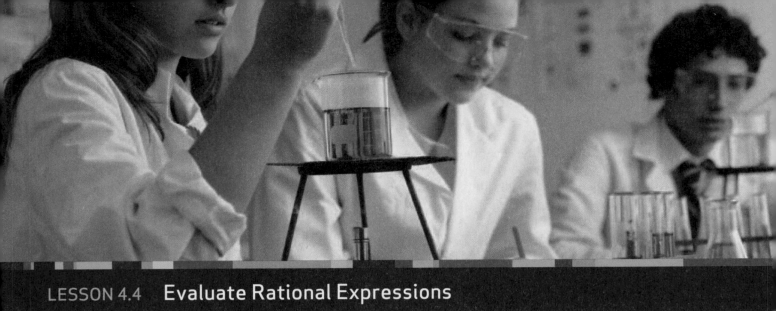

LESSON 4.4   Evaluate Rational Expressions

■ LESSON OBJECTIVES

- Evaluate rational expressions
- Simplify rational expressions
- Add, subtract, multiply, and divide rational expressions

■ CORE SKILLS & PRACTICES

- Evaluate Expressions
- Perform Operations

Key Terms

rational expression
a ratio of two polynomials

**restricted value
(of a rational expression)**
a value of the variable for which the denominator of the rational expression is equal to 0

Vocabulary

polynomial
an expression made up of numbers and variables with whole-numbered exponents and any sum, difference, or product of them

prime number
a whole number > 1 whose only two factors are 1 and itself

reciprocals
two numbers or expressions whose product is 1

least common denominator (LCD)
the least common multiple of two or more denominators

Key Concept

A rational expression is a ratio of two polynomials. Rational expressions are similar to fractions and can be simplified, multiplied, divided, added, and subtracted using methods similar to those for fractions.

Simplifying Rational Expressions

A **polynomial** is an expression containing one or more terms made up of numbers and/or variables with whole-number exponents. A **rational expression** is a ratio of two polynomials. The word "rational" stems from the word "ratio," indicating a comparison of a numerator and a denominator. Rational expressions appear often in math and science. For example, in physics, rational expressions can be used to describe the motion of objects along a curved or circular path.

Rational Expressions

Below are examples and non-examples of rational expressions. Note that the numerator and the denominator in the rational expressions are both polynomials. Remember that terms in a polynomial cannot have negative exponents, division by a variable, variable exponents, or variables under a radical.

Examples of Rational Expressions		
$\dfrac{5}{x}$	$\dfrac{n+1}{n-1}$	$\dfrac{r^2+5r}{r^2+7r+10}$

Non-Examples of Rational Expressions			
$\dfrac{5}{2^x}$	$\dfrac{3-\sqrt{n}}{2n}$	$\dfrac{5}{1+\dfrac{1}{y}}$	$\dfrac{2x^{-2}+3}{x}$

A rational expression is undefined when the denominator is equal to 0. A **restricted value** of a rational expression is any value of the variable for which the denominator is equal to 0.

Evaluate Rational Expressions

Example 1: Find Restricted Values

Find the restricted value(s) for each rational expression.

a. $\frac{5}{x}$ b. $\frac{r^2 + 5r}{r^2 + 7r + 10}$

Step 1 Set the denominator equal to 0.

a. $x = 0$ b. $r^2 + 7r + 10 = 0$

Step 2 Solve the equation. The solution or solutions are the restricted values.

a. $x = 0$ b. $(r + 2)(r + 5) = 0$
  $r + 2 = 0$ or $r + 5 = 0$
  $r = -2$ or $r = -5$

The restricted value is 0.

The restricted values are -2 and -5.

Simplifying Rational Expressions

A rational expression is simplified when its numerator and denominator have no common factors other than 1. To simplify a rational expression, you may need to factor the numerator, denominator, or both to recognize common factors that can be divided out. The original expression and the simplified expression, under the same restrictions, are equivalent.

Example 2: Simplify a Rational Expression

Simplify the rational expression $\frac{x^2 + x}{x^2 + 3x + 2}$

Step 1 Factor the numerator and the denominator.

$$\frac{x^2 + x}{x^2 + 3x + 2} = \frac{x(x + 1)}{(x + 1)(x + 2)}$$

Step 2 Identify restricted values.

Restricted values: $x = -1$ and $x = -2$

Step 3 Cancel out common factors in the numerator 1 and denominator. Because a quantity divided by itself equals 1, replace factors that are divided out with 1.

$$\frac{x\overset{1}{\cancel{(x + 1)}}}{\underset{1}{\cancel{(x + 1)}}(x + 2)}$$

Step 4 Write the simplified expression and indicate the restricted values.

$\frac{x}{x + 2}$; $x \neq -1$, $x \neq -2$

CORE SKILL

Evaluate Expressions

You can evaluate rational expressions the same way you evaluate other algebraic expressions. Just replace the variable for whichever value you want to substitute in. Then simplify both the numerator and denominator using the appropriate order of operations. Finally, reduce the fraction by dividing out any common factors in the numerator and denominator.

Now you try. Evaluate the rational expression $\frac{-3g^2 - 1}{-4g + 2}$ when $g = -3$.

CALCULATOR SKILL

You can use a calculator to evaluate rational expressions, but remember to use parentheses around the numerator and the denominator. For example, to evaluate $\frac{x - 3}{2x + 1}$ when $x = 2$, you must enter $(2 - 3) \div (2 \times 2 + 1)$. Try entering the expression with and without parentheses to see that the calculator displays different answers. Always include the parentheses to be sure your answer is correct.

◣ Think about Math

Directions: Simplify the rational expressions and state the restricted values.

1. $\frac{x^3 - 3x^2}{x^2 + 2x - 15}$ **2.** $\frac{5x}{5x + 10}$ **3.** $\frac{x^2 - 3x - 18}{x^2 - x - 12}$

Multiplying and Dividing Rational Expressions

The formula $d = rt$ describes the distance traveled at a constant rate in a time period. Rational expressions are often used to represent rates, because rates are ratios of one quantity to another. To use the formula $d = rt$ when r is a rational expression, multiply and divide with rational expressions.

Evaluate Rational Expressions

To multiply fractions, you
multiply the numerators,
multiply the denominators,
and then simplify your answer.
However, it may be easier to
divide out common factors
before you multiply. Begin by
rewriting each fraction so that
its numerator and denominator
are products of prime numbers.
A **prime number** is a whole
number greater than 1 whose
only two factors are 1 and itself.

Rewrite the fractions below
so that their numerators and
denominators are products
of prime numbers to find the
product. $\frac{22}{27} \times \frac{63}{66}$.

Multiplying Fractions

Multiplying rational expressions is similar to multiplying fractions.

Example 3: Multiply Fractions

Find the product $\frac{2}{3} \times \frac{3}{4}$.

Step 1 Multiply numerators and denominators. $\frac{2}{3} \times \frac{3}{4} = \frac{2 \times 3}{3 \times 4} = \frac{6}{12}$

Step 2 Simplify the answer by dividing out $\frac{6}{12} = \frac{1 \times \cancel{6}}{2 \times \cancel{6}} = \frac{1}{2}$
common factors.

Multiplying Rational Expressions

When multiplying rational expressions, factor each numerator and
denominator and divide out common factors before multiplying.

Example 4: Multiply Rational Expressions

Find the product $\frac{2x}{x+4} \times \frac{x^2 + 5x + 4}{4x}$.

Step 1 Factor the numerators and the denominators.

$$\frac{2x}{x+4} = \frac{2 \times x}{(x+4)} \qquad \frac{x^2 + 5x + 4}{4x} = \frac{(x+1)(x+4)}{2 \times 2 \times x}$$

Step 2 Divide out common factors. As with
fractions, you can divide a factor in
either numerator by a factor in
either denominator.

$$\frac{\cancel{2} \times \cancel{x}}{\cancel{(x+4)}} \times \frac{(x+1)\cancel{(x+4)}}{2 \times 2 \times \cancel{x}}$$

Step 3 Multiply the remaining factors. $\frac{1 \times (x+1) \times 1}{1 \times 2 \times 1} = \frac{x+1}{2}$

Dividing Fractions

Two numbers or expressions are reciprocals if their product is 1. Remember
that dividing is the same as multiplying by the reciprocal. Dividing rational
expressions is similar to dividing fractions.

Example 5: Divide Fractions

Find the quotient $\frac{8}{11} \div \frac{4}{33}$.

Step 1 Rewrite division as multiplication by the $\frac{8}{11} \div \frac{4}{33} = \frac{8}{11} \times \frac{33}{4}$
reciprocal. (The reciprocal of $\frac{4}{33}$ is $\frac{33}{4}$).

Step 2 Rewrite each fraction so that its numerator and denominator
are products of prime numbers.

$$\frac{8}{11} = \frac{2 \times 2 \times 2}{11} \qquad \frac{33}{4} = \frac{3 \times 11}{2 \times 2}$$

Step 3 Divide out common factors.

$$\frac{\cancel{2} \times 2 \times 2}{\cancel{11}} \times \frac{3 \times \cancel{11}}{\cancel{2} \times 2}$$

Step 4 Multiply the remaining factors and simplify the answer.

$$\frac{1 \times 2 \times 3 \times 1}{1 \times 1} = \frac{6}{1} = 6$$

Dividing Rational Expressions

To divide rational expressions, first rewrite the division as multiplication by the reciprocal. Then follow the steps for multiplying rational expressions.

Example 6: Divide Rational Expressions

Find the quotient $\frac{3x - 9}{3x} \div \frac{x^2 - 9}{9x + 3}$.

Step 1 Rewrite division as multiplication by the reciprocal.

$$\frac{3x - 9}{3x} \times \frac{9x + 3}{x^2 - 9} \quad \longleftarrow \quad \boxed{\text{To write the reciprocal, interchange the numerator and the denominator.}}$$

Step 2 Factor the numerators and the denominators.

$$\frac{3x - 9}{3x} = \frac{3(x - 3)}{3 \times x} \qquad \frac{9x + 3}{x^2 - 9} = \frac{3(3x + 1)}{(x - 3)(x + 3)}$$

Step 3 Divide out common factors.

$$\frac{3(\cancel{x - 3})}{\cancel{3} \times x} \times \frac{\cancel{3}(3x + 1)}{(\cancel{x - 3})(x + 3)}$$

Step 4 Multiply the remaining factors.

$$\frac{3 \times 1 \times 1 \times (3x + 1)}{1 \times x \times 1 \times (x + 3)} = \frac{3(3x + 1)}{x(x + 3)}$$

Think about Math

Directions: Perform the operations.

1. $\frac{6x}{x^2 - 3x - 18} \times \frac{x + 3}{2x^2 + 8x}$

2. $\frac{x^2 - 16}{21x} \div \frac{4x + 16}{7x^2}$

Adding and Subtracting Rational Expressions

Rational expressions are often used in work problems. If you know the amount of time it takes for several individual people to complete a task when working alone, you can add rational expressions to determine how long it will take them to complete the task working together.

Adding with Like Denominators

Adding rational expressions with like denominators is similar to adding fractions with like denominators. To add fractions with like denominators, add the numerators and keep the like denominator. Simplify the answer if necessary.

$$\frac{11}{17} + \frac{2}{17} = \frac{11 + 2}{17} = \frac{13}{17} \qquad \frac{3}{8} + \frac{5}{8} = \frac{3 + 5}{8} = \frac{8}{8} = 1$$

Example 7: Add Rational Expressions with Like Denominators

Find the sum $\frac{3x - 1}{x + 4} + \frac{2x + 4}{x + 4}$.

Step 1 Add the numerators and keep the like denominator.

$$\frac{(3x - 1) + (2x + 4)}{x + 4}$$

Step 2 Combine like terms in the numerator. Simplify the answer, if necessary.

$$\frac{3x - 1 + 2x + 4}{x + 4} = \frac{3x + 2x - 1 + 4}{x + 4} = \frac{5x + 3}{x + 4}$$

Evaluate Rational Expressions

WORKPLACE SKILL

Plan and Organize

A production manager is considering hiring a new employee. The new employee would help another employee complete production jobs. The present employee can complete a production job working alone in x hours. The manager estimates that the new employee will be able to complete a production job working alone in $x + 5$ hours. To help decide whether she should hire the new employee, the manager must determine how long it will take the two employees working together to complete a production job. Add the rational expressions below to determine the fraction of a production job that the two employees will complete per hour when working together.

The present employee completes $\frac{1}{x}$ job per hour.

The new employee will complete $\frac{1}{x+5}$ job per hour.

Adding with Unlike Denominators

One way to add fractions with unlike denominators is to use the **least common denominator (LCD)**. The LCD is the least common multiple of two or more denominators.

Example 8: Add Fractions with Unlike Denominators

Find the sum $\frac{5}{12} + \frac{7}{18}$.

Step 1 Factor each denominator into a product of prime numbers. Write the products using exponents.

$$12 = 2 \times 2 \times 3 = 2^2 \times 3 \qquad 18 = 2 \times 3 \times 3 = 2 \times 3^2$$

Step 2 For every prime-number factor in the denominators, identify its greatest power. Multiply these powers to find the LCD.

$$\text{LCD} = 2^2 \times 3^2$$
$$= 4 \times 9 = 36$$

> The prime-number factors in the denominators are 2 and 3. The greatest power of 2 that appears in either denominator is 2^2 and the greatest power of 3 that appears in either denominator is 3^2.

Step 3 Rewrite each fraction as an equivalent fraction whose denominator is the LCD.

$$\overset{\times 3}{\frac{5}{12} = \frac{15}{36}}_{\times 3} \qquad \overset{\times 2}{\frac{7}{18} = \frac{14}{36}}_{\times 2}$$

Step 4 Add the fractions. Simplify the answer, if necessary.

$$\frac{15}{36} + \frac{14}{36} = \frac{15+14}{36} = \frac{29}{36}$$

To add rational expressions with unlike denominators, first identify the LCD. Then use the LCD to write rational expressions with like denominators.

Example 9: Add Rational Expressions with Unlike Denominators

Find the sum $\frac{2}{x+3} + \frac{2x}{x-1}$.

Step 1 Factor all denominators, if possible. In this case the denominators cannot be factored. Proceed to Step 2.

Step 2 Identify the LCD. Because the denominators cannot be factored, the LCD is the product of the denominators.

$$\text{LCD} = (x+3)(x-1)$$

Step 3 Rewrite each rational expression as an equivalent expression whose denominator is the LCD.

$$\overset{\times (x-1)}{\frac{2}{x+3} = \frac{2(x-1)}{(x+3)(x-1)}}_{\times (x-1)} \qquad \overset{\times (x+3)}{\frac{2x}{x-1} = \frac{2x(x+3)}{(x+3)(x-1)}}_{\times (x+3)}$$

Step 4 Add the rational expressions. Simplify the answer, if necessary.

$$\frac{2(x-1)}{(x+3)(x-1)} + \frac{2x(x+3)}{(x+3)(x-1)} = \frac{2(x-1) + 2x(x+3)}{(x+3)(x-1)}$$

$$\boxed{\text{Distributive Property}} \rightarrow = \frac{2x - 2 + 2x^2 + 6x}{(x+3)(x-1)}$$

$$\boxed{\text{Combine like terms.}} \rightarrow = \frac{2x^2 + 8x - 2}{(x+3)(x-1)}$$

Evaluate Rational Expressions

Subtracting with Like Denominators

Subtracting rational expressions is similar to adding rational expressions.

> **Example 10:** Subtract with Like Denominators
>
> Find the difference $\dfrac{2x-3}{x+2} - \dfrac{x+5}{x+2}$.
>
> **Step 1** Subtract the numerators. Keep the like denominator.
>
> $$\frac{(2x-3)-(x+5)}{x+2}$$
>
> **Step 2** Combine like terms in the numerator. Simplify the answer, if necessary.
>
> $$\frac{(2x-3)-(x+5)}{x+2} = \frac{2x-3-x-5}{x+2} = \frac{x-8}{x+2}$$

Subtracting with Unlike Denominators

To subtract rational expressions with unlike denominators, first find a common denominator.

> **Example 11:** Subtract with Unlike Denominators
>
> Find the difference $\dfrac{3x}{x^2-1} - \dfrac{x+2}{x-1}$.
>
> **Step 1** Factor all denominators.
>
> $$x^2 - 1 = (x-1)(x+1) \qquad x - 1$$
>
> The denominator $x - 1$ cannot be factored.
>
> **Step 2** Identify the LCD. To find the LCD, multiply every factor in the denominators.
>
> $$\text{LCD} = (x-1)(x+1)$$
>
> The factors in the denominators are $x - 1$ and $x + 1$.
>
> **Step 3** Rewrite each rational expression as an equivalent expression whose denominator is the LCD.
>
> $\times (x+1)$
>
> $$\frac{3x}{x^2-1} = \frac{3x}{(x-1)(x+1)} \qquad \frac{x+2}{x-1} = \frac{(x+2)(x+1)}{(x-1)(x+1)}$$
>
> $\times (x+1)$
>
> **Step 4** Subtract the rational expressions with like denominators by subtracting the numerators and keeping the like denominator. Simplify the answer, if necessary.
>
> $$\frac{3x}{(x-1)(x+1)} - \frac{(x+2)(x+1)}{(x-1)(x+1)} = \frac{3x-(x+2)(x+1)}{(x-1)(x+1)}$$
>
> Multiply $(x+2)(x+1)$ in the numerator
>
> $$= \frac{3x-(x^2+3x+2)}{(x-1)(x+1)}$$
>
> $$= \frac{3x-x^2-3x-2}{(x-1)(x+1)}$$
>
> Combine like terms.
>
> $$= \frac{(x^2-2)}{(x-1)(x+1)}$$

Vocabulary Review

Directions: Write the missing term in the blank.

LCD   prime number   restricted values
polynomial   rational expression   reciprocals

1. An algebraic expression with one or more terms in which each variable is raised to a whole-number exponent is called a _____.

2. If two numbers or expressions are multiplied and yield a product of 1, the numbers or expressions are _____.

3. A ratio of two polynomials is called a _____.

4. The least common multiple of two or more denominators is the _____.

5. _____ make the denominator of a rational expression equal to 0.

6. A whole number greater than 1 whose only two factors are 1 and itself is a _____.

Skill Review

Directions: Read each problem and complete the task.

1. Determine which of the following are rational expressions:
$$\frac{r+1}{r-4}, \frac{4}{3x}, \frac{8^{-2}+3x}{4x}, \frac{n^2-81}{n-1}, \frac{9-\sqrt{n}}{5}$$

2. Find the restricted value(s) for each rational expression.

   a. $\frac{x+1}{x-1}$

   b. $\frac{7}{x}$

   c. $\frac{x^2-64}{x^2-x-12}$

3. Which shows the rational expression $\frac{x+2}{x^2+5x+6}$ correctly simplified with its restricted values?

   A. $\frac{1}{x+2}; x \neq -2$

   B. $\frac{1}{x+2}; x \neq -2; x \neq -3$

   C. $\frac{1}{x+3}; x \neq -3$

   D. $\frac{1}{x+3}; x \neq -2; x \neq -3$

Directions: Perform each operation.

4. $\frac{2x}{5x} \times \frac{x+1}{x}$

5. $\frac{7r}{8r+1} + \frac{7}{8r+1}$

6. $\frac{4n-3}{n-3} - \frac{1}{n-3}$

7. A rectangular plot of land has area $x^3 + 8x^2 + 15x$ and length $x + 3$. Which expression represents the width of the plot of land?

   A. $x + 5$
   B. $x - 5$
   C. $x(x + 5)$
   D. $5(x - 5)$

Skill Practice

Directions: Read each problem and complete the task.

1. When the rational expressions below are added and the sum is simplified, which term appears in the numerator?

   $14 + 2n^2 - 1 + n - 3n - 4$

   A. -16
   B. $13n^2$
   C. $21n$
   D. $-20n$

2. A student subtracted two rational expressions and arrived at the incorrect answer below. Explain the student's error. What is the correct answer?

   $$\frac{7+x}{x} - \frac{x+1}{x+2}$$

   Step 1: $\frac{(7+x)(x+2) - x(x+1)}{x(x+2)}$

   Step 2: $7\frac{x+14+x^2+2x-x^2-1x}{x^2+2x}$

   Step 3: $\frac{2x^2+8x+14}{x^2+2x}$

3. Write two rational expressions whose sum is $\frac{x-1}{x+4}$.

4. Identify the missing numerator.

   $$\frac{x+2}{x-5} \div \frac{?}{x^2-25} = x+5$$

5. Simplify the rational expression $\frac{a^2 - b^2}{2a - 2b}$.

6. It takes Rider 3 hours longer than Morgan to complete a repair job.

   a. Let m = the length of time it takes Morgan to complete the job. Write an expression that represents the amount of time it takes Rider to complete the job.

   b. The fraction of the job that Morgan completes in one hour is $\frac{1}{m}$. Write an expression that represents the fraction of the job that Rider completes in one hour.

   c. Add your answers to parts b and c. This sum represents the fraction of the job that Morgan and Rider complete in one hour when they are working together.

   d. Suppose that Morgan completes the job in 3 hours. What fraction of the job will Morgan and Rider complete in one hour working together? How long will it take Morgan and Rider to complete the job working together?

Directions: Choose the best answer to each question.

1. Nathan lays carpet with a length of $x + 3$ and a width of $2x^2 + 5$. So, the area of his room is _____ .

2. Ellen uses the _____ to rewrite $(x - 5)(2x + 6) = 0$ as $x - 5 = 0$ and $2x + 6 = 0$.

3. Which polynomial is not in standard form?
   A. $5x(7 + 3x)$
   B. $5x^2 + 12x - 3$
   C. $3 - 14x + 4x^2$
   D. $3x^2 + 6x + 4 + 2x$

4. Find the quotient of $\frac{x + 2}{3x^2} \div \frac{x^2 + x - 2}{15x}$.

   A. $\frac{x^2 - x}{5}$
   B. $\frac{x^3 + 3x^2 - 4}{45x^3}$
   C. $\frac{5}{x(x - 1)}$
   D. $\frac{5}{x^2 + 10}$

5. Factor the expression $2x^2 + 4x - 30$.
   A. $(x + 10)(x - 6)$
   B. $2(x + 5)(x - 3)$
   C. $(2x + 5)(x - 3)$
   D. $(2x + 10)(2x - 6)$

6. Leah is solving the equation $\frac{3m + 24}{m^2 + 6m - 16}$. She knows that there are restrictions to the possible values of m. The restricted values are _____ .

7. The GCF (greatest common factor) of the expression $12x^3y - 8x^2y^3$ is _____ .

8. Factor the expression $16xy^2 + 24x^2y^3$.
   A. $40x^3y^5$
   B. $8(xy^2 + 3x^2y^3)$
   C. $8xy(2 + 3xy)$
   D. $8xy^2(2 + 3xy)$

9. Theresa's company uses the expression $2x^2 + 16x + 50$ to represent their earnings based on the number of items ordered, x. What is the value when 25 items are ordered?
   A. 70
   B. 500
   C. 1,075
   D. 1,700

10. Find the solution(s) for the equation: $x^2 - 4x + 4 = 16$.
    A. $x = 6$
    B. $x = 6; x = -2$
    C. $x = 18$
    D. $x = -6; x = 2$

11. A rectangular plot of land has an area of $x^2 + x - 12$ and a width of $x^2 + 4x$. So the length of the plot is _____ .

12. Anthony has 64 square tiles and would like to arrange them in a square. He uses the equation $x^2 = 64$ to model the situation. How many tiles wide will his arrangement be?
    A. –8
    B. 0
    C. 8
    D. 32

Directions: Choose the best answer to each question.

13. Factor the expression $x^2 + 2x - 48$.

   A. $(x - 8)(x + 6)$
   B. $(x - 12)(x + 4)$
   C. $(x + 8)(x - 6)$
   D. $(x + 12)(x - 4)$

14. Amber wrote the equation $5x^3 + 4x^2 - 3x + 10$ and said that the _____ was 5.

15. Mila studies plants that grow both above and below sea level. She expresses the depths and heights that a certain plant can grow with the equation $x^2 + 16x + 20 = 0$ where x represents the number of feet above or below sea level. What are the depths that the plant can grow?

   A. −2 ft; −14 ft
   B. −2 ft
   C. −8 ft
   D. 8 ft, 2 ft

16. The sum of $\frac{x}{x-3}$ and $\frac{6}{x+3}$ is equal to

_____.

17. Jerrod uses the equation $x^2 - x - 30 = 0$ to represent the incomes and outflows in his budget. What are the values of x?

   A. $x = 10, x = -3$
   B. $x = -10, x = 3$
   C. $x = 6, x = -5$
   D. $x = -6, x = 5$

Check Your Understanding

On the following chart, circle the items you missed. The last column shows pages you can review to study the content covered in the question. Review those lessons in which you missed half or more of the questions.

Lesson	Item Number(s)			Review Page(s)
	Procedural	Conceptual	Problem Solving	
4.1 Evaluate Polynomials	1	3, 7, 14	9	118–123
4.2 Factor Polynomials	5, 8, 13	2		124–129
4.3 Solve Quadratic Equations	10		12, 15, 17	130–137
4.4 Evaluate Rational Expressions	4, 16	6	11	138–145

Chapter 5

Linear Equations in the Coordinate Plane

When you look at a map, you will sometimes use the latitude and longitude lines to help you identify where a town or geographical feature is located. These lines help you orient yourself. Similarly in math, the coordinate plane gives you information on where points, lines, and functions are graphed and how they relate to each other. This information can be calculated but is often easier to see when graphed on the coordinate plane.

Paul Burns/Blend Images LLC

Lesson 5.1

Interpret Slope

Understanding the coordinate plane is key to graphing equations of all types. The graph of a line on the coordinate plane can tell you information like the slope, or steepness, and what points it passes through. Learn how to calculate slope and interpret its real-world meaning.

Lesson 5.2

Write the Equation of a Line

People give directions in different ways. Some may use the names of the roads, others may use route numbers, others may describe the way using landmarks; however, they all describe the same path. Similarly, you can write the equation of a line in multiple ways. Learn how to write the equation of a line using different features like slope, intercepts, and points on the line.

Lesson 5.3

Graph Linear Equations

When solving real-life problems involving linear equations, you want to understand how the two variables relate to each other. While the equation can quickly tell you the slope, or even the intercept, the graph immediately shows you visually the relationship between the two. Learn how to graph linear equations on the coordinate plane.

Lesson 5.4

Solve Systems of Linear Equations

Sometimes you can solve a simple problem by modeling it with a linear equation. Often, however, you have multiple situations that need to be factored in, and one linear equation cannot describe the whole scenario. Learn how to solve a system of linear equations and model real-world situations.

Goal Setting

Think about the last time you made plans with a friend. How did you find a date that worked for both of you? Were there times you were free that your friend was not? How many attempts did it take to find a time? Did you find more than one time that would work?

When you solve a system of linear equations you are looking for a solution that works for both equations. What information do you know about each equation? How can you find a solution for both equations? Can you have more than one solution?

LESSON OBJECTIVES

- Determine the slope of a line from a graph, equation, or table
- Interpret unit rate as the slope in a proportional relationship of real-world and mathematical problems

CORE SKILLS & PRACTICES

- Make Use of Structure
- Use Ratio Reasoning

Key Terms

proportional relationship
an equation of the form $y = kx$ (when $k \neq 0$)

slope
the ratio of vertical change to horizontal change

unit rate
a rate that compares to one unit, such as miles per gallon

Vocabulary

coordinate plane
a grid formed by the intersection of a horizontal number line and a vertical number line

ordered pair
a pair of numbers (x, y) that is used to describe the location of a point on the coordinate plane

quadrant
one of the four regions of the coordinate plane formed by the intersection of the x- and y-axes

Key Concept

Slope, a measure of the steepness of a line, is the ratio of vertical change to horizontal change (or rise over run). For lines that represent proportional relationships, the slope of the line is equal to the unit rate.

Points and Lines in the Coordinate Plane

In some cities, streets that run east and west are named with numbers, and streets that run north and south are named with letters. Due to this convenient naming system, it is easy for people who are unfamiliar with the city to identify their current locations and navigate to a specific address.

Points in Four Quadrants

The **coordinate plane** is formed by a horizontal number line (the x-axis) and a vertical number line (the y-axis) that intersect at 0. The axes divide the coordinate plane into four regions called **quadrants**.

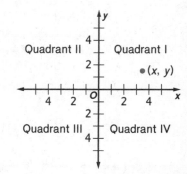

Points in the plane are described with two numbers, or an **ordered pair** (x, y). The first number in an ordered pair is the x-coordinate and the second number is the y-coordinate.

Plotting Points

Use x- and y-coordinates to plot a point in the plane. Start at $(0, 0)$, move horizontally based on the x-coordinate, and then move vertically based on the y-coordinate.

Example 1: Plotting a Point

Plot the point $(-3, 4)$.

Step 1 Start at $(0, 0)$.

Step 2 The x-coordinate is -3, so move 3 units left.

Step 3 The y-coordinate is 4, so move 4 units up.

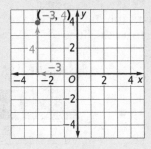

Multiple Representations of Lines

There are several different ways to represent a line in the coordinate plane by describing the points that form the line.

Equations

The verbal description "y is 3 more than twice x" can be written as the equation $y = 3 + 2x$, or $y = 2x + 3$.

Verbal Description

y	is	3 more than	twice x
↓	↓	↓	↓
y	=	$3\ +$	$2x$

Tables

To make a table of values for the equation $y = 2x + 3$, choose several values to substitute for x and then find the corresponding values of y.

x	$y = 2x + 3$
0	3
1	5
2	7

Make Use of Structure

Points in Quadrant I are plotted by starting at $(0, 0)$, moving to the right, and then moving up. These movements correspond to an ordered pair with a positive x-coordinate and a positive y-coordinate. Therefore, all points in Quadrant I have positive coordinates.

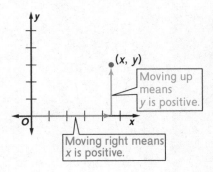

What is true about the signs of the coordinates of points in Quadrants II, III, and IV?

Graphs

To graph the line, plot the points (x, y) from the table and connect the points to form the line.

Table

x	y
0	3
1	5
2	7

Graph

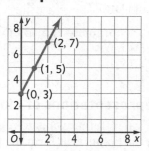

Think about Math

Directions: Answer the following questions.

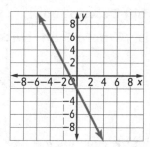

1. Which ordered pair(s) represent solutions to the equation of the line?

   A. $(-4, 6)$
   B. $(-2, 4)$
   C. $(2, -6)$
   D. $(4, -4)$

2. Which ordered pair or pairs could be in a table of values for the line?

   A. $(-6, 6)$
   B. $(0, -2)$
   C. $(3, -8)$
   D. $(6, -6)$

The Slope of a Line

The pitch of a roof is a ratio that describes how steep the roof is. A pitch of 4:12, for example, means that the roof rises 4 inches vertically for every 12 inches of horizontal run. You can also use the ratio of vertical change to horizontal change to describe the steepness of a line in the coordinate plane. This ratio is called slope.

Defining Slope

Nonvertical lines in the plane have a measure of steepness, called the **slope** of the line, which is the ratio of the vertical change (rise) to the horizontal change (run).

Positive	Lines that rise from left to right have positive slopes.	Slope $= \frac{\text{rise}}{\text{run}} = \frac{2}{3}$ **Positive Slope**
Negative	Lines that fall from left to right have negative slopes.	Slope $= \frac{\text{rise}}{\text{run}} = \frac{-4}{2} = -2$ **Negative Slope**
Zero	Because horizontal lines have no vertical change, the rise is 0. This means that the slope of a horizontal line is 0.	Slope $= \frac{\text{rise}}{\text{run}} = \frac{0}{3} = 0$ **Zero Slope**

The Slope Formula

The slope of a line is the same between any two points on the line. You can use the coordinates of two points on a line to calculate the slope. Subtract the y-coordinates to find the rise, and subtract the x-coordinates to find the run.

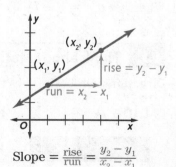

$$\text{Slope} = \frac{\text{rise}}{\text{run}} = \frac{y_2 - y_1}{x_2 - x_1}$$

When using a calculator to determine the slope of the line, parentheses are needed to make sure the order of operations are performed correctly. On the TI-30XS MultiView™ calculator, the parentheses buttons, (and), help keep the numerator together and denominator together. For example, calculating the slope between the points (3, 5) and (4, 1), you must press the calculator buttons in this order:

(, 1 , − , 5 ,) ,

÷ , (, 4 , − , 3 ,

) , and enter. This should give a slope of −4. Recheck the buttons you pushed if the answer is different.

Health Literacy

Resting heart rate is a measure of your heart's efficiency. The lower your resting heart rate, the less exertion on your heart as it pumps blood through your body. The resting heart rate of an average adult is between 60 and 100 beats per minute.

Four people calculated their resting heart rates.

- Person A's heart beat 6 times in 5 seconds and 24 times in 20 seconds.
- Person C wrote the equation $y = 48x$ to describe his resting heart rate, where y is the number of heartbeats in x minutes.
- The heart rates of Person B and Person D are described in the table and graph below, where x is minutes and y is heartbeats.

Person B

x	y
$\frac{1}{6}$	16
$\frac{1}{2}$	48

Person D

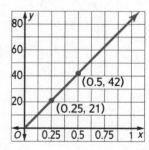

Who has the greatest resting heart rate?

To find the slope of a line from its graph, follow these steps.

Example 2: Finding the Slope

Step 1 Identify two points on the line.

Step 2 Substitute the coordinates of the points into the slope formula. Be sure to subtract the coordinates in the same order.

Step 3 Evaluate to determine the slope of the line.

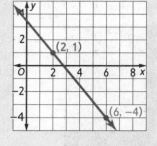

$$\text{Slope} = \frac{\text{rise}}{\text{run}} = \frac{y_2 - y_1}{x_2 - x_1} = \frac{-4 - 1}{6 - 2} = \frac{-5}{4} = -\frac{5}{4}$$

Slopes from Equations and Tables

To find the slope of a line from a table, choose two points from the table and use the slope formula.

$$\text{Slope} = \frac{\text{rise}}{\text{run}} = \frac{y_2 - y_1}{x_2 - x_1}$$

x	y
−4	0
2	3
6	5

$(-4, 0)$

$(6, 5)$

$$\text{Slope} = \frac{5 - 0}{6 - (-4)} = \frac{1}{2}$$

To find the slope of a line from an equation, use the equation to find two points on the line.

Choose a value for one variable and solve for the value of the other variable. Then use the slope formula.

$12x + 3y = 6$

When $x = 0$, $y = 2 \rightarrow (0, 2)$ $\text{Slope} = \frac{\text{rise}}{\text{run}} = \frac{y_2 - y_1}{x_2 - x_1}$

When $y = 0$, $x = \frac{1}{2} \rightarrow (\frac{1}{2}, 0)$ $\text{Slope} = \frac{0 - 2}{\frac{1}{2} - 0} = \frac{-2}{\frac{1}{2}} = -2 \times 2 = -4$

⌐ Think about Math

Directions: Order the lines from least slope to greatest slope.

Line A

x	y
8	10
24	14

Line B

$y = -2x + 1$

Line C

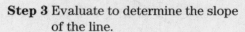

Slope as a Unit Rate

In real-world proportional relationships, the slope of a line does more than just describe its steepness. Knowing the slope can also provide context and help you better understand the relationship between variables, such as wages earned and miles traveled.

Proportional Relationships

Two variables x and y have a **proportional relationship** if there exists a nonzero number k such that $y = kx$. The constant k is called the constant of proportionality. It represents a unit rate, which is a ratio that compares a quantity to one unit, such as miles per gallon.

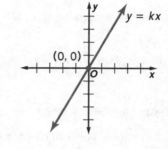

When $x = 0$, $y = k(0) = 0$. Therefore, a line that represents a proportional relationship passes through $(0, 0)$.

The slope of a line that represents a proportional relationship is k, the constant of proportionality.

Connecting Slope and Unit Rate

The graph shows the prices for different quantities of cherries.

For a point (x, y) on this line, x pounds of cherries cost y dollars.

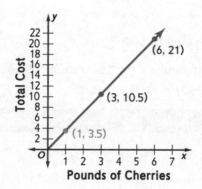

Using any two points, you can find that the slope of this line is 3.5. This is a proportional relationship with constant of proportionality 3.5. The unit rate can be found by dividing the cost by the number of pounds, which is $3.50 per pound. All three quantities (slope, unit rate, and constant of proportionality) are the same.

Notice that the point whose x-coordinate is 1 gives the unit cost per pound, $3.50.

The graph of a proportional relationship passes through the point $(1, k)$, where k is the constant of proportionality, unit rate, or slope of the line.

CORE SKILL

Use Ratio Reasoning

Use ratio reasoning to solve problems involving proportional relationships in which you must make a comparison. For example, when given two proportional relationships relating speed, time, and distance, find the unit rate for each relationship by finding the slope. You can then use this rate to compare times and distances.

Two cyclists traveling at constant speeds are tracking their distance traveled for periods of time. For Cyclist A, the equation $y = 0.3x$ describes miles traveled y in x minutes. For Cyclist B, this relationship is described in the table, where x is minutes and y is miles.

Cyclist B

x	y
0	0
1	0.2
2	0.4
3	0.6

After 30 minutes, how much farther will Cyclist A have traveled than Cyclist B?

Remember that when $x = 0$ and $y = 0$, there is a proportional relationship. Therefore, you can find the unit rate by finding the y value when $x = 1$. So the unit rate for Cyclist B is $0.2x$.

Vocabulary Review

Directions: Write the missing term in the blank.

> **coordinate plane**   **ordered pair**   **proportional relationship**
> **quadrant**   **slope**   **unit rate**

1. A(n) _____ is a rate that compares a quantity to one unit, such as miles per gallon.

2. The _____ is formed by the intersection of a horizontal number line and a vertical number line.

3. One of the four regions of the coordinate plane formed by the intersection of the x- and y-axes is called a(n) _____.

4. An equation of the form $y = kx$ for some nonzero k describes a(n) _____.

5. _____ is the constant ratio of vertical change to horizontal change for a line.

6. A pair of numbers (x, y) that is used to describe the location of a point in the coordinate plane is a(n) _____.

Skill Review

Directions: Read each problem and complete the task.

1. Write each ordered pair in the table to indicate the quadrant in which it is located.

   $(1, -15), (11, 2), (-10, 12), (-1, 8), (-20, -1), (7, -18)$

Quadrant I	Quadrant II	Quadrant III	Quadrant IV

2. Explain the relationship between the graph of a line and the solutions to the equation of the line. Use the graph shown as an example in your explanation.

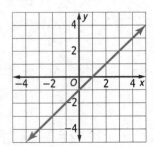

3. What is the slope of the line?

   A. −2
   B. −0.5
   C. 0.5
   D. 2

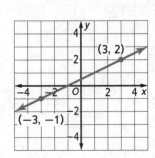

4. Which describes the unit rate associated with the table?

Time in Minutes, x	Words Read, y
2	160
4	320
6	480

A. The person's reading rate is 80 words per minute.
B. The person's reading rate is 82 words per minute.
C. The person's reading rate is 84 words per minute.
D. The person's reading rate is 86 words per minute.

5. The table shows a proportional relationship between time worked and money earned. Write the missing value in the table.

Time (hours), x	Money Earned ($), y
1	
2	21
4	42

Skill Practice

Directions: Read each problem and complete the task.

1. Give an example of a point in Quadrant III.

2. Which point or points lie on the line described by the equation $y = 3x + 4$?

A. $(-1, 1)$
B. $(0, -4)$
C. $(1, 7)$
D. $(3, 13)$

3. Lincoln says that the slope of the line described by the table is $\frac{1}{2}$. What error did Lincoln make? What is the correct slope?

x	y
2	6
4	10
6	14

4. On the graph below, plot and label the point that represents the unit rate.

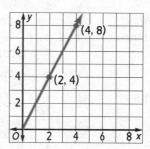

5. The table and graph show the costs in dollars y of x pounds of apples at two local grocery stores. At which store does it cost less to buy 5 pounds of apples? Describe a way you could answer this question without calculating the cost of 5 pounds of apples at either store.

Store A

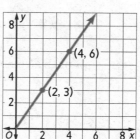

Store B

Apples (lbs), x	Price ($), y
1	1.70
3	5.10

LESSON OBJECTIVES

- Write the equation of a line given the slope and a point
- Write the equation of a line passing through two given distinct points
- Write the equation of a line from a graph or a table

CORE SKILLS & PRACTICES

- Build Solution Pathways
- Model with Mathematics

Key Terms

standard form of a linear equation
$Ax + By = C$, where A is a whole number and both A and B cannot be 0

y-intercept
the y-coordinate of the point where a line crosses the y-axis

slope-intercept form
$y = mx + b$, where m is the slope of the line and b is the y-intercept

point-slope form
an equation that allows points on a line to be calculated if one point and the slope is known

Vocabulary

coefficient
a number that is multiplied by a variable

slope
the ratio of rise to run

158   Lesson 5.2

Key Concept

The equation of a line can be written in many different ways. You can use given information about the line to determine the best way to write the equation.

Using Slope and y-Intercept

A cab company charges an initial fee and then an additional charge for each mile traveled. To find the total cost of a cab ride, you can use the same formula used to write the equation of a line. If you were to graph this information, your graph would be a line that describes the relationship between the distance traveled and the total cost of the cab ride.

Standard Form

A linear equation can be written in several ways. The **standard form of a linear equation** is $Ax + By = C$. In standard form, A (the coefficient of x) must be a whole number, and A and B cannot both be equal to 0. A **coefficient** is a number placed before a variable that is multiplied by a variable.

To write an equation in standard form, perform operations on both sides until the equation is in the form $Ax + By = C$, where A is a whole number.

Example 1: Standard Form of a Linear Equation

Write the equation $4y = 2x - 5$ in standard form.

Step 1 Subtract $2x$ from both sides.

$$4y = 2x - 5$$
$$\underline{-2x \qquad -2x}$$
$$-2x + 4y = -5$$

Step 2 Multiply both sides by -1.

$$-1(-2x + 4y) = -1(-5)$$
$$2x - 4y = 5$$

For this equation, $A = 2$, $B = -4$, and $C = 5$.

Slope-Intercept Form

When the **slope**, which is the ratio of rise to run, and y-intercept are known, you can write the equation of the line in **slope-intercept form**, $y = mx + b$, where m is the slope of the line and b is the y-intercept. The **y-intercept** is the y-coordinate of the point where a line crosses the y-axis.

Point-Slope Form

Point-slope form is another way to write the equation of a line. You can write a linear equation in point-slope form if you know the slope m and a point $(4, -3)$ on the line.

Point-Slope Form: $y - y_1 = m(x - x_1)$

To write this equation in slope-intercept form, solve for y.

Subtract 3 from both sides:
$y + 3 - 3 = x - 4 - 3$
$y = x - 7$

Now you know that the y-intercept is -7. The graph of this equation crosses the y-axis at $(0, -7)$.

CORE SKILL

Build Solution Pathways

When solving a mathematics problem, you need to consider the given information as well as the desired answer. This will help you determine a method to find the solution.

Find the equation of a line with slope 2 that passes through the point $(-3, 4)$. Write the equation in slope-intercept form. Because you are given the slope and a point, you can use either the slope-intercept form or the point-slope form to write the equation. However, the directions ask for the equation in slope-intercept form, so it likely will be simpler to start with slope-intercept form.

Think about Math

Directions: Use the following information to answer the questions.

A line has slope −3 and passes through the point (1, 2).

1. Write the equation of this line in point-slope form.
2. What are the values of A, B, and C when this equation is written in standard form?
3. Write the equation of the line in slope-intercept form.

Using Two Distinct Points

Joe is inspecting a wheelchair ramp built by a homeowner. He has measured the points where the wheelchair ramp begins and ends. In order to meet city building codes, the wheelchair ramp must have a slope of one inch in height for every one foot of horizontal distance. Joe must determine whether this ramp meets building codes.

Write the Equation Given Two Points

If you know two points on a line, you can write the equation of the line using either slope-intercept form or point-slope form. The first step is to use the slope formula to find the slope of the line.

Example 5: Write an Equation Given Two Points

A line contains the points (4, −4) and (3, 0). Write the equation of this line in slope-intercept form.

Step 1 Find the slope. Substitute the given values into the slope formula.
$$m = \frac{-4 - 0}{4 - 3} = \frac{-4}{1} = -4$$

Step 2 Now you know the slope and two points on the line. You can use either slope-intercept form or point-slope form to write the equation. Whichever form you use, choose only one of the given points to substitute.

$$y = mx + b$$

Choose slope-intercept form.

$$0 = -4(3) + b$$

Choose (3, 0):
$m = -4, x = 3, y = 0$

$$0 = -12 + b$$
$$12 = b \qquad \text{Solve for } b.$$

Step 3 Write the equation.
$$m = -4, b = 12$$
$$y = -4x + 12$$

Find Points from a Graph

You can write the equation of a line by using a graph.

Example 6: Write an Equation Given a Graph

Write the equation of the line shown in standard form.

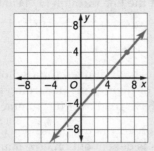

Step 1 Identify the coordinates of two points on the line. Then find the slope by substituting the coordinates into the slope formula.

Two points on this line are $(7, 4)$ and $(2, -2)$.
Let $(7, 4) = (x_1, y_1)$ and $(2, -2) = (x_1, y_1)$

$$m = \frac{y_2 - y_1}{x_2 - x_1} = \frac{4 - (-2)}{7 - 2} = \frac{6}{5}$$

Step 2 Use any method to write the equation. In this example, point slope form is used.

$$y - y_1 = m(x - x_1)$$
$$y - (-2) = \frac{6}{5}(x - 2)$$
$$y + 2 = \frac{6}{5}x - \frac{12}{5}$$
$$y = \frac{6}{5}x - \frac{22}{5}$$

Step 3 Write the equation in standard form.

$$y = \frac{6}{5}x - \frac{22}{5}$$
$$5y = 6x - 22$$
$$-6x + 5y = -22$$
$$-1(-6x + 5y) = -1(-22)$$
$$6x - 5y = 22$$

Model with Mathematics

The graph shows the estimated resting heart rate in beats per minute based on a person's age. Using this information, write the equation of the given line and then use it to find the resting heart rate of a person who is 20 years old.

Average Resting Heart Rate

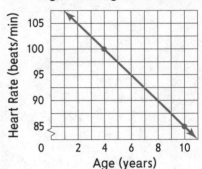

◢ Think about Math

Directions: Answer the following questions.

1. The prices for boat rental are represented on a graph. The line starts at (0, 0) and passes through (3, 60). What is the slope?

2. What is the equation of the line through (1, 4) and (2, 1)?

Using Tables

Several high-school students have applied for a scholarship awarded by the local library. One measure of their effort in earning the scholarship includes the number of books they have read over the summer.

One student decides to graph how many books she can read over the summer, and finds that this is dependent on how many weeks she has free over the break. She then converts her graph to a table.

Summer Reading

Number of Weeks	Number of Books Read
1	3
2	6
3	9

Make a Table from a Graph

Make a table to represent the line graphed.

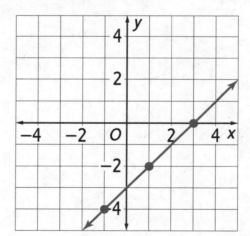

Example 7: Make a Table from a Graph

Step 1 Look at the line that is graphed and identify a few points. This line contains the points (3, 0), (1, −2), and (−1, −4).

Step 2 For each point, write the x-coordinate and the y-coordinate in the appropriate column of the table.

x	y
3	0
1	−2
−1	−4

Think about Math

Directions: Determine which of these points are on the line. If the point is on the line, mark and label the point on the line.

(0, 2), (−2, 4), (−2, 2), (1, 5), (−4, 1), (4, 5)

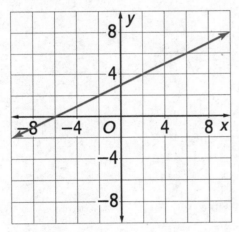

Understand Data in Different Formats

Sometimes you will need to compare similar data in different formats. A manufacturing company has collected the data shown in the graph and table that describes the relationship between time in hours and number of items each factory produces. How does the slope of a line showing this relationship for Factory 1 compare to the slope of the line showing this relationship for Factory 2? Which factory will produce the most items in an 8-hour workday? How do you know?

Factory 1

Hours	Items Assembled
1	1500
3	4500
4	6000

Factory 2

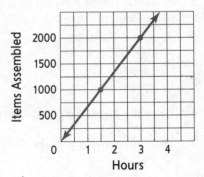

Write the Equation of a Line

Vocabulary Review

Directions: Draw a line to match the term to its definition.

standard form of a
linear equation

the y-coordinate of the point where a line
crosses the y-axis

y-intercept

$y - y_1 = m(x - x_1)$

slope-intercept form

the ratio of rise to run

point-slope form

$Ax + By = C$

coefficient

a number that is multiplied by a variable

slope

$y = mx + b$, where m is the slope of the line and
b is the y-intercept

Skill Review

Directions: Read each problem and complete the task.

1. A line has slope 4 and passes through the point $(-4, 6)$.
    What is the equation of this line in point-slope form?

    A. $y + 4 = x + 4$
    B. $y + 4 = 6x + 4$
    C. $y - 6 = x + 4$
    D. $y - 6 = 4x + 16$

2. What is the equation of the line written in standard form?

    $3y = 4x + 2$

    A. $4x - 3y = -2$
    B. $-3x + 4y = 2$
    C. $-2x + 3y = 4$
    D. $-4x + 2y = 3$

3. A line has slope -3 and passes through $(0, 0)$. What is the equation
    of the line in slope-intercept form?

    A. $y = -3x + 3$
    B. $y = 3x + 3$
    C. $y = 3x$
    D. $y = -3x$

4. Write the equation of the line through $(0, 2)$ and $(1, 4)$ in slope-intercept form.

5. Write the equation of the line through $(-4, 3)$ and $(-1, -1)$ in point-slope form.

6. Represent the information in the graph in a table. Identify at least three points.

Bicycle Rental

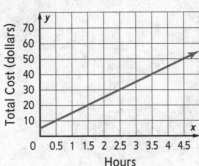

Skill Practice

Directions: Read each problem and complete the task.

1. Emily's commission at a T-shirt shop depends on how many T-shirts she sells each day. She earns $10 for the first T-shirt she sells and $2 for each T-shirt after that. The line that represents her commission has a slope of 2 and passes through (5, 18). What is the equation of the line? Use the equation to find out how much Emily earns in commission for selling 20 T-shirts.

2. A slide is planned for a playground. When represented on a graph, the slide begins at (5, 4) and ends at (0, 0). What is the slope? Write the equation of the line in slope-intercept form.

3. The graph shows the cost to rent a boat at a marina. Find the equation of the line. Then find the cost of renting a boat for 8 hours.

Boat Rental

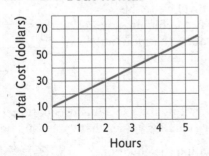

4. A line on a graph represents a ramp that extends from the back of a moving truck to the ground. The line has slope $-\frac{1}{2}$ and passes through (8, 0). The y-intercept represents the height of the back of the moving truck. How tall is the back of the moving truck?

A. 2 feet
B. 4 feet
C. 6 feet
D. 8 feet

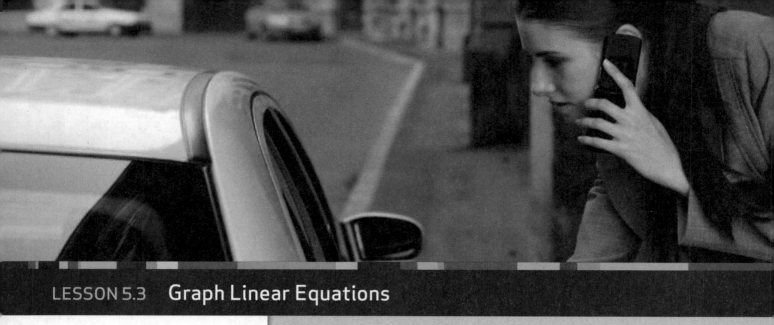

LESSON OBJECTIVES

- Complete a table of x- and y-values for a linear equation
- Use x- and y-values to graph a linear equation
- Graph linear equations to solve real-world problems

CORE SKILLS & PRACTICES

- Solve Linear Equations
- Interpret Graphs

Key Terms

ordered pair
a pair of numbers (x, y) that is used to describe the location of a point in the coordinate plane

slope
the ratio of rise to run

y-intercept
The y-coordinate of a point where a graph crosses the y-axis

Vocabulary

slope-intercept form
$y = mx + b$, where m is the slope of the line and b is the y-intercept

x-value
the horizontal value in an ordered pair

y-value
the vertical value in an ordered pair

Key Concept

You can visualize how two variables in an equation are related by graphing the equation. Solutions of a linear equation can be plotted as ordered pairs on the coordinate plane. You can also use the special forms of linear equations to graph them.

Using Ordered Pairs

Graphing equations is one way to see the equation visually. Each graph is made up of many points, called ordered pairs, which record both the horizontal and vertical direction from the origin. The first value in an ordered pair is the x-value, which is the horizontal value along the x-axis. The second value is the y-value, which is the vertical value along the y-axis. You can use ordered pairs to graph points of relationships between two different objects, such as time vs. distance, products sold vs. profit, etc.

Making a Table

Consider this simple linear equation, $y = 5x$. This equation could represent the total cost in cents, y, for x minutes of cell phone data usage, at a cost of 5 cents per minute.

To find ordered pairs, make a table of the solutions. In the left column, you will write the x-values that you choose. In the right column, you will write the corresponding y-values that you find by evaluating the equation for each x-value.

x	y = 5x
0	
1	
2	

Finding x- and y-values

You can choose any values to substitute for x in a linear equation. However, it makes the most sense to choose smaller values that can easily be graphed.

The y-values are found by evaluating the equation at the x-values you've chosen, so you may want to choose x-values that make it easier to calculate y-values. You need at least two points to graph a line, so choose at least two to three x-values. Let's start with 0, 1, and 2. Calculate the y-values for the equation by substituting each x-value into the equation.

x	y = 5x
0	5(0) = 0
1	5(1) = 5
2	5(2) = 10

Graphing a Line

Now that you have ordered pairs for points on the line $y = 5x$, graph the points and connect them to form a line.

The three ordered pairs using the x- and y-values are $(0, 0)$, $(1, 5)$, and $(2, 10)$.

Connect the points you plotted to form a line. The line represents the linear equation. The solutions of the linear equation are all of the points on the line. Be sure the coordinate axes are labeled and the scale makes sense.

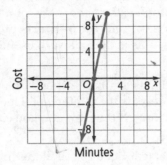

Think about Math

Directions: Find the y-value for each x-value. Plot the points to graph the equation.

x	$y = -4x + 5$
1	
2	
3	

Solve Linear Equations

An integral part of algebra and problem solving involves solving linear equations. To graph a linear equation, you will often have to solve first for y. To solve a linear equation, isolate the variable on one side of the equation. To do this, perform the same operation on both sides of the equation.

Solve $-2x - 8 = 4$ for x.

CALCULATOR SKILL

When making a table of a function, your calculator can be a helpful way to solve for different values of x. You can set a number as a variable in your TI-30XS MultiView™ calculator and then enter the equation using the variable. This can be especially helpful when you want to substitute a negative value for x. To store a number as a variable, type the number you want and then press 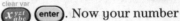 . Now your number is stored as a variable and you can type it directly in your calculator. If you wanted to find $y = 9 - 2x$, when x is the number you stored, you can type

   .

Find $y = 5x + 3$ for $x = -4$ by storing the variable and by entering it into your calculator. What are the advantages of storing it as a variable?

You can learn a lot about an equation by looking at its graph. Learning how to interpret graphs can help you when graphing linear equations. You will be able to quickly recognize whether you have correctly graphed the equation. You can identify whether a line has a positive or negative slope and a positive or negative y-intercept by looking at its position on a graph.

The slope of a line can be either positive or negative. A line with a positive slope goes up from left to right. A line with a negative slope goes down from left to right. A line with a positive y-intercept will cross the y-axis above zero and a line with a negative y-intercept will cross the y-axis below zero.

On a blank graph, draw a line with a positive slope and a negative y-intercept. Write the equation of the line. Then draw a line with a negative slope and a positive y-intercept. Write the equation of the line.

Using Slope-Intercept Form

When a linear equation is in slope-intercept form, it is easy to graph. Many people use linear equations and graphs to build things, create budgets, and monitor fast-changing data. A fast, easy way to graph an equation is an important time-saving tool.

Writing an Equation in Slope-Intercept Form

The **slope-intercept form** of an equation gives you the slope of the line and the y-intercept. The **slope** is the "steepness" of a line. On a graph, it is measured as the ratio of rise to run. The **y-intercept** is the point at which the line crosses the y-axis. This makes it a very useful form for graphing an equation.

The slope-intercept equation is $y = mx + b$. In this equation, m represents the slope of the line and b represents the y-intercept. Linear equations are often written in slope-intercept form. If they are not, you can rewrite them.

Example 1: Convert a Linear Equation to Slope-Intercept Form

Step 1 Solve for y.

$$5x + 6y = 12$$

Step 2 Reorder the terms. In this equation, first subtract $5x$ from both sides, then divide both sides by 6.

$$6y = 12 - 5x$$
$$y = 2 - \frac{5}{6}x$$

Step 3 Arrange the terms to be in slope-intercept format.

$$y = -\frac{5}{6}x + 2$$

Graphing an Equation

If you know the slope of a line and its *y*-intercept, you can graph it.

Example 2: Use Slope and y-intercept to Graph an Equation.

Step 1 Find the *y*-intercept. From the equation $y = 2x + 5$, we know that the *y*-intercept is 5. This means that one point on the line is (0, 5).

Step 2 Use the slope to find a second point on the line. From the equation $y = 2x + 5$, we know that the slope is 2. It is helpful to think of this slope as a fraction over 1, where the numerator is the distance you move on the *y*-axis and the denominator is the distance you move on the *x*-axis. A slope of 2 means move 2 units up and 1 unit to the right from a point on the line to identify another point on the line. If the slope were negative, you would move 2 units down and 1 unit to the left.

$$y = 2x + 5$$

Step 3 The point that is 2 units up and 1 unit right from (0, 5) is (1, 7). After plotting the second point, draw a line between these two points that extends beyond the points in both directions.

$$y = 2x + 5$$

Think about Math

Directions: Answer the following questions.

Taxi Fare

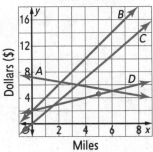

1. Taxi fare with Green Taxis is $2.00 plus $0.50 per mile. The equation to represent total taxi fare is $y = 0.5x + 2$. Which line on the graph matches this equation?

   A. Line A   B. Line B
   C. Line C   D. Line D

2. What is the total fare for a 5-mile trip with Green Taxis?

Evaluate the Answer

When you answer test questions, it is a good idea to spend a few seconds evaluating your answer. By going back and checking your answer, you may catch a mistake you made while you have the opportunity to correct it. When you evaluate an answer, you are making sure the answer makes sense.

A student was asked to graph the equation $y = -\frac{1}{2}x + 2$. The student's graph is shown below. Does the answer make sense? Why or why not?

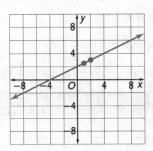

Vocabulary Review

Directions: Write the missing term in the blank.

y-intercept   slope   ordered pair
x-value   *y*-value   slope-intercept form

1. The _____ of a line is its ratio of rise to run.

2. The point at which a line crosses the *y*-axis is the _____ .

3. The _____ is the horizontal value in an ordered pair.

4. A(n) _____ is the horizontal and vertical values that denote the position of a point on a line on a graph

5. A linear equation written as $y = mx + b$ is written in _____ .

6. The vertical value in an ordered pair is the _____ .

Skill Review

Directions: Read each problem and complete the task.

1. Find the *y*-value for each *x*-value. Plot the points to graph the equation.

x	y = 3x − 2
0	
1	
2	

2. A store is running a special on cereal. The cost per box of cereal decreases the more boxes you buy. If price is the y-value, and number of boxes is the *x*-value, which best describes the slope of the line and *y*-intercept?

  A.   The slope is negative and the *y*-intercept is negative.
  B.   The slope is negative and the *y*-intercept is positive.
  C.   The slope is positive and the *y*-intercept is positive.
  D.   The slope is positive and the *y*-intercept is negative.

3. Marla is 20 miles from home at her uncle's house. She continues driving away from home toward her grandmother's house at a constant speed of 60 mph.

   a. Graph an equation to represent her total miles from home.

   b. If Marla drives for two hours, how far away from home will she be when she is at her grandmother's house?

      A. 140 miles

      B. 120 miles

      C. 60 miles

      D. 20 miles

4. Solve for x.

   $-4x - 3 = 13$

5. Kellan brings x soccer balls to practice. His coach brings 12 soccer balls to practice. If $2x + 4$ equals the number of balls his coach brought to practice, how many balls did Kellan bring to practice?

Skill Practice

Directions: Read each problem and complete the task.

1. A kennel charges an upfront fee of $12. Each day of boarding is $8 for cats and $12 for dogs. Graph the equations $y = 8x + 12$ and $y = 12x + 12$.

   a. What is the cost for boarding a cat for a week? What is the cost of boarding a dog for a week?

   b. For how many dogs and how many cats is the total price per day the same?

      A. 2 dogs and 3 cats

      B. 3 dogs and 2 cats

      C. 2 dogs and 2 cats

      D. 3 dogs and 4 cats

2. Ellie wrote a paper for class that she is typing into the computer. She types at a rate of 120 words per minute. The equation to represent how long it will take her to type papers of different lengths is $y = 120x$.

   a. Graph the equation.

   b. How long will it take Ellie to type a 3,000-word paper?

3. The equation $y = -3x - 2$ represents Jaden's movement of his pieces across a game board.

   a. Plot the point for the y-intercept and a second point with $x = 1$ to graph the line.

   b. If Jaden began at the y-intercept and moved 1 unit to the right with each turn, how many units down from his starting point was he after 3 turns?

■ LESSON OBJECTIVES

- Solve a system of linear equations algebraically and graphically
- Solve problems leading to a system of linear equations

■ CORE SKILLS & PRACTICES

- Represent Real-World Problems
- Solve Pairs of Linear Equations

Key Terms

system of linear equations
a set of two or more linear equations with two or more variables

independent system
a system that has one solution

inconsistent system
a system that has no solutions

dependent system
a system that has an infinite number of solutions

Vocabulary

substitution method
a method of solving a system of equations by solving one equation for one variable and substituting the resulting expression into the other equation

elimination method
a method of solving a system of equations by adding or subtracting equations to eliminate one of the variables

Key Concept

Just like a solution of an equation is a value that makes the equation true, a solution of a system of equations is a set of values that makes all of the equations in the system true. You can solve systems of linear equations graphically by finding the point at which the graphs of the equations intersect. You can also solve systems algebraically, by using the substitution or the elimination method.

The Graphing Method

When planning a cookout for a large group of friends, you must determine what to buy and in what quantities. How many people will want hot dogs or hamburgers? How many buns come in a package? You can use linear equations to represent these questions and then graph the equations on a coordinate plane. The intersection of the lines is the solution and will help you make decisions about cookout supplies.

Systems of Linear Equations

A **system of linear equations** is a set of two or more linear equations with two or more variables. The system shown is a system of two linear equations with two variables.

$$y - x = 1$$
$$x + y = 3$$

A solution of a system of two linear equations with two variables is an ordered pair that makes *both* equations true. The ordered pair $(4, 5)$ is *not* the solution of this system. It makes the first equation true, but not the second one. The ordered pair $(5, -2)$ is *not* the solution of the system. It makes the second equation true, but not the first one. The ordered pair $(1, 2)$ is the solution of the system. It makes both equations true.

Solving by Graphing

Remember that the graph of an equation contains all ordered pairs that make the equation true. One way, then, to solve a system of equations is to graph the equations and find the intersection point. Because the intersection point lies on both graphs, it will make both equations true.

Example 1: Solve a System by Graphing

Solve the system by graphing.

$$3x + 3y = 9 \qquad 4x + 2y = 8$$

Step 1 In order to make it easier to graph, write each equation in slope-intercept form by solving for y.

$3x + 3y = 9$	$4x + 2y = 8$
$3y = -3x + 9$	$2y = -4x + 8$
$y = -x + 3$	$y = -2x + 4$
slope: -1	slope: -2
y-intercept: 3	y-intercept: 4

Step 2 Use the slopes and y-intercepts to graph both equations. Identify the point where the lines intersect. This point is the solution.

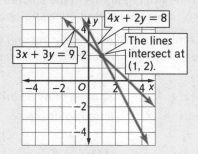

$4x + 2y = 8$

$3x + 3y = 9$

The lines intersect at (1, 2).

Step 3 Check the solution by substituting into *both* original equations. The solution of the system is $(1, 2)$.

Check
$3(1) + 3(2) = 9$ ✔
$4(1) + 2(2) = 8$ ✔

A system of equations that has exactly one solution, like the system in Example 1, is called an **independent system**.

CORE SKILL

Represent Real-World Problems

Systems of linear equations can be used to represent real-world problems. To write a system of linear equations to represent a real-world problem, read the problem carefully to identify the given information, determine the unknown values that will be represented with variables, and decide which operations are necessary. Write a system of linear equations to represent the problem.

Tickets for a play cost either $10 or $15. A total of 200 tickets are sold. The total amount of money paid for tickets is $2,600. Write a pair of equations that can be written to model the situation.

Inconsistent and Dependent Systems

Not all systems of equations have exactly one solution. For a system of two linear equations in two variables, there are two other possibilities.

The graphs of the two equations could be parallel lines. Because the lines do not intersect, such a system has no solutions. A system with no solutions is called an **inconsistent system**.

$$y = -x + 5 \qquad \text{This system has no solutions}$$
$$x + y = 3$$

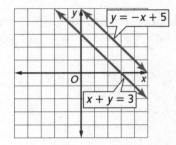

The graphs of the two equations could be the same line. The graph will look like just one line. However, both equations in the system are represented by the same line, so the "two" lines intersect at every point on the line.

This means that every point on the line is a solution and the system has infinitely many solutions. A system with infinitely many solutions is called a **dependent system**.

$$y = \frac{2}{3}x + 4 \qquad \text{This system has infinitely}$$
$$3y - 2x = 12 \qquad \text{many solutions.}$$

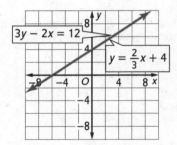

Think about Math

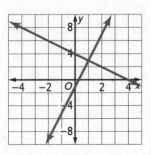

Directions: Answer the following question.

What is the solution of the system shown in the graph?

A. (0, 4)
B. (1, 4)
C. (0, 0)
D. (1, 3)

The Substitution Method

When cooking, you might substitute one ingredient for another ingredient. You make substitutions in math as well. Sometimes you are given a value to substitute for a variable in an equation. By substituting the value, you can solve the equation. You can also use substitution to solve a system of equations.

Solving by Substitution

Sometimes it may be difficult to solve a system by graphing because you cannot read the exact coordinates of the intersection point. There are also algebraic methods to solve a system. One algebraic method is the **substitution method**.

In the substitution method, the first step is to solve one of the equations for one of the variables. You then substitute into the other equation.

Example 2: Solve a System by Substitution

Solve the system by substitution.
$$-3x + 2y = -12$$
$$x - y = 2$$

Step 1 Choose an equation to solve for one of the variables. You may choose either equation and solve for either variable. In this case, the simplest choice is to solve the second equation for x.

$x - y = 2$
$x = 2 + y$　　Add y to both sides.

Step 2 Substitute into the *other* equation. Now you have one equation that contains one variable. Solve for the variable. In this case, substitute $2 + y$ for x in the first equation. Then solve for y.

$-3(2 + y) + 2y = -12$
$-6 - 3y + 2y = -12$　　Distributive Property
$-6 - y = -12$　　Combine like terms.
$-y = -6$　　Add 6 to both sides.
$y = 6$　　Multiply both sides by -1.

Step 3 Substitute the value you found in Step 2 into either original equation to find the value of the other variable. Here, substitute $y = 6$ into the second equation and solve for x.

$x - y = 2$　　Choose the second equation.
$x - 6 = 2$　　Substitute $y = 6$.
$x = 8$　　Add 6 to both sides.

Step 4 Write the solution as an ordered pair. Check by substituting into both original equations. The solution is $x = 8$ and $y = 6$, or (8, 6).

Check　　$-3(8) + 2(6) = -12$ ✔
　　　　$8 - 6 = 2$ ✔

Think about Math

Directions: Answer the following question.

1. Caden spent $600 on supplies to start a document-shredding business from his home. He estimates it will cost $2.00 per box of documents to shred. He plans to charge $10 per box of documents. How many boxes of documents will he need to shred to break even?

The Elimination Method

Many of us have "To Do" lists—lists of tasks that need to be done. When you complete a task and cross it off your list, you eliminate it. As you eliminate tasks, your list becomes more and more manageable. The same is true when you eliminate a variable from a system of linear equations. The system becomes easier to solve!

Multiplying One Equation

Another algebraic method to solve systems of equations is the **elimination method**. In this method, you eliminate one variable by adding or subtracting equations.

Example 3: Eliminate by Multiplying One Equation

Solve the system by elimination.

$3x + y = 11$
$4x + y = 14$

Step 1 Multiply one or both of the equations by a constant so that the two x-terms or the two y-terms will have opposite coefficients. If either of these equations is multiplied by -1, the y-terms will have opposite coefficients.

$4x + y = 14$	Choose the second equation.
$-1(4x + y) = (-1)\,14$	Multiply both sides by -1.
$-4x - y = -14$	Simplify.

Step 2 Add the equations so that terms are eliminated. Add the new equation from Step 1 to the first equation.

$$\begin{array}{r} 3x + y = 11 \\ +\ -4x - y = -14 \\ \hline -x \quad\ = -3 \end{array}$$

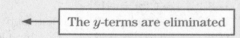 The y-terms are eliminated

Now you have one equation that contains one variable. Solve for the variable.

$-x = -3$	
$x = 3$	To solve for x and eliminate the negative, multiply both sides by -1.

Step 3 Substitute the value you found in Step 2 into either original equation to find the value of the other variable.

$3x + y = 11$	Choose the first equation.
$3(3) + y = 11$	Substitute $x = 3$.
$9 + y = 11$	Simplify.
$y = 2$	Subtract 9 from both sides.

Step 4 Write the solution as an ordered pair. Check by substituting into both original equations. The solution is $x = 3$ and $y = 2$, or $(3, 2)$.

Check
$3(3) + 2 = 11$ ✓
$4(3) + 2 = 14$ ✓

Multiplying Both Equations

You may have to multiply both equations by a constant before you can add to eliminate terms.

Example 4: Eliminate by Multiplying Both Equations

Solve the system by elimination.

$5x + 2y = -4$
$-2x + 6y = 5$

Step 1 Multiply one or both of the equations by a constant so that the two x-terms or the two y-terms will have opposite coefficients. If the first equation is multiplied by 2 and the second is multiplied by 5, the x-terms will have opposite coefficients.

Multiply the first equation by 2 and simplify:
$2(5x + 2y) = 2(-4)$
$10x + 4y = -8$

Multiply the second equation by 5 and simplify:
$5(-2x + 6y) = 5(5)$
$-10x + 30y = 25$

Step 2 Add the equations from Step 1 so that terms are eliminated.

$$\begin{array}{r} 10x + 4y = -8 \\ + \ -10x + 30y = 25 \\ \hline 34y = 17 \end{array}$$

← The x-terms are eliminated

Now you have one equation that contains one variable.

Solve for y.
$34y = 17$
$y = 0.5$ \qquad Divide both sides by 34.

Step 3 Substitute the value you found in Step 2 into either equation to find the value of the other variable. Substitute $y = 0.5$ into either equation and solve for y.

$5x + 2y = -4$	Choose the first equation.
$5x + 2(0.5) = -4$	Substitute $y = 0.5$.
$5x + 1 = -4$	Simplify.
$5x = -5$	Subtract 1 from both sides.
$x = -1$	Divide both sides by 5.

Step 4 Write the solution as an ordered pair. The solution is $x = -1$ and $y = 0.5$, or $(-1, 0.5)$. Check by substituting into both original equations.

Check \qquad $5(-1) + 2(0.5) = -4$ ✔
\qquad\qquad $-2(-1) + 6(0.5) = 5$ ✔

Think about Math

Directions: Answer the following question.

1. Elias purchased rectangular and square patio blocks to build a patio. He paid $2 for each square block and $3 for each rectangular block. The total cost was $350, and he bought 150 blocks in all. How many square blocks and how many rectangular blocks did Elias purchase?

Solve Pairs of Linear Equations

You have learned three methods to solve a system of equations: graphing, substitution, and elimination. When solving a system, you should try to choose the method that is most efficient for that particular system.

- If both equations are solved for y and the numbers are relatively small, graphing may be a good choice.

- If one of the equations is solved for a variable or can easily be solved for a variable, substitution may be the best method.

- In all other cases, elimination may be best.

For each system of equations, describe the solution method that would work best.

$-5x + 4y = 12$
$2x - y = 5$

$y = x + 2$
$y = -x - 1$

$y = 2x + 8$
$4x + y = 5$

Vocabulary Review

Directions: Write the missing term in the blank.

system of linear equations	**independent system**	**inconsistent system**
dependent system	**substitution method**	**elimination method**

1. A system that has exactly one solution is called a(n) _____.

2. The _____ is a method of solving a system of equations by solving one equation for one variable and substituting the resulting expression into the other equation.

3. A(n) _____ has an infinite number of solutions.

4. A set of two of more linear equations with two or more variables is a(n) _____.

5. A(n) _____ has no solutions.

6. The _____ is a method of solving a system of equations by adding or subtracting equations to eliminate one of the variables.

Skill Review

Directions: Read each problem and complete the task.

1. Identify the type of system shown in the graph.

$$x + y = 2$$
$$2x + y = 5$$

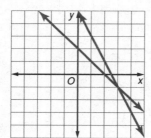

2. Which statement or statements are true of the system of linear equations?

$$-3x + 2y = -12$$
$$x - y = 2$$

A. The system is independent.
B. The system is dependent.
C. The solution of the system is (8, 6).
D. The solution of the system is (4, −8).

3. Find the values of two numbers if their sum is 5 and their difference is 1.

a. Write a system of equations to solve the problem.

b. What are the values of x and y?

4. A quiz has 20 questions worth a total of 40 points. Each multiple-choice question is worth 1 point, and each short-answer question is worth 5 points. Which system of equations can be used to represent this situation?

A. $x + y = 20$
  $x + 5y = 40$

B. $5x + y = 20$
  $x + y = 40$

C. $x + y = 20$
  $5x + 5y = 40$

D. $x - y = 20$
  $x + y = 40$

5. What is the solution of the system shown in the graph?

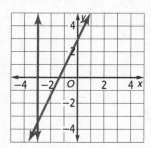

6. Amy and Dan are looking for a photographer for their wedding. Photographer A charges $400 for the photo shoot and $6 for each print, no matter the size. Photographer B charges $500 for the photo shoot and $4 for each print.

   a. For how many prints are the total prices of the two photographers the same?

   b. Which photographer will be the best value if they expect to order 100 prints?

Skill Practice

Directions: Read each problem and complete the task.

1. A theater is selling tickets to a musical. On the first day of ticket sales, they sold 25 senior tickets and 14 child tickets for a total of $362. On the second day, they sold 5 senior tickets and 7 child tickets for a total of $106. What is the price each for one senior ticket and one child ticket?

   A. $8 for a child and $12 for a senior
   B. $10 for a child and $10 for a senior
   C. $8 for a child and $10 for a senior
   D. $10 for a child and $8 for a senior

2. Taya is a certified lifeguard. She spent $90 to get her certification. She gets paid a base salary at the pool of $40 per week plus $10 for each swimming class she assists. How many classes does she need to teach in the first week to earn back what she spent on her certification?

3. Iris flew two separate airlines to reach her destination. With the first airline, she had to pay $26 to check her bag, plus $3 for every pound that her bag was over the weight limit. On the second flight, she had to pay $5 per pound that her bag was over the weight limit, in addition to the checked bag fee of $22. The two airlines have the same weight limit for checked luggage. Iris's luggage fees were the same for both flights.

   a. Graph the two equations to find out how many pounds over the weight limit Iris's bag was.

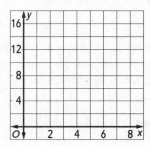

   b. What was Iris's checked luggage fee for each flight?

4. Eric is comparing phone plans. Information about each plan is shown in the table.

	Monthly Fee	Price per Minute of Talk or Text
Plan A	$22	$0.05
Plan B	$18	$0.07

   a. For what number of minutes is the total monthly cost of the two plans the same?

   A. 400
   B. 200
   C. 100
   D. 20

   b. If Eric spends an average of 150 minutes talking and texting each month, which phone plan is a better option for him?

Directions: Choose the best answer to each question.

1. Bailey's Barbeque sells adult meals at one price and children's meals at another price. At lunch they sold 15 adult meals and 5 children's meals for a total of $90. At dinner they sold 12 adult meals and 6 children's meals for a total of $78. What is the price for each adult meal and the price for each children's meal?

   A. Adult meal: $4, Child meal: $6
   B. Adult meal: $5, Child meal: $3
   C. Adult meal: $6, Child meal: $1
   D. Adult meal: $3, Child meal: $5

2. A line contains the points (4, 7) and (8, 19). What is the equation of this line in slope-intercept form?

   A. $4x + 7y = 19$
   B. $y = 4x + 7$
   C. $7x + 8y = 19$
   D. $y = 3x - 5$

3. Apples cost $3 per pound. When you graph this rate as a line on a coordinate plane, (1, 3) is a point on the graph. Another point on the line is _____.

4. Abby graphs these two equations: $y = 2x + 2$ and $x + 2y = 4$. The solution to the system of equations is represented by the point _____.

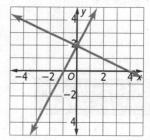

5. Greg graphed a line shown below. Which is a point on the line?

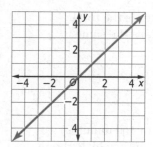

   A. $(-2, 2)$
   B. $(6, 7)$
   C. $(7, 7)$
   D. $(0, 5)$

6. Sophia is solving this system of equations with the elimination method: $2x + 5y = 10$ and $x - 5y = 11$. What are the values of x and y?

   A. $x = 7, y = -\frac{4}{5}$
   B. $x = 7, y = \frac{4}{5}$
   C. $x = \frac{4}{5}, y = -7$
   D. $x = -\frac{4}{5}, y = 7$

7. Jen graphs a line that has the points (4, 28) and (7, 43). What is the slope of the line she graphed?

   A. 71
   B. 15
   C. 5
   D. $\frac{1}{5}$

8. The equation $10x + 2y = 6$ can be rewritten in point-slope form as _____.

Directions: Use Graph A for Problems 9–11.

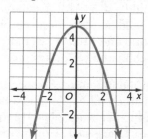

Graph A

9. The y-intercept is 5 and the x-intercepts are _____.

10. The increasing interval is $x < 0$ and the decreasing interval is _____.

11. The _____ is 5 and there is no relative minimum.

12. Eva plotted a point on a coordinate plane. To plot the point, she started at the point (0, 0), then moved two units right and 3 units down. What is the coordinate pair for the point?

    A. $(2, -3)$
    B. $(3, -2)$
    C. $(-2, 3)$
    D. $(2, 3)$

13. Heather graphed a line with the points (6, 4) and (9, 5). Then she made a table with the x- and y-coordinates. What are the x- and y-coordinates for another entry in the table?

x-coordinate	y-coordinate
6	4
9	5

A. (12, 6)
B. (15, 8)
C. (6, 12)
D. (10, 7)

14. Crystal is using the substitution method to solve this system of linear equations: $5x + 2y = 18$ and $x + 6y = 5$. Which of the equations below could she use to solve the problem?

A. $5x + 2\left(\frac{5}{6} - x\right) = 18$

B. $5\left(\frac{5}{6} - x\right) + 2y = 18$

C. $5(x + 6y) + 2y = 18$
D. $5(5 - 6y) + 2y = 18$

15. Which situation is represented by the graph?

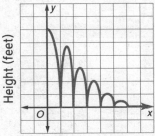

Time (seconds)

A. The height of a child as he grows into an adult.
B. A rubber ball dropped from 4 feet and bounces several times before rolling to a stop.
C. The speed of a car when driving, stopping at a stop sign, then continuing for 10 minutes, and stopping at its destination.
D. The depth of the ocean floor as the distance from the shore increases.

16. Ella records the inputs and outputs of a function in a table. The table represents the equation _____.

x	y
0	2
1	7
2	12

Check Your Understanding

On the following chart, circle the items you missed. The last column shows pages you can review to study the content covered in the question. Review those lessons in which you missed half or more of the questions.

Lesson	Item Number(s)			Review Page(s)
	Procedural	Conceptual	Problem Solving	
5.1 Interpret Slope	16	3	7, 12	150–157
5.2 Write the Equation of a Line	2	8		158–165
5.3 Graph Linear Equations	5	9, 10, 11, 15	13	166–171
5.4 Solve Systems of Linear Equations	4	6, 14	1	172–179

Chapter 6

Functions

Factories produce diverse items from cars to loaves of bread. A production line on a factory floor begins with raw materials and ends with a fully packaged product, ready for shipping to the customer. In a similar way, a function has an input, and certain operations are performed on this input value to yield an output value. In this chapter you will understand how to work with functions, graphing, and identifying key features.

Neil Beer/Getty Images

Lesson 6.1
Identify a Function

Every square is a rectangle, but not every rectangle is a square. Similarly every function is an equation, but not every equation defines a function. Learn how to identify a function using the vertical line test.

Lesson 6.2
Identify Linear and Quadratic Functions

You learned methods for factoring quadratic equations to find solutions. Those solutions correspond to x-intercepts on the graph of quadratic functions. Learn how to identify linear and quadratic functions by analyzing consecutive differences.

Lesson 6.3
Identify Key Features of a Graph

What are some important features of a function? How can you identify those features in a graph? What do those features translate to in the equation of the function? Learn to identify key features of a function in tables, graphs, and equations.

Lesson 6.4
Compare Functions

When solving a real-world problem, what is your first step? Do you make a table, graph, or equation? How do you decide which type will be most appropriate? Learn techniques for using multiple representations to solve real-world problems.

Goal Setting

Think about solving a linear equation. What steps do you follow? Did you graph the equation or solve the equation algebraically? What information is easy to tell from an equation that is not immediately found on a graph? What information is easy to find from a graph?

How do you solve a quadratic equation algebraically? How might graphing the quadratic equation help you solve the problem?

LESSON OBJECTIVES

- Recognize a function as a table of values, a graph, an equation, and in the context of a scenario
- Evaluate linear and quadratic functions
- Plot points in a coordinate plane

CORE SKILLS & PRACTICES

- Use Math Tools Appropriately
- Solve Real-World Problems

Key Terms

function
a rule that assigns exactly one output to each input

linear function
a function that can be written in the form $f(x) = mx + b$, where m and b are constants, whose graph is a non-vertical line

quadratic function
a polynomial that has 2 as its highest power of x

Vocabulary

domain
the set of inputs of a function

one-to-one function
a function for which every value in the range has exactly one element assigned to it from the domain

range
the set of outputs of a function

Key Concept

A function assigns exactly one output for each input. The inputs of a function are a given set, and the outputs for this function create another set. The outputs are what the function did to the set of inputs. A good way to identify a function is to use the Vertical Line Test.

Functions

Functions are used in physics, environmental science, biology, economics, business, and finance. They are used in every field of study. Interpreting the output of functions tells us a lot about our world!

Function and Its Purpose

A **function** is a rule that assigns exactly one output to each input. The inputs and the outputs can be represented by sets. An input is an element from a set called the **domain**. An output is an element from a set called the **range**. We can represent a function using the symbol $f(x)$.

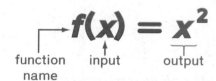

$$f(x) = x^2$$

function name   input   output

Functions allow people to see unique relationships both numerically and graphically.

Tables of Values

We can use tables of values to represent functions. A table represents a function if there is exactly one value, $f(x)$, in the range for each value, x, in the domain.

The table shows a function in which every domain value is multiplied by 2. The results are the range.

Domain	Range
x	**f(x)**
1	2
2	4
3	6
4	8

$f(x) = 2x$

Identify a Function

Example 1: Identifying a Function from a Table

Tell whether each table represents a function.

a.

Domain	Range
−1	4
0	0
1	4
2	8

b.

Domain	Range
0	0
1	1
4	−2
4	2

Check whether each domain value has exactly one range value.

a. Each domain value has exactly one range value. Even though 4 appears twice in the range, the domain value −1 has only one range value, and 1 has only one range value.

b. The domain value 4 has two range values, −2 and 2. This table does not represent a function.

The table in Example 1a represents a function, but it does not represent a **one-to-one function**. A one-to-one function is a function in which every domain value has a different range value.

Graphs

We can represent a function as a graph on a coordinate plane. The elements of the domain are represented by x-coordinates and the elements of the range are represented by y-coordinates. A graph represents a function if no two points have the same x-coordinate and different y-coordinates.

Example 2: Identifying a Function from a Graph

Tell whether each graph represents a function.

a.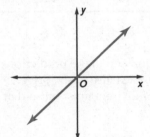

b.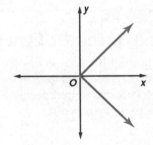

a. This graph represents a function because no two points have the same x-coordinate and different y-coordinates.

b. This graph does not represent a function because there are many points with the same x-coordinate and different y-coordinates.

Use Math Tools Appropriately

You can use a tool called the Vertical Line Test to determine whether a graph represents a function.

For a graph to represent a function, no two points can have the same x-coordinate and different y-coordinates. If two points did have the same x-coordinate but different y-coordinates, these two points would lie on a vertical line. For example, the points $(4, 2)$ and $(4, −3)$ lie on the vertical line $x = 4$. Therefore, a graph represents a function if there is no vertical line that intersects the graph at more than one point.

Use the Vertical Line Test to determine whether each graph represents a function.

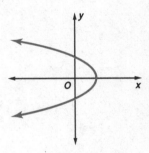

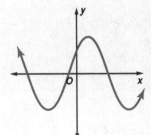

Equations

We can use equations to represent functions. Equations describe the rules for inputs to outputs mathematically. An example of an equation that represents a function is $f(x) = 2x + 3$.

> **Example 3: Writing an Equation for a Function**
>
> Suppose every item in a store is priced at $5.00. Write a function to represent the cost of x items. What is the cost of 4 items? If someone spent $30.00, how many items did this person buy?
>
> **Step 1** Because the cost of each item is $5.00, the cost of x items is $5x$. The function is $f(x) = 5x$.
>
> **Step 2** To find the cost of 4 items, evaluate $f(x)$ when $x = 4$: $f(4) = 5(4) = 20$. The cost of 4 items is $20.00.
>
> **Step 3** Someone spent $30.00 and bought x items. Substitute 30 for $f(x)$ and solve for x: $30 = 5x$, so $x = 6$. This person bought 6 items.

◣ Think about Math

Directions: Tell whether the table and the graph represent functions. Explain your answers.

1.

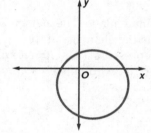

2.

Domain	Range
1	1
2	1
3	1

Linear and Quadratic Functions

No one can predict the future, but with functions you can make predictions based on certain rules. Businesses use functions to measure costs, price, and revenue. This can help determine how many of an item must be sold at a certain price to make a profit.

Evaluate Linear Functions

A **linear function** expresses a linear equation using function notation. A linear function can be written in the form $f(x) = mx + b$. The graph of a linear function is a non-vertical line. Vertical lines do not represent functions because every point on a vertical line has the same x-coordinate.

> **Example 4: Evaluating a Linear Function**
>
> Find the value of $f(x) = \frac{1}{2}x + 3$ when $x = 4$.
>
> **Step 1** Substitute 4 for x. $\qquad\qquad\qquad f(4) = \frac{1}{2}(4) + 3$
>
> **Step 2** Simplify by multiplying and then adding. $\qquad = 2 + 3$
> $\qquad\qquad\qquad\qquad\qquad\qquad\qquad\qquad\qquad = 5$
>
> When $x = 4, f(x) = 5$.

Evaluate Quadratic Functions

A **quadratic function** expresses a quadratic equation using function notation. A quadratic function can be written in the form $f(x) = ax^2 + bx + c$, where $a \neq 0$. Evaluating a quadratic function is similar to evaluating a linear function. In a quadratic function, however, there will be an exponent that must be evaluated before any other operations are performed.

> **Example 5: Evaluating a Quadratic Function**
>
> Find the value of $f(x) = x^2 + 4x - 3$ when $x = -2$.
>
> **Step 1** Substitute -2 for x. $\qquad\qquad f(-2) = (-2)^2 + 4(-2) - 3$
>
> **Step 2** Simplify. First evaluate the $\qquad = 4 - 8 - 3$
> exponent. Then multiply. $\qquad\qquad = -4 - 3$
> Finally, add and subtract $\qquad\qquad = -7$
> from left to right.
>
> When $x = -2, f(x) = -7$.

Think about Math

1. Find the value of the function $f(x) = -7x + 4$ when $x = -2$, $x = 1$, and $x = 4$.
2. Find the value of the function $f(x) = 2x^2 + 2$ when $x = -2$, $x = 0$, and $x = \frac{1}{2}$.

21ST CENTURY SKILL

Business Literacy

A business uses the function $C(x) = x^2 - 6x + 8$ to determine the cost to produce x tennis rackets. How much does it cost to produce 20 tennis rackets?

Solve Real-World Problems

The function $H(t) = -16t^2 + 256$ is used in physics to describe the height H in feet of an object t seconds after it has been dropped from a height of 256 feet.

Evaluate this function when $t = 0$, $t = 1$, $t = 2$, $t = 3$, and $t = 4$. What do your answers represent? Graph the corresponding points in the coordinate plane and connect them to form the graph of the function.

Functions in the Coordinate Plane

We graph functions in the coordinate plane to learn more about them. We can observe a graph's characteristics to better understand the behavior of the function. We can look at rates of change and tell how much a racecar driver is accelerating or decelerating through a curve.

Plot Points on the Coordinate Plane

When function notation was introduced, we replaced y with $f(x)$.

$$y = x + 1$$
$$f(x) = x + 1$$
$$\text{So, } y = f(x).$$

On a coordinate plane, we plot points (x, y). For a function, each point will be $(x, f(x))$. Again, y is replaced with $f(x)$.

We can evaluate a function for any value of x in the domain and graph the corresponding point on the coordinate plane.

Example 6: Using Functions to Graph Points

a. For the function $f(x) = x + 1$, graph the point whose x-coordinate is 2.

Step 1 Find the value of the function when $x = 2$.

$f(2) = 2 + 1 = 3$

Step 2 When $x = 2$, $f(x) = 3$. Graph the point $(2, 3)$.

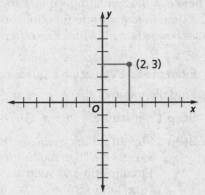

b. For the function $f(x) = 2x - 5$, graph the point whose x-coordinate is 3.

Step 1 Find the value of the function when $x = 3$.

$f(3) = 2(3) - 5 = 6 - 5 = 1$

Step 2 When $x = 3$, $f(x) = 1$. Graph the point $(3, 1)$.

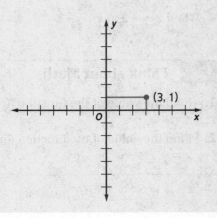

Piecewise Functions

A piecewise function consists of two or more parts. It applies two or more algebraic rules to different parts of the domain. Even though it consists of more than one rule, it is treated as one function. This creates graphs with unique characteristics. We still have to make sure all the combined parts assign exactly one range value for each domain value.

Example 7: Evaluating and Graphing a Piecewise Function

Evaluate the piecewise function when $x = -2$, $x = -1$, $x = 1$, $x = 2$, and $x = 3$. Then graph the function.

$$f(x) = \begin{cases} x \text{ when } x < 1 \\ 3 \text{ when } x = 1 \\ -x \text{ when } x > 1 \end{cases}$$

Step 1 Evaluate the function when $x = -2$.
$-2 < 1$, so use the first part of the function, $f(x) = x$.

$f(-2) = -2$

Step 2 Evaluate the function when $x = -1$.
$-1 < 1$, so use the first part of the function, $f(x) = x$.

$f(-1) = -1$

Step 3 Evaluate the function when $x = 1$.
Use the second part of the function, $f(x) = 3$.

$f(1) = 3$

Step 4 Evaluate the function when $x = 2$.
$2 > 1$, so use the third part of the function, $f(x) = -x$.

$f(2) = -2$

Step 5 Evaluate the function when $x = 3$.
$3 > 1$, so use the third part of the function, $f(x) = -x$.

$f(3) = -3$

Step 6 Graph the points $(-2, -2)$, $(-1, -1)$, $(1, 3)$, $(2, -2)$, and $(3, -3)$. Connect the points to form the graph.

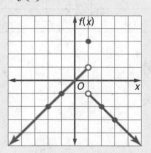

Think about Math

Evaluate the piecewise function when $x = -2$, $x = -1$, $x = 1$, $x = 2$, and $x = 3$. Then graph the function.

$$f(x) = \begin{cases} -2x \text{ when } x < 0 \\ 2x \text{ when } x \geq 0 \end{cases}$$

Vocabulary Review

Directions: Write the missing term in the blank.

domain	quadratic function	function
linear function	one-to-one function	range

1. The set of outputs of a function is the _____.

2. In a _____, each input has exactly one output.

3. A _____ can be written in the form $f(x) = ax^2 + bx + c$, where $a \neq 0$.

4. The set of inputs of a function is the _____.

5. A _____ can be written in the form $f(x) = mx + b$.

6. In a _____, each input has a different output.

Skill Review

Directions: Read each problem and complete the task.

1. Which table or tables represent functions? Which function or functions are one-to one?

Table A

Domain	Range
−3	3
−2	2
−1	1
0	0

Table B

Domain	Range
3	0
6	1
9	0
12	1

Table C

Domain	Range
−1	1
0	2
−1	3
−2	4

Table D

Domain	Range
0.2	0.04
0.5	0.25
0.8	0.64
1.1	1.21

2. Which graph does not represent a function?

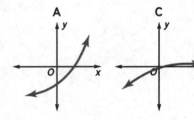

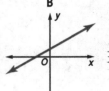

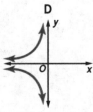

3. The one-to-one function $f(x) = 6.5x$ represents the cost to download x books from a Web site.

   a. What is the cost per book?

   b. Find the cost of downloading 4, 5, 6, and 7 books.

4. What is the value of the function $f(x) = x^2 - 2x + 7$ when $x = -3$?

   A. 22

   B. 10

   C. 7

   D. 4

5. Evaluate the function $f(x) = -x^2 - x + 2$ when $x = -2$, $x = -1$, $x = 0$, $x = 1$, and $x = 2$. Then graph the corresponding points.

6. Graph the piecewise function.
$$f(x) = \begin{cases} x - 1 \text{ when } x < -1 \\ 0 \text{ when } x = -1 \\ -x + 1 \text{ when } x > -1 \end{cases}$$

Skill Practice

Directions: Read each problem and complete the task.

1. Identify the table that does not represent a function. Use the definition of a function to explain why this table does not represent a function.

Table A

Domain	−1	0	1	2
Range	−1	0	1	2

Table B

Domain	−1	0	1	2
Range	0	0	0	0

Table C

Domain	−1	0	−1	2
Range	−1	0	1	2

Table D

Domain	−1	0	1	2
Range	−1	0	−1	−2

2. Explain why the Vertical Line Test can be used to determine whether a graph represents a function.

3. Which situation could be represented by the function $f(x) = 8x$?

   A. Lindsey bought 8 pieces of candy for $1.00.

   B. Pat earns $8.00 per hour.

   C. Ken worked for 8 hours.

   D. Julian sold 8 more raffle tickets than his cousin sold.

4. Lou ran at a speed of 8 miles per hour.

   a. Write a function to represent the number of miles that Lou would run in x hours if he maintained this speed.

   b. If Lou could run at this speed for 2 hours, what distance would he run?

   c. Is the function a one-to-one function? Explain.

   d. At this speed, how many minutes will it take Lou to run one mile?

   e. Is the function linear or quadratic? Explain.

5. The function $H(t) = -16t^2 + 400$ gives the height H in feet of a ball t seconds after it has been dropped from a height of 400 feet. What is the value of this function when $t = 5$? What is the meaning of this value in the context of the problem?

6. Ricardo said that the piecewise relationship below is a function. Explain to Ricardo why he is incorrect. How could you modify the relationship so that it is a piecewise function?
$$f(x) = \begin{cases} x + 1 \text{ when } x \geq 1 \\ x - 1 \text{ when } x \leq 1 \end{cases}$$

Identify Linear and Quadratic Functions

LESSON OBJECTIVES

- Evaluate linear and quadratic functions in the form of a table or graph
- Recognize linear and quadratic functions in the form of a table or graph

CORE SKILLS & PRACTICES

- Critique the Reasoning of Others

Key Terms

common difference
the amount that is the same between consecutive differences

consecutive difference
the subtraction between the next and current terms in a table

Vocabulary

linear function
a function that represents a line

quadratic function
a polynomial function that has 2 as its highest power of x

coordinate
the pairs (x, y) graphed on a plane

Key Concept

Linear and quadratic functions express a relationship between two variables—one independent and the other dependent. As the independent variable changes, the dependent variable of linear functions changes at a constant rate while the dependent variable of quadratic functions does not change at a constant rate.

Evaluating Linear and Quadratic Functions

Linear and quadratic equations help model interest rates and gravitational forces, respectively. Plugging values into these equations helps determine the amount of interest and the amount of gravity.

Linear Functions

A **linear function** represents a line. Algebraically, the equation of a line is written as $f(x) = mx + b$.

The table shows values of the linear function $f(x) = 2x - 3$ for several values of x.

x	$f(x) = 2x - 3$	Consecutive Difference
−3	$2(-3) - 3 = -9$	$(-7) - (-9) = 2$
−2	$2(-2) - 3 = -7$	$(-5) - (-7) = 2$
−1	$2(-1) - 3 = -5$	$(-3) - (-5) = 2$
0	$2(0) - 3 = -3$	$(-1) - (-3) = 2$
1	$2(1) - 3 = -1$	$1 - (-1) = 2$
2	$2(2) - 3 = 1$	$3 - 1 = 2$
3	$2(3) - 3 = 3$	

Notice in the table that the x-values differ by 1. In a table with this quality, if you take any two y-values next to each other and subtract them, you find their **consecutive differences**. Notice in the table above that the consecutive differences are the same. For this reason, the difference is called a **common difference**.

Think about Math

Directions: Answer the following questions.

1. What is the common consecutive difference for the function $f(x) = -3x + 2$?

2. What is the common consecutive difference for the function $f(x) = x - 4$?

Quadratic Functions

A **quadratic function** is a polynomial function that has 2 as its highest power of x. The graph of a quadratic function is a parabola.

Just like with linear functions, you can evaluate quadratic functions and make a table of values.

The table below shows that, unlike linear functions, quadratic functions do not have common consecutive differences.

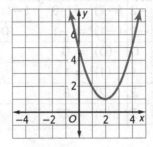

x	$f(x) = x^2 - 4x + 5$	Consecutive Difference
−3	$(-3)^2 - 4(-3) + 5 = 26$	$17 - 26 = -9$
−2	$(-2)^2 - 4(-2) + 5 = 17$	$10 - 17 = -7$
−1	$(-1)^2 - 4(-1) + 5 = 10$	$5 - 10 = -5$
0	$(0)^2 - 4(0) + 5 = 5$	$2 - 5 = -3$
1	$(1)^2 - 4(1) + 5 = 2$	$1 - 2 = -1$
2	$(2)^2 - 4(2) + 5 = 1$	$2 - 1 = 1$
3	$(3)^2 - 4(3) + 5 = 2$	

However, the consecutive differences of the quadratic function are *exactly* the y-values found in the previous linear example. So quadratic functions have common *second* consecutive differences.

x	$f(x) = x^2 - 4x + 5$	Consecutive Difference	2nd Consecutive Difference
−3	$(-3)^2 - 4(-3) + 5 = 26$	$17 - 26 = -9$	$(-7) - (-9) = 2$
−2	$(-2)^2 - 4(-2) + 5 = 17$	$10 - 17 = -7$	$(-5) - (-7) = 2$
−1	$(-1)^2 - 4(-1) + 5 = 10$	$5 - 10 = -5$	$(-3) - (-5) = 2$
0	$(0)^2 - 4(0) + 5 = 5$	$2 - 5 = -3$	$(-1) - (-3) = 2$
1	$(1)^2 - 4(1) + 5 = 2$	$1 - 2 = -1$	$1 - (-1) = 2$
2	$(2)^2 - 4(2) + 5 = 1$	$2 - 1 = 1$	
3	$(3)^2 - 4(3) + 5 = 2$		

TEST-TAKING SKILL

Eliminate Unnecessary Information

Sometimes on a test you are given extra information not needed to solve the problem. Read each test question carefully and determine what information is needed to solve the problem.

The table below shows the first, second, and third differences for a function $f(x)$.

x	f(x)	Consecutive Differences		
		1st	2nd	3rd
8	2	1	1	0
9	3	2	1	0
10	5	3	1	0
11	8	4	1	
12	12	5		
13	17			

Is $f(x)$ a quadratic function? What information in the table is not needed to answer this question?

Examples of Non-Linear/Quadratic Functions

All polynomial functions eventually turn out to have common consecutive differences. For example, the function $f(x) = x^3$ has common third consecutive differences.

x	$f(x) = x^3$	Consecutive Difference	2nd Consecutive Difference	3rd Consecutive Difference
−3	$(−3)^3 = −27$	$(−8) − (−27) = 19$	$7 − 19 = −12$	$(−6) − (−12) = 6$
−2	$(−2)^3 = −8$	$(−1) − (−8) = 7$	$1 − 7 = −6$	$0 − (−6) = 6$
−1	$(−1)^3 = −1$	$0 − (−1) = 1$	$1 − 1 = 0$	$6 − 0 = 6$
0	$(0)^3 = 0$	$1 − 0 = 1$	$7 − 1 = 6$	$12 − 6 = 6$
1	$(1)^3 = 1$	$8 − 1 = 7$	$19 − 7 = 12$	
2	$(2)^3 = 8$	$27 − 8 = 19$		
3	$(3)^3 = 27$			

Some functions never turn out to have common consecutive differences. For example, the function $f(x) = 2^x$ has no common consecutive differences. Notice how the second consecutive differences are the same as the first consecutive differences.

x	$f(x) = 2^x$	Consecutive Difference	2nd Consecutive Difference
0	$2^0 = 1$	$2 − 1 = 1$	$2 − 1 = 1$
1	$2^1 = 2$	$4 − 2 = 2$	$4 − 2 = 2$
2	$2^2 = 4$	$8 − 4 = 4$	$8 − 4 = 4$
3	$2^3 = 8$	$16 − 8 = 8$	$16 − 8 = 8$
4	$2^4 = 16$	$32 − 16 = 16$	$32 − 16 = 16$
5	$2^5 = 32$	$64 − 32 = 32$	
6	$2^6 = 64$		

Recognizing Linear and Quadratic Functions

Some computer scientists make and break computer codes. These codes are needed to secure national secrets, as well as your banking information. One of the earliest codes in history was the Caesar Cipher, which uses a linear function to move each letter to another letter.

Linear Functions

The function below appears to be linear. To be sure, check by finding consecutive differences.

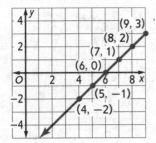

Example 1: Identifying Linear Functions

Step 1 Record the **coordinates,** or the pairs (x, y) graphed on a plane, of several points from the graph in a table. Make sure that the x-values change by 1.

Step 2 Find the consecutive differences of the y-values.

x	f(x)	Consecutive Differences
4	−2	1
5	−1	1
6	0	1
7	1	1
8	2	1
9	3	

Because the function has common first consecutive differences, it is linear.

CALCULATOR SKILL

Many 4-function calculators do not have the ability to enter in tables. The TI-30XS MultiView™ calculator can not only enter a table of values, it can generate the table for you. Press the (table) key, enter the function you want to find a table of values for, choose a start value for x and a step value (the increment between each x value). A reasonable start value for x may be -3 and a reasonable step value for x may be 1. Using the down arrow, you can scroll throughout the table to see the table of values.

Quadratic Functions

The function below appears to be quadratic. To be sure, check by finding second consecutive differences.

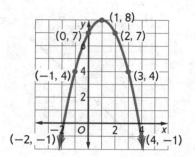

Example 2: Identifying Quadratic Functions

Step 1 Record the coordinates of several points from the graph in a table. Make sure the x-values change by 1.

Step 2 Find the first consecutive differences.

Step 3 Find the second consecutive differences.

x	f(x)	Consecutive Differences	2nd Consecutive Differences
−2	−1	5	−2
−1	4	3	−2
0	7	1	−2
1	8	−1	−2
2	7	−3	−2
3	4	−5	
4	−1		

Because the function has common second consecutive differences, it is quadratic.

Examples of Non-Linear/Quadratic Functions

Sometimes the graph of a function may appear to be linear or quadratic, but in fact it is not.

Identify Linear and Quadratic Functions

Example 3: Showing that a Function is Not Quadratic

Is the function in the graph quadratic?

The function appears to be quadratic. To be sure, check by finding second consecutive differences.

Step 1 Record the coordinates of several points from the graph in a table. Make sure the x-values change by 1.

Step 2 Find the first consecutive differences.

Step 3 Find the second consecutive differences.

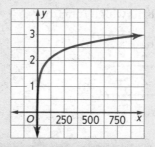

x	$f(x)$	Consecutive Differences	2nd Consecutive Differences
−3	81	−65	50
−2	16	−15	14
−1	1	−1	2
0	0	1	14
1	1	15	50
2	16	65	
3	81		

The second consecutive differences show that the function is not quadratic.

Example 4: Showing that a Function is Not Linear

Is the function in the graph linear?

The function appears to be linear. To be sure, check by finding consecutive differences.

Step 1 Record the coordinates of several points from the graph in a table.

Step 2 Find the first consecutive differences. It appears that this function has common consecutive differences and is therefore linear. However, this is not the case. The function is not linear. What went wrong? Look at the table in this example. The x-values do not differ by 1 and so the differences between the values of $f(x)$ are not consecutive differences.

x	$f(x)$	Consecutive Differences?
1	0	1
10	1	1
100	2	1
1000	3	1
10000	4	

Making sure the x-values differ by 1 is very important when trying to determine whether a graph is linear or quadratic.

Critique the Reasoning of Others

Mario makes the table of values below and calculates the first differences as shown. He claims that the function $f(x)$ is linear because there are common first consecutive differences. Do you agree with Mario? Why or why not?

x	$f(x)$	Consecutive Differences
3	1	6
4	7	6
5	1	6
6	7	6
7	1	6
8	7	

Vocabulary Review

Directions: Write the missing term in the blank.

common difference	**consecutive difference**	**coordinate**
linear function	**quadratic function**	

1. A function that represents a line is a _____.

2. When differences are the same, they are called _____.

3. A polynomial function that has 2 as its highest power of x is a _____.

4. A point graphed on a plane is also called a _____.

5. The difference between terms that follow each other in a table is a _____.

Skill Review

Directions: Read each problem and complete the task.

1. Which table of values corresponds to the function $f(x) = -5x - 3$?

A.

x	f(x)
−3	−18
−2	−13
−1	−8
0	−3
1	2
2	7
3	12

B.

x	f(x)
−3	12
−2	7
−1	2
0	−3
1	−8
2	−13
3	−18

C.

x	f(x)
−3	15
−2	10
−1	5
0	0
1	−5
2	−10
3	−15

D.

x	f(x)
−3	−18
−2	−13
−1	−8
0	−3
1	−8
2	−13
3	−18

2. The table of values below corresponds to which function?

x	−3	−2	−1	0	1	2	3
f(x)	−3	2	5	6	5	2	−3

A. $f(x) = x^2 - 6$

B. $f(x) = -x^2 + 6$

C. $f(x) = -x - 6$

D. $f(x) = x - 6$

3. Use second differences to show that the function $f(x) = 3x^2 - 1$ is quadratic.

Skill Practice

Directions: Read each problem and complete the task.

1. Heidi says that the function $f(x) = -4x + 7$ is linear because it is in the form $f(x) = mx + b$. Neal made the table of values below for the function and concluded that the function is not linear because the first consecutive differences are not common. Which student is correct? Describe the other student's error.

x	f(x)	Consecutive Differences
−2	15	−4
−1	11	−8
1	3	−8
3	−5	−4
4	−9	

2. You have seen that linear functions have common first consecutive differences and that quadratic functions have common second consecutive differences. You have also seen that $f(x) = x^3$ has common third consecutive differences. Make a conjecture about $f(x) = x^4$ and common consecutive differences. Then test your conjecture by making a table of values for $f(x) = x^4$.

3. Enter values into the tables so that Table A represents a linear function and Table B represents a quadratic function. Show that your answers are correct.

Table A   Table B

x	f(x)

x	f(x)

4. Use first consecutive differences to show that the function is linear.

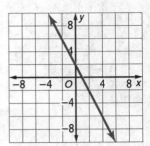

5. Use second consecutive differences to show that the function is quadratic.

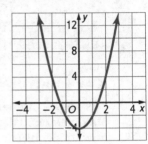

LESSON OBJECTIVES

- Identify key features of a graph
- Draw a graph when given its key features
- Graph a real-world relationship by identifying key features

CORE SKILLS & PRACTICES

- Make Use of Structure
- Gather Information

Key Terms

end behavior
describes the appearance of a graph as it extends in both directions away from zero

relative maximum/minimum
the y-coordinate of any point that is the highest/lowest point for some section of the graph

Vocabulary

line symmetry
a figure displays this when there is a line that divides the figure into two halves that are the mirror images of each other

rotational symmetry
a figure displays this when it can be rotated less than 360° around a point to coincide with itself

x-intercept
the x-coordinate of a point where a graph crosses the x-axis

y-intercept
the y-coordinate of a point where a graph crosses the y-axis

Key Concept

You can sketch graphs if you know or can determine their key features.

Key Features

Businesses often use information from graphs to make important decisions. Making a good decision requires knowing how to read a graph correctly and how to identify and interpret its most important features.

Intercepts and Intervals

An **x-intercept** is the x-coordinate of a point where a graph crosses the x-axis. A **y-intercept** is the y-coordinate of a point where a graph crosses the y-axis. Intercepts can sometimes be determined by examining a graph.

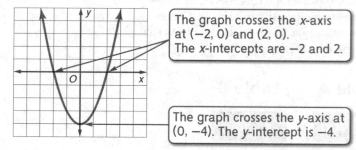

The graph crosses the x-axis at $(-2, 0)$ and $(2, 0)$. The x-intercepts are -2 and 2.

The graph crosses the y-axis at $(0, -4)$. The y-intercept is -4.

This graph crosses the x-axis at $(-2, 0)$ and $(2, 0)$, so the x-intercepts are -2 and 2. The graph crosses the y-axis at $(0, -4)$, so the y-intercept is -4.

Not all graphs have x- and y-intercepts.

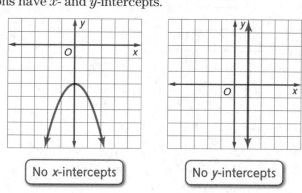

No x-intercepts

No y-intercepts

200   Lesson 6.3

Identify Key Features of a Graph

A *positive interval* describes the values of x for which the graph is above the x-axis. This graph is above the x-axis when $x < -2$ and $x > 2$.

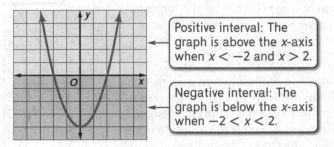

Positive interval: The graph is above the x-axis when $x < -2$ and $x > 2$.

Negative interval: The graph is below the x-axis when $-2 < x < 2$.

A *negative interval* describes the values of x for which the graph is below the x-axis. This graph is below the x-axis when $-2 < x < 2$.

Some graphs are always above the x-axis, and some graphs are always below the x-axis.

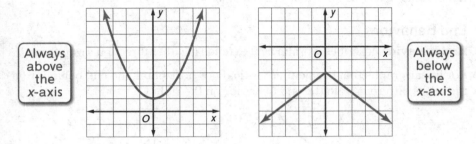

Always above the x-axis

Always below the x-axis

Increasing and Decreasing

A *decreasing interval* describes the values of x for which the graph falls from left to right. On a decreasing interval, the values of y decrease as the values of x increase. This graph falls from left to right when $x < 0$.

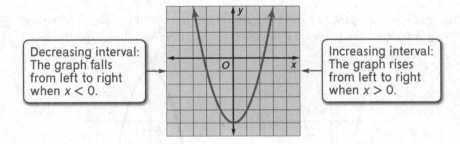

Decreasing interval: The graph falls from left to right when $x < 0$.

Increasing interval: The graph rises from left to right when $x > 0$.

An *increasing interval* describes the values of x for which the graph rises from left to right. On an increasing interval, the values of y increase as the values of x increase. This graph rises from left to right when $x > 0$.

Some graphs are always rising or always falling for all values of x.

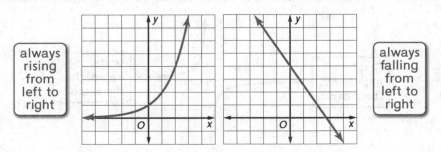

always rising from left to right

always falling from left to right

Make Use of Structure

You can use the structure of a linear graph to determine general information about key features of linear graphs. For example, a linear graph could be vertical, horizontal, or neither. If the graph is not vertical or horizontal, it will rise from left to right or it will fall from left to right.

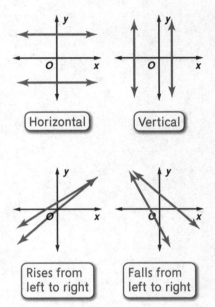

Horizontal

Vertical

Rises from left to right

Falls from left to right

From these possibilities, you can see that all linear graphs have one or two intercepts.

Now you try. Think about the structure of a quadratic graph. Visualize or sketch the different possibilities. What are the possible numbers of intercepts for a quadratic graph?

Identify Key Features of a Graph

Relative Minimums and Maximums

A **relative maximum/minimum** is a *y*-coordinate of any point that is the highest/lowest point for some section of the graph. Relative maximums/minimums occur at "hills"/"valleys" in the graph.

A relative maximum or minimum may or may not occur at the highest or lowest point on the entire graph. Relative maximums and minimums can exist even if the graph overall does not have a highest or lowest point, as long as there are one or more sections of the graph that have a highest or lowest point.

Graphs that are always rising or always falling do not have relative minimums or maximums. There are no "hills" or "valleys."

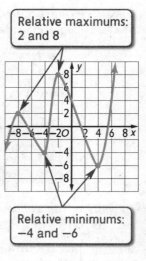

Relative maximums: 2 and 8

Relative minimums: −4 and −6

End Behavior

End behavior describes a graph as it extends in either direction away from 0.

For linear graphs that are neither vertical nor horizontal, the end behavior on the left is different from the end behavior on the right.

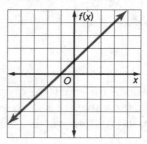
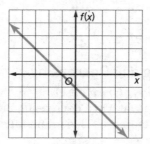

For quadratic graphs, the end behavior is the same on both sides. Whether the graph extends up or down depends on the direction the graph opens.

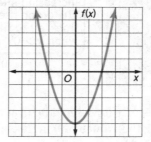
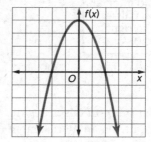

Some graphs have "ends" that do not extend up or down indefinitely. For example, this graph has "ends" that approach, but never reach, the *x*-axis.

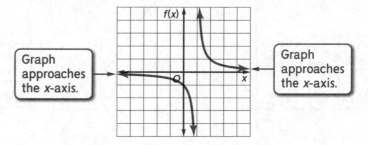

Graph approaches the *x*-axis.

Graph approaches the *x*-axis.

Identify Key Features of a Graph

Symmetry

A graph has **line symmetry** if there is a line that divides the figure into two halves that are mirror images of each other. For example, a quadratic graph is symmetrical about a vertical line.

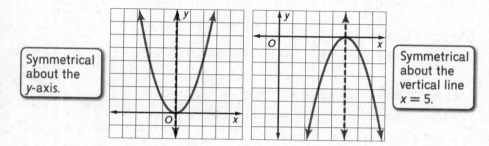

Symmetrical about the y-axis.

Symmetrical about the vertical line $x = 5$.

A graph has **rotational symmetry** if it can be rotated less than 360° around a point to coincide with itself.

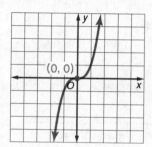

This graph is symmetrical around the point (0, 0). The graph will coincide with itself after a rotation of 180° around (0, 0).

Some graphs have no symmetry.

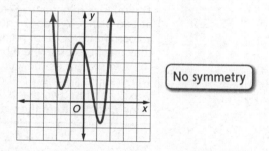

No symmetry

Periodic Graphs

Some graphs are periodic. This means that one piece of the graph repeats over equal intervals.

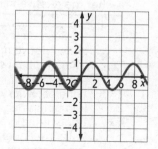

This graph is formed by repeating the highlighted section over and over again in both directions.

CORE SKILL

Gather Information

When you are asked to identify the graph of a function, one method is to substitute values for x and generate ordered pairs. However, it may require less work to use the function rule to find information about the key features of the graph. Then you can match the key features to the correct graph.

Suppose you were given several graphs and asked to identify the graph of $f(x) = x^2 + 5x + 6$.

- Find the y-intercept by substituting 0 for x in the function rule.

$$f(0) = 0^2 + 5(0) + 6 = 0 + 0 + 6 = 6$$

- Find the x-intercepts by substituting 0 for $f(x)$ and factoring to solve the quadratic equation.

$$0 = x^2 + 5x + 6 = (x + 2)(x + 3)$$
$$x = -2 \qquad x = -3$$

Think about a graph that shows a y-intercept of 6 and x-intercepts of -2 and -3. What key features might you use to identify the graph of $4x + 2y = 12$?

Use Key Features to Draw a Graph

Forensic artists make sketches of people based on witnesses' descriptions of physical features—hair and eye color, jaw line, eyebrow thickness and shape, and so on. Similarly, when you are given a description of the key features of a graph, you can make a sketch of the graph.

Sketch a Graph

You can sketch a graph when you know some of its key features.

Example 1: Sketching a Graph Using Key Features

Sketch a graph with the following features.

- The y-intercept is 4.
- The x-intercepts are -2, -1, 1, and 2.
- There is one relative maximum, 4. It occurs at one point.
- There is one relative minimum, -2. It occurs at two points.
- The graph is symmetrical about the y-axis.
- The end behavior on the left is the same as the end behavior on the right.

Step 1 Graph the intercepts, the relative maximum, and the relative minimum.

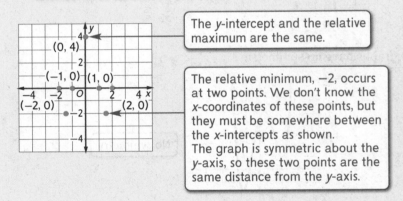

The y-intercept and the relative maximum are the same.

The relative minimum, -2, occurs at two points. We don't know the x-coordinates of these points, but they must be somewhere between the x-intercepts as shown. The graph is symmetric about the y-axis, so these two points are the same distance from the y-axis.

Step 2 Sketch the graph through the points. Make sure that the graph has all of the key features listed above.

2. The graph rises from the relative minimum to the relative maximum.

3. The graph falls from the relative maximum until it reaches the relative minimum.

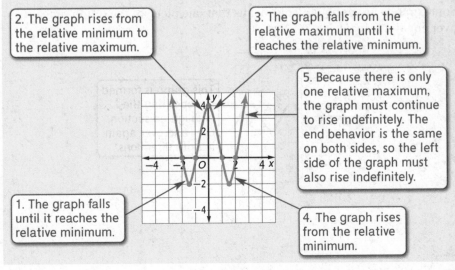

5. Because there is only one relative maximum, the graph must continue to rise indefinitely. The end behavior is the same on both sides, so the left side of the graph must also rise indefinitely.

1. The graph falls until it reaches the relative minimum.

4. The graph rises from the relative minimum.

Real-World Graphs

You can make a graph to describe a real-world situation. Think about the key features of a graph and how they translate to the situation.

Example 2: Graphing a Real-World Situation

Chloe leaves home driving at a speed of 30 miles per hour. She drives at this speed for 8 minutes, stops at a red light for 2 minutes, and then drives at 30 miles per hour for 10 more minutes until she arrives at work. Make a graph to represent the situation.

Step 1 Assign a variable to each quantity. Let x = time and y = distance traveled.

Step 2 Identify the intercepts. When x (time) is 0, y (distance traveled) is also 0. Both the x- and y-intercepts are 0.

Step 3 Identify increasing and decreasing intervals. Distance traveled increases when Chloe is driving – during the first 8 minutes and the last 10 minutes. Chloe drove for a total of $8 + 2 + 10 = 20$ minutes, so the graph will rise for $0 < x < 8$ and $10 < x \le 20$. Distance traveled does not decrease, so there are no decreasing intervals. Distance does not change during the 2 minutes that Chloe is stopped at the red light, so the graph will neither rise nor fall when x is between 8 and 10.

Step 4 Use key features and the given information to sketch the graph. Because Chloe was driving at a constant rate (30 miles per hour), the beginning and end of her trip will be linear.

The part of her trip stopped at the red light will be constant.

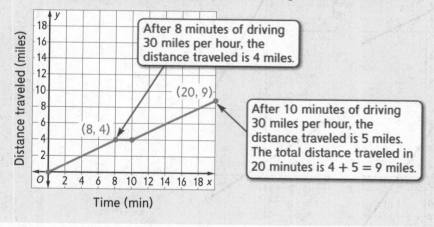

Think about Math

Directions: Sketch a graph with the following key features.

- There are three x-intercepts.
- There is a relative maximum.
- The end behavior on the left is different from the end behavior on the right.
- The graph has no symmetry.

When given a function rule, it is usually fairly straightforward to find y-intercept(s)—simply substitute 0 for x. However, finding x-intercept(s) can be more involved. If the function is quadratic, you may need to factor or use the quadratic formula. When using a calculator and the quadratic formula to find x-intercepts, it may be useful to first calculate $b^2 - 4ac$ and write the value on paper. Then enter (−b + "value")/2a and (−b − "value")/2a. The parentheses are important because your calculator uses the order of operations. If you do not include the parentheses, the calculator will compute the division before the addition/subtraction.

Use a calculator and the quadratic formula to find the x-intercepts of $f(x) = 6x^2 - 7x - 3$.

Vocabulary Review

Directions: Fill in each blank with a word from the list below.

end behavior	**line symmetry**	**relative maximum/minimum**
rotational symmetry	**x-intercept**	**y-intercept**

1. A(n) _____ is the y-coordinate of a point where a graph crosses the y-axis.

2. A(n) _____ is the y-coordinate of any point that is the highest/lowest point for some section of a graph.

3. A(n) _____ is the x-coordinate of a point where a graph crosses the x-axis.

4. A figure has _____ if there is a line that divides the figure into two halves that are mirror images of each other.

5. _____ describes the appearance of a graph as it extends in both directions away from 0.

6. A figure has _____ if it can be rotated less than 360° around a point to coincide with itself.

Skill Review

Directions: Read each problem and complete the task.

1. Which is the graph of $2x + 2y = 6$? Explain how you can use key features to identify the correct graph.

A 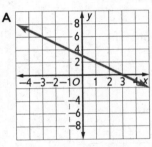 B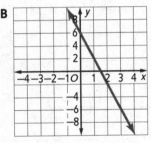

C 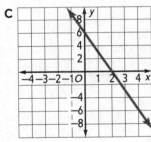 D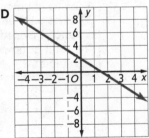

2. Sketch a graph with the following key features.
   - The x-intercepts are −3, −1, 1, and 3.
   - The y-intercept is −9.
   - There is one relative minimum, −9. It occurs at one point.
   - There is one relative maximum, 16. It occurs at two points.
   - The graph extends down indefinitely in both directions.
   - The graph is symmetrical about the y-axis.

Identify Key Features of a Graph

3. Which situation is best represented by the graph?

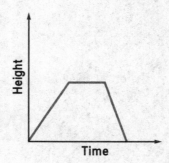

A. A rubber ball is dropped from a height of 5 feet. It bounces several times before rolling to a stop on the ground.

B. An elevator begins at the ground floor of an office building. It goes up three floors and remains on the third floor for several minutes. Then it goes down to the second floor, where a passenger gets in and goes up to the fifth floor.

C. Leanne hikes up to a mountain peak at a speed of 2.5 miles per hour. When she reaches the top, she rests for a while, and then hikes back down at a speed of 4 miles per hour.

D. A diver is on a board one meter above the ground. He dives into a pool that is 12 feet deep and then swims up to the water's surface.

Skill Practice

Directions: Read each problem and complete the task.

1. Describe the key features of the graph.

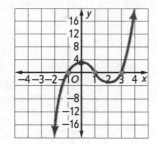

Identify the:

- x- and y-intercepts
- positive and negative intervals
- increasing and decreasing intervals
- relative minimum(s) and maximum(s)

2. Sketch a quadratic graph that matches each description. If a graph is not possible, explain why.

  a. The graph has no relative minimum.

  b. The graph has no relative minimums or maximums.

  c. The graph has no symmetry.

3. Write a real-world situation about a vehicle or an object whose speed changes over time. Make a graph to represent your situation. Identify the key features of your graph and describe their meanings in the context of your situation.

4. Use key features to sketch the graph of $f(x) = x^2 - 9$. Describe the key features you used.

5. Drake claims that the only key features needed to sketch a linear graph are the intercepts. Do you agree with Drake? Explain why or why not.

LESSON 6.4 Compare Functions

LESSON OBJECTIVES

- Compare proportional relationships represented in different ways
- Compare linear functions represented in different ways
- Compare quadratic functions represented in different ways

CORE SKILLS & PRACTICES

- Use Ratio and Rate Reasoning
- Make Sense of Problems

Key Terms

proportional relationship
a relationship between two quantities x and y such that the ratio of y to x is always equal to a nonzero constant k

Vocabulary

slope
the ratio of rise to run

y-intercept
the y-coordinate of a point where a graph crosses the y-axis

quadratic function
a function that can be written in the form $y = ax^2 + bx + c$, where $a \neq 0$

Key Concept

Functions can be represented in many ways—graphs, tables, equations, verbal descriptions, and so on. To compare two or more functions represented in different ways, you will have to use the information given in each representation to determine key features that can be compared.

Compare Proportional Relationships

Many real-world relationships are proportional. For example, an employee's pay is proportional to the number of hours the employee works. Being able to compare proportional relationships will allow you to evaluate situations and make decisions.

Distance-Rate-Time

Two variable quantities x and y are in a **proportional relationship** if the ratio of y to x is always equal to a nonzero constant k.

$$\frac{y}{k} = x$$

Solving the equation above for y gives $y = kx$. The linear equation $y = kx$ is a line with slope k that passes through $(0, 0)$. Remember that the **slope** of a line is the ratio of rise to run.

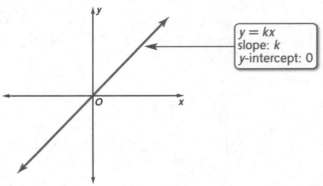

$y = kx$
slope: k
y-intercept: 0

When a person or an object is traveling at a constant speed (r), the relationship between distance (d) and time (t) is a proportional relationship described by the equation $d = rt$. The speed (r) is the rate of change or slope.

Example 1: Comparing Distance Relationships

Alison and Lia competed in a race. The graph and equation describe their race performances. Who ran faster, Alison or Lia?

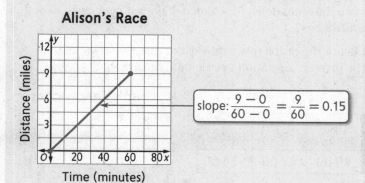

Alison's Race

Lia's Race

$d = 0.1t$, where d represents distance in miles and t represents time in minutes

$$\text{slope: } \frac{9-0}{60-0} = \frac{9}{60} = 0.15$$

Step 1 Find Alison's speed. Alison's speed is equal to the slope of the graph, 0.15. This means that Alison's speed was 0.15 miles per minute.

Step 2 Find Lia's speed. Lia's speed is represented by the coefficient of t, 0.1. Lia's speed was 0.1 miles per minute.

Step 3 Compare the speeds. $0.15 > 0.1$, so Alison ran faster.

Hourly Pay Rates

When a person is paid a fixed amount per hour, the relationship between time worked and total pay is a proportional relationship. The amount earned per hour is the rate of change.

Example 2: Comparing Hourly Pay Rates

The table and the equation describe Kyle's and Reese's pay at their jobs. Who earns more per hour, Kyle or Reese?

Reese's Pay

$p = 10h$, where p represents total pay and h represents hours worked

Kyle's Pay

Hours Worked	0	1	2	3	4	5
Total Pay	0	20	40	60	80	100

Step 1 Find the amount that Kyle earns per hour. You can see from the table that when the number of hours worked increases by 1, the total pay increases by $20. Kyle earns $20 per hour.

Step 2 Find the amount that Reese earns per hour. This amount is represented by the coefficient of h, 10. Reese earns $10 per hour.

Step 3 Compare. $20 > 10$, so Kyle earns more money per hour.

Use Ratio and Rate Reasoning

Sometimes when comparing proportional relationships, you will have to convert units.

Dean earns $75 per day and he works 7 days per week. The graph below describes Dean's spending.

Dean's Spending

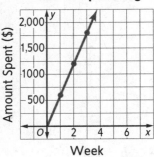

To determine whether Dean's pay supports his spending, you will have to compare the amount he earns to the amount he spends. Notice that Dean's spending is described weekly while his pay is described daily. You will have to use reasoning to compare weekly amounts to daily amounts.

Use the graph to determine how much Dean spends per week.

How can you determine how much Dean earns per week?

Does Dean's pay support his spending? Explain.

Cost

When you buy more than one of the same item, the relationship between the number of items purchased and total cost is a proportional relationship. The cost per item is the rate of change.

Example 3: Comparing Cost

The table and the graph give information about the cost of turkey sandwiches at two different delis. At which deli does it cost more to buy three turkey sandwiches?

Neither the table nor the graph gives the cost of three sandwiches, so you must use the given information to calculate the answer.

Dawn's Deli

Number of Sandwiches	0	2	4
Total Cost	$0.00	$7.00	$14.00

Sam's Sandwiches

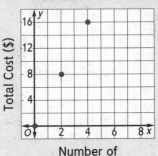

Number of Sandwiches

Step 1 Find the cost of a turkey sandwich at Dawn's Deli.
Two sandwiches cost $7.00.
Therefore, one sandwich costs $7.00 ÷ 2 = $3.50.

Step 2 Find the cost of a turkey sandwich at Sam's Sandwiches.
Two sandwiches cost $8.00.
Therefore, one sandwich costs $8.00 ÷ 2 = $4.00.

Step 3 Compare the costs. $3.50 < $4.00, so a turkey sandwich costs more at Sam's Sandwiches.
Therefore, as long as the price stays constant, three turkey sandwiches cost more at Sam's Sandwiches.

Think about Math

Three proportional relationships are represented by the table, the equation, and the graph. Order the relationships from greatest rate of change to least rate of change.

Relationship A

x	y
0	0
5	2
10	4

Relationship B

$y = 1.5x$

Relationship C

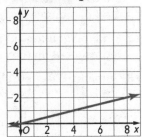

Compare Linear Functions

When choosing a monthly plan for text messages, comparing the monthly charge and rate per text message for each plan can help you determine which plan is best for you. Plans such as these can often be modeled by linear functions.

Compare Slopes

When given information about two linear functions, you can determine which has the greater slope.

Example 4: Compare Slopes

The linear function $g(x)$ passes through the points $(1, 12)$ and $(-2, 0)$. The linear function $f(x)$ is shown in the graph. Which function has the greater slope?

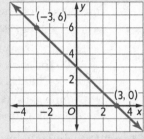

Step 1 Use two points on the graph to find the slope of $f(x)$:

$$m = \frac{6-0}{-3-3} = \frac{6}{-6} = -1$$

Step 2 Use the given points to find the slope of $g(x)$: $m = \frac{0-12}{-2-1} = \frac{-12}{-3} = 4$

Step 3 Compare. $-1 < 4$, so $g(x)$ has the greater slope.

Compare y-intercepts

A **y-intercept** is the y-coordinate of a point where a graph crosses the y-axis. When given information about two linear functions, you can determine which has a greater y-intercept.

Example 5: Compare y-intercepts

Two linear functions are described in the table and graph. Which function has the greater y-intercept?

Graph of g(x)

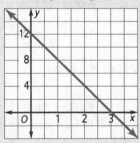

x	−3	−2	−1	0	1	2	3
f(x)	15	12	9	6	3	0	−3

Step 1 Identify the y-intercept of $f(x)$. The y-intercept is the y-coordinate of the point whose x-coordinate is 0. You can see from the fourth row of the table that the y-intercept is 6.

Step 2 Identify the y-intercept of $g(x)$. The graph of $g(x)$ crosses the y-axis at $(0, 12)$, so the y-intercept is 12.

Step 3 Compare. $6 < 12$, so $g(x)$ has the greater y-intercept.

Linear functions can model cost situations in which there is an initial cost as well as a cost per item. In these situations, the initial cost is represented by the y-intercept and the cost per item is represented by the slope.

Health Literacy

Yesterday, Max spent one hour bowling and burned about 224 calories. Today Max will either swim or jump rope. The total number of calories burned during exercise for yesterday and today will depend on the amount of time Max spends swimming or jumping rope, as shown in the table and the graph.

Swimming

Hours Spent Swimming Today	Total Calories Burned Yesterday and Today
0	224
0.5	447
1	670

Jumping Rope

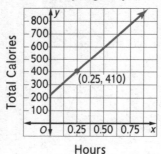

Hours

Max wants to burn a total of at least 400 calories for yesterday and today. If Max decides to swim today, how long must he swim to meet his goal? If Max decides to jump rope today, how long must he jump rope to meet his goal?

Example 6: Compare Costs

Luisa wants to join an online book club that allows her to download books to her tablet. For Book Club A, there is a one-time membership fee of $15.99, and each book downloaded costs $3.00. The costs for Book Club B are described in the graph. Which book club costs more per download? Which book club has the greater membership fee?

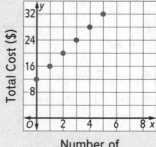

Book Club B

Total Cost ($) / Number of Books Downloaded

Step 1 Find and compare the cost per download for each club.

- The cost per download for Club A is $3.00.
- From the graph, you can see that the total cost increases by $4.00 for each book downloaded. The cost per download for Book Club B is $4.00.
- Compare. $3.00 < $4.00, so the cost per download is greater for Book Club B.

Step 2 Find and compare the membership fee for each club.

- The membership fee for Club A is $15.99.
- The membership fee is the cost when 0 books are downloaded. From the graph, you can see that the membership fee for Book Club B is $12.00.
- Compare. $12.00 < $15.99, so the membership fee is greater for Book Club A.

Compare Quadratic Functions

Quadratic functions, functions that can be written in the form $y = ax^2 + bx + c$ where $a \neq 0$, are commonly used to model the motion of objects—objects that are dropped, thrown, kicked, and so on. You can compare the motion of two different objects by comparing the quadratic functions that model their motion.

Compare Zeros

Example 7: Compare Zeros

A red golf ball and a blue golf ball were hit at the same time from a platform 48 feet above the ground. The equation and the table describe the motion of each golf ball. Which golf ball reached the ground first?

Red Golf Ball

$y = -16x^2 + 52x + 48$, where y represents the height above the ground in feet and x represents the time in seconds after the ball is hit.

Blue Golf Ball

Time After the Ball is Hit (seconds)	0	1	2	3
Height Above the Ground (feet)	48	64	48	0

Step 1 Determine when the red golf ball reached the ground. When the ball reaches the ground, the height $y = 0$. Substitute 0 for y and solve the quadratic equation.

$$0 = -16x^2 + 52x + 48$$
$$= (-2x + 8)(8x + 6)$$

Factor and use the Zero Product Property.

$$-2x + 8 = 8 \quad \text{or} \quad 8x + 6 = 0$$
$$\boxed{x = 4} \quad \text{or} \quad x = -\frac{3}{4}$$

Step 2 Determine when the blue golf ball reached the ground. According to the table, the blue golf ball reached the ground (height = 0) after 3 seconds.

Step 3 Compare. $3 < 4$, so the blue golf ball reached the ground first.

Because x represents time, the negative solution does not make sense. The red golf ball reached the ground after 4 seconds.

Compare Maximums

You can use quadratic functions to determine the maximum height reached by an object. You can compare quadratic functions to determine which of two objects reached a greater height.

> **Example 8:** Compare Maximums
>
> The graph and the table describe two of Liam's kicks in yesterday's soccer game. For which kick did the soccer ball reach a greater height?
>
> **Liam's First Kick**
>
>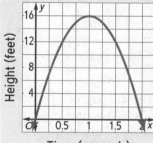
>
> Time (seconds)
>
> **Liam's Second Kick**
>
> | Time After the Ball is Kicked (seconds) | 0 | 0.5 | 1 | 1.5 | 2 |
> |---|---|---|---|---|---|
> | Height Above the Ground (feet) | 0 | 13 | 18 | 15 | 4 |
>
> **Step 1** Examine the graph. The greatest height reached by the soccer ball is represented by the maximum—about 16 feet.
>
> **Step 2** Examine the table. You can see that the soccer ball reached a height of 18 feet after 1 second.
>
> **Step 3** Compare. We cannot be sure that the ball's maximum height in the second kick was 18 feet, but 18 is greater than the maximum height in the first kick. Therefore, the soccer ball reached a greater height in Liam's second kick.

CORE PRACTICE

Make Sense of Problems

To make sense of problems, look at all of the given information. Identify what you are asked to find and develop a plan to determine the answer.

How would you determine if a function had the same x-intercepts as the function shown?

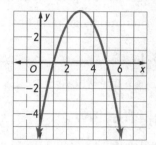

Compare Functions

Vocabulary Review

Directions: Fill in the blanks with one of the terms below. Terms may be used more than once.

slope   *y*-intercepts   quadratic function   proportional relationship

1. A linear function is a function whose graph is a line. Features of linear functions that can be

   compared include _____, which measures the steepness of a line,

   and _____, which indicate where a line crosses the *y*-axis.

2. A function that can be written in the form $y = ax^2 + bx + c$ where a ≠ 0 is a

   _____. Features of this type of function that can be compared include

   minimums, maximums, and intercepts.

3. In a _____, the ratio of *y* to *x* is equal to a nonzero constant *k*. The graph

   is a line whose _____ is *k*.

Skill Review

Directions: Read each problem and complete the task.

The heights of two model rockets that were launched at the same time are described in the table and graph below. Use this information for 1 and 2.

Rocket A

Time After Launch (seconds)	Height Above Ground (feet)
0	25
0.5	37
1	41
1.5	37
2	25
2.5	5

Rocket B

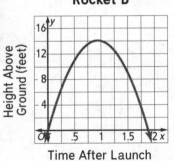

Time After Launch (seconds)

1. Jonas says that he cannot determine which rocket reached the ground first because the table does not contain any information about the time that Rocket A reached the ground. Is Jonas correct? Explain why or why not.

2. Which statement is correct?
   A. Rocket A was launched from 25 feet above the ground and Rocket B from the ground.
   B. Both rockets were launched from the ground.
   C. Rocket A was launched 5 feet above the ground and Rocket B from 25 feet above the ground.
   D. Both were launched from 25 feet above the ground.

3. A car is traveling at a rate of 60 miles per hour. A truck is traveling as shown in the graph. Which statement is correct?

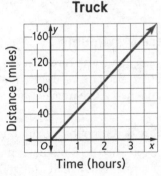

Truck

   A. After 4 hours, the car will have traveled twice as far as the truck.
   B. The truck's rate of speed is less than $\frac{1}{2}$ the car's rate of speed.
   C. It will take the truck 4 hours to travel the same distance that the car travels in 3 hours.
   D. The car's rate of speed is $\frac{3}{4}$ the truck's rate of speed.

4. A work day is 8 hours. Sandra's total pay p when she works h hours is described by the equation $p = 21.5h$. Boris's pay is described in the table. Who earns more money per day? How much more?

Hours Worked	Total Pay
0	0
2	45
4	90

Compare Functions

Skill Practice

Directions: Read each problem and complete the task.

1. Lila works 5 days per week. Her weekly pay and spending are described in the table and graph below. Which statement or statements are correct?

Lila's Pay

Days Worked	0	1	2	3	4	5
Total Pay	$0	$100	$200	$300	$400	$500

Lila's Spending

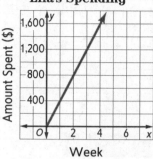

A. The amount that Lila spends each week is more than the amount she earns each week.
B. Lila must work 4 days to earn as much as she spends in one week.
C. After working 4 weeks, Lila could deposit $500 in a savings account.
D. If Lila reduces her weekly spending by $50, she can save $150 per week.

2. Order the functions from least y-intercept to greatest y-intercept.

Function A
$f(x) = x^2 + 5x - 1$

Function B

x	g(x)
0	−7
1	1
4	1

Function C

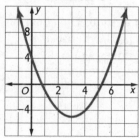

The total cost for a taxi ride usually consists of an initial charge plus a charge per mile. Taxi fares in four different cities are described below. Use this information for 3–5.

Taxi Fares – New York City

Miles	Total Cost
0	$2.50
2	$7.50
4	$12.50
6	$17.50

Taxi Fares – Chicago

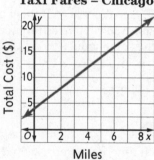

The graph of Miami's taxi fares contains the points (1, 4.9) and (5, 14.5). The first coordinate in the ordered pairs represents distance in miles and the second coordinate represents cost.

In Dallas, the cost c to travel m miles in a taxi is $c = 1.8m + 2.25$.

3. Which city has the greatest initial charge? What is this initial charge?

4. Which city has the least charge per mile? What is this charge per mile?

5. Which shows the cities listed in order from least to greatest based on the total cost of a 15-mile taxi ride?

A. Chicago, Dallas, New York City, Miami
B. Dallas, Chicago, Miami, New York City
C. New York City, Miami, Chicago, Dallas
D. Miami, New York City, Dallas, Chicago

Directions: Choose the best answer to each question.

1. Which table of values corresponds to the function $f(x) = 2x^2 - 4x + 10$?

A.
x	f(x)
−2	26
−1	16
0	10
1	8
2	10

B.
x	f(x)
26	1
16	2
10	3
8	4
10	5

C.
x	f(x)
−2	10
−1	8
0	10
1	8
2	10

D.
x	f(x)
26	−2
16	−1
10	0
8	1
10	2

2. The function shown with the table is _____ because the 2nd consecutive differences are the same value.

x	f(x)
−2	2
−1	1
0	2
1	5
2	10

3. Which function is represented by the graph?

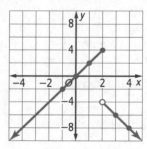

A. $f(x) = \begin{cases} -3x; x \geq 2 \\ 3x; \quad x < 2 \end{cases}$

B. $f(x) = -3x$

C. $f(x) = 3x$

D. $f(x) = \begin{cases} -3x; x > 2 \\ 3x; \quad x \leq 2 \end{cases}$

4. The graph below represents the situation of someone _____.

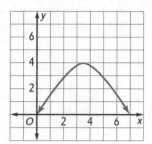

5. A function is a rule that assigns exactly one output to each input. The set of inputs is called the _____. The set of outputs is called the range.

Directions: Use the graph below to answer questions 6–8.

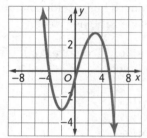

6. The y-intercept is -0.75 and the x-intercepts are _____.

7. The increasing interval is _____ and the decreasing intervals are $x < 2$ and $x > 3$.

8. The _____ is 3 and the relative minimum is -3.

9. Ellen is paid \$18 an hour at her job. The equation $\frac{1}{15}p = h$ shows how Jake is paid, where p is the pay and h is the number of hours worked. So, _____ is paid more per hour.

Directions: Choose the best answer to each question.

10. Which table of values corresponds to the function $f(x) = 12x - 20$?

A.

x	f(x)
−44	−2
−32	−1
−20	0
−8	1
4	2

B.

x	f(x)
−2	−44
−1	−32
0	−20
1	−8
2	4

C.

x	f(x)
0	20
2	4
3	16
4	28
5	80

D.

x	f(x)
−2	−4
−1	8
0	20
1	32
2	44

11. The input represents the number of products a company makes for a month. The output of a function represents the profit a company will make that month. How much profit will the company make if they make 100 products?

$f(x) = 2x^2 + 3x - 80$

A. 420
B. 620
C. 20,220
D. 20,380

12. Which pair of points forms a line with a greater slope than the slope of the line shown?

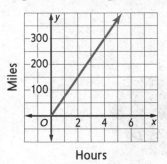

Hours

A. (4, 153); (8, 313)
B. (2, 14); (4, 24)
C. (3, 25); (6, 55)
D. (2, 170); (5, 410)

13. Ava threw a ball in the air and graphed its heights. Bill also threw a ball in the air and recorded its heights in a table. Which is true?

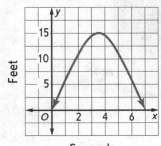

Seconds

time(s)	height(s)
2	10
3	15
4	18
5	15
6	10

A. Ava threw her ball higher than Bill threw his.
B. Bill threw his ball higher than Ava threw hers.
C. Ava and Bill threw their balls at the same height.
D. Bill didn't throw his ball as high as Ava threw hers.

Check Your Understanding

On the following chart, circle the items you missed. The last column shows pages you can review to study the content covered in the question. Review those lessons in which you missed half or more of the questions.

Lesson	Item Number(s)			Review Page(s)
	Procedural	Conceptual	Problem Solving	
6.1 Identify a Function	3	4, 5	11	184–191
6.2 Identify Linear and Quadratic Functions	2			192–199
6.3 Identify Key Features of a Graph		6, 7, 8	12	200–207
6.4 Compare Functions	1, 10		9, 13	208–215

Chapter 7

Geometry and Measurement

Owning a home can mean a lot of home improvement projects. Outside you might put in fencing, or add flower beds. Inside you might want to change the flooring or repaint the walls. All of these projects will involve finding the perimeter or area of spaces. In this chapter you will learn how to calculate perimeter, area, circumference, and volume of geometric figures.

Lesson 7.1
Compute Perimeter and Area of Polygons

When doing home renovations you often want to redo your floors. In order to buy the right amount of materials, you need to know the area of the floor. You also need to know the perimeter so you can buy enough trim. Learn how to calculate perimeter and area of triangles, rectangles, and other polygons.

Lesson 7.2
Compute Circumference and Area of Circles

Circumference and area of circles are calculated with formulas using the number π (pi). Pi is a special number that equals approximately 3.14. Learn about pi and how to calculate circumference and area.

Lesson 7.3
Compute Surface Area and Volume

Perimeter and area only apply to 2-dimensional figures. When you have a 3-dimensional figure you calculate surface area and volume instead. Learn how surface area is an extension of finding area, and how to use formulas to calculate volume.

Lesson 7.4
Compute Perimeter, Area, Surface Area, and Volume of Composite Figures

A composite figure is made of two or more 2-dimensional shapes. A composite solid is made of two or more 3-dimensional solids. In order to calculate the dimensions of a composite figure or object, break the figure or object into its component pieces. Then find the dimensions of each piece and add them together. Learn how to calculate measurements of composite figures.

Goal Setting

Construction workers have to measure and determine the dimensions of the spaces they work in. How could this chapter help them in their job? What other kind of jobs use measurement and geometry?

Where in your daily life do you encounter polygons, circles, and composite figures? When would you need to compute the dimensions of these shapes? How could you apply the formulas you learn in this chapter in your life?

LESSON 7.1 Compute Perimeter and Area of Polygons

LESSON OBJECTIVES

- Compute the perimeter of a polygon
- Use geometric formulas to find the area of a polygon
- Determine a side length of a polygon when given the perimeter or area

CORE SKILLS & PRACTICES

- Calculate Area
- Perform Operations

Key Terms

area
the number of non-overlapping square units needed to exactly cover the entire inside of a polygon

perimeter
the distance around the outside of a polygon

polygon
a closed figure in a plane that is formed by three or more segments

Vocabulary

hypotenuse
the longest side of a right triangle that is also opposite the right angle

parallelogram
a four-sided polygon whose opposite sides are parallel

trapezoid
a four-sided polygon with exactly one pair of parallel opposite sides

Key Concept

Formulas can be used to find the perimeter and area of polygons.

Rectangles

Rectangles are everywhere. For example, floors, walls, and ceilings are often rectangles. When building or decorating a room, you can find the perimeter and area of rectangles to determine the quantities of supplies needed.

Perimeter

A **polygon** is a closed figure in a plane that is formed by three or more segments. These segments are the sides of the polygon.

The **perimeter** of a polygon is the distance around the outside of the polygon. The letter P is often used to represent perimeter.

To find the perimeter of a rectangle, add the four side lengths. Another way is to find two times the length, ℓ, plus two times the width, w.

Polygons

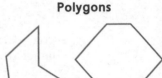

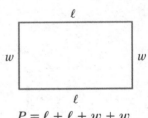

$$P = \ell + \ell + w + w$$
$$P = 2\ell + 2w$$

Example 1: Perimeter of a Rectangle

Find the perimeter of the rectangle.

$P = 2\ell + 2w$
$ = 2(5) + 2(3)$
$ = 10 + 6$
$ = 16$ inches

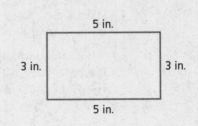

Compute Perimeter and Area of Polygons

Area

The **area** of a polygon is the number of non-overlapping square units needed to exactly cover the entire inside of the polygon. The letter A is often used to represent area.

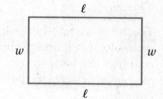

To find the area of a rectangle, multiply the length, ℓ, times the width, w.

Example 2: Area of a Rectangle

Find the area of the rectangle.

$A = \ell w$
$\quad = 5(3)$
$\quad = 15$ square inches

5 in.

3 in.

Missing Side Lengths

If you know the perimeter or area and either the length, ℓ, or the width, w, you can determine the other dimension of the rectangle.

Example 3: Width of a Rectangle

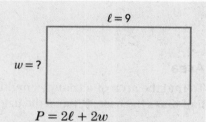

The perimeter of the rectangle is 28 units. What is the width?

Step 1 Substitute the known values into the formula for perimeter.

$$P = 2\ell + 2w$$
$$28 = 2(9) + 2w$$

Step 2 Simplify and solve the equation for w. The width is 5 units.

$$28 = 18 + 2w$$
$$10 = 2w$$
$$5 = w$$

⌐ Think about Math

Directions: Answer the following questions.

1. What is the length of a rectangle whose area is 111.6 m² and whose width is 9 m?

   A. 12.4 m
   B. 102.6 m
   C. 120.6 m
   D. 1004 m

2. A rectangle has a perimeter of 36.8 yd and a width of 9.6 yd. What is the area of the rectangle?

   A. 3.8 yd²
   B. 8.8 yd²
   C. 17.6 yd²
   D. 84.48 yd²

CORE SKILL

Calculate Area

You may have to calculate the area with sides that are not whole numbers. For example, to find the area of a rectangle whose length is 11.2 meters and whose width is 7.5 meters, you will have to multiply decimals. What is the area of this rectangle? How can you use estimation to determine whether your answer is reasonable?

Triangles

Triangles are another polygon that you can see almost everywhere. For example, many bridges have triangles included in their construction because triangles are very sturdy shapes. When determining the size of a bridge, it may be necessary to find the area and/or perimeter of the triangles it would contain.

Perimeter

To find the perimeter of a triangle, add the three side lengths together.

Example 4: Perimeter of a Triangle

Find the perimeter of the triangle.

$$P = a + b + c$$
$$= 12 + 12 + 10$$
$$= 34 \text{ feet}$$

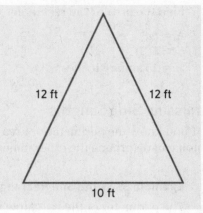

12 ft   12 ft

10 ft

Area

To find the area of a triangle, multiply one-half times the base, b, times the height, h.

The base can be any side of the triangle. The height is a line segment that is perpendicular to the base and extends to the vertex of the triangle opposite the base. The location of the height depends on the type of the triangle.

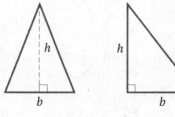

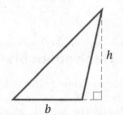

Example 5: Area of a Triangle

Find the area of each triangle.

Each triangle has a base of 6 meters and a height of 8 meters. The area of each triangle is

$$A = \tfrac{1}{2}bh$$
$$= \tfrac{1}{2} \times 6 \times 8$$
$$= 3 \times 8$$
$$= 24 \text{ m}^2$$

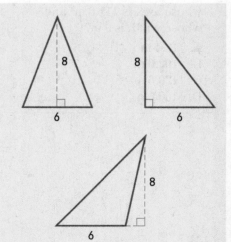

Pythagorean Theorem

The Pythagorean theorem describes the relationship between the legs of a right triangle and the **hypotenuse**, which is the longest side of a right triangle. The legs form the right angle and the hypotenuse is opposite the right angle.

Sometimes, the Pythagorean theorem can be used to determine a missing length that is needed to find the area of a triangle.

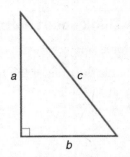

Pythagorean Theorem:
$a^2 + b^2 = c^2$

Example 6: Pythagorean Theorem

Calculate the area of the triangle.

Step 1 Find the height of the triangle. Notice that the height is the leg of a right triangle whose other leg is 18 cm and whose hypotenuse is 30 cm. Use the Pythagorean Theorem to calculate the height.

$a^2 + b^2 = c^2$
$h^2 + 18^2 = 30^2$
$h^2 + 324 = 900$
$h^2 = 576$
$h = 24 \text{ cm}$

Step 2 Use the value of h from Step 1 and the area formula to find the area of the original triangle. Notice that the base of the original triangle is $18 + 18 = 36$ cm.

$A = \frac{1}{2}bh$
$= \frac{1}{2}(36)(24)$
$= 18(24)$
$= 432 \text{ cm}^2$

Evaluate the Answer

Sometimes you can check your answer by solving the problem in a different way.

One way to find the area of the triangle below is to use the formula $A = \frac{1}{2}bh$.

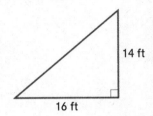

$A = \frac{1}{2}bh = \frac{1}{2}(16)(14)$
$= 8(14) = 112 \text{ ft}^2$

Notice that the triangle can be doubled to form a rectangle. Another way to find the area of this triangle is to first find the area of the rectangle and then divide by 2.

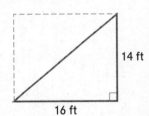

Area of rectangle $= \ell w$
$= 16(14)$
$= 224 \text{ ft}^2$

Area of triangle $= 224 \div 2$
$= 112 \text{ ft}^2$

Notice that both methods give the same answer.

Find the area of the triangle shown below. Then check your answer by solving in a different way to see if you arrive at the same answer.

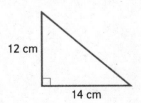

◢ Think about Math

Directions: Answer the following question.

1. What is the area of the triangle?
   A. 40 in.^{2}
   B. 80 in.^{2}
   C. 85 in.^{2}
   D. 170 in.^{2}

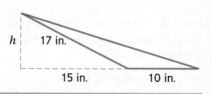

Parallelograms and Trapezoids

Parallelograms and trapezoids are other polygons that you may see in the real world. For example, some rooftops are shaped like trapezoids or parallelograms. To determine how much material would be needed to cover the roof, you would need to find its area.

Parallelograms

A **parallelogram** is a four-sided polygon whose opposite sides are parallel.

The area of a parallelogram is found by multiplying the base of the parallelogram by the height. The base can be any side of the parallelogram. The height is perpendicular to the base and extends to the side opposite the base.

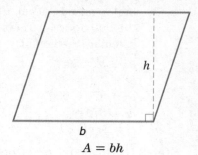

$$A = bh$$

Example 7: Area of a Parallelogram

Find the area of the parallelogram.

$$\begin{aligned} A &= bh \\ &= 15(12) \\ &= 180 \text{ in.}^2 \end{aligned}$$

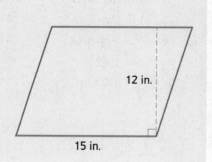

Trapezoids

A **trapezoid** is a four-sided polygon with exactly one pair of parallel opposite sides. A trapezoid has two parallel bases, b_1 and b_2. The height of a trapezoid is the distance between the bases and is perpendicular to the bases.

The area of a trapezoid is computed by multiplying one-half times the height times the sum of the bases.

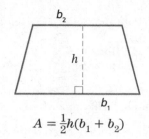

$$A = \tfrac{1}{2}h(b_1 + b_2)$$

Example 8: Area of a Trapezoid

Calculate the area of the trapezoid.

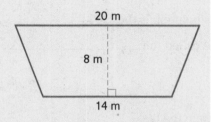

$A = \tfrac{1}{2}h(b_1 + b_2)$

$= \tfrac{1}{2}(8)(14 + 20)$

$= \tfrac{1}{2}(8)(34)$

$= 4(34)$

$= 136 \text{ m}^2$

Think about Math

Directions: Answer the following questions.

1. The area of a parallelogram is 1,732 square inches and its base is 62 inches. What is the approximate height of the parallelogram?

   A. 28 in.
   B. 56 in.
   C. 118 in.
   D. 242 in.

2. The area of a trapezoid is 210 square centimeters. Its height is 15 centimeters and one of its bases measures 12 centimeters. What is the length of the other base?

   A. 8 cm
   B. 16 cm
   C. 32 cm
   D. 180 cm

CORE SKILL

Perform Operations

To use the formula for area of a trapezoid, you must know the order in which to perform the operations. For a trapezoid whose height is 6 units and whose bases are 10 units and 15 units, the area is $\tfrac{1}{2}(6)(10 + 15)$ square units. To simplify this expression, use the order of operations.

Step 1 Add inside the parentheses.

$\tfrac{1}{2}(6)(10 + 15) = \tfrac{1}{2}(6)(25)$

Step 2 Multiply from left to right.

$\tfrac{1}{2}(6)(25) = 3(25) = 75 \text{ sq units}$

Find the area of a trapezoid whose height is 6 cm and whose bases are 12 cm and 8 cm.

Vocabulary Review

Directions: Write the missing term in the blank.

| perimeter | area | polygon |
| hypotenuse | parallelogram | trapezoid |

1. A _____ is a four-sided polygon with exactly one pair of parallel opposite sides.

2. The number of non-overlapping square units needed to cover the entire inside of a polygon is the

   _____ of the polygon.

3. A closed figure in a plane that is formed by three or more segments is a _____.

4. A four-sided polygon whose opposite sides are parallel is a _____.

5. The distance around the outside of a polygon is the _____ of the polygon.

6. The longest side of a right triangle is called the _____.

Skill Review

Directions: Read each problem and complete the task.

1. Kevin is getting ready to seed a plot of ground that is shaped like a trapezoid with the dimensions shown.

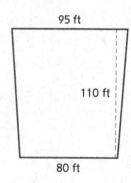

95 ft

110 ft

80 ft

Seeding ground costs $0.35 per square foot. What will it cost Kevin to seed this plot of ground?

A. $3,080.00
B. $3,368.75
C. $6,737.50
D. $9,625.00

2. What is the area of the triangle shown?

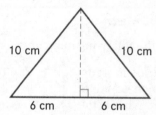

10 cm   10 cm

6 cm   6 cm

A. 32 cm²
B. 24 cm²
C. 48 cm²
D. 120 cm²

Compute Perimeter and Area of Polygons

3. Explain the difference between perimeter and area of a polygon. Give examples of both in your explanation.

4. A sheet of legal-sized paper is 14 inches long and 8.5 inches wide. What is the perimeter of a sheet of legal-sized paper?

   A. 22.5 in.
   B. 45 in.
   C. 59.5 in.
   D. 119 in.

5. Charlie creates and sells jewelry. He has designed a triangular pendant with two 2-inch sides and one 1.5-inch side. Charlie will use wire to form the triangle. He wants to make several of these pendants to sell at his store. How many pendants can Charlie make using 5 feet of wire?

6. Keisha and Howie are cutting parallelograms for an art project. When figuring out how many parallelograms to cut out, they must determine the area of each parallelogram. Keisha says the area is $1\frac{1}{2}$ in². Howie says the area is $2\frac{1}{2}$ in². Who is correct? Explain the other student's error.

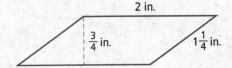

Skill Practice

Directions: Read each problem and complete the task.

1. What is the area of the parallelogram shown?

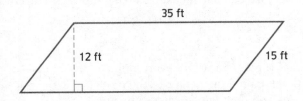

   A. 100 ft²
   B. 180 ft²
   C. 420 ft²
   D. 525 ft²

2. What are the length and the width of a rectangle whose area is 147 square inches and whose perimeter is 104 inches?

   A. L = 6 inches and W = 24 inches
   B. L = 7 inches and W = 21 inches
   C. L and W = 36.75 inches
   D. L = 3 inches and W = 49 inches

3. Ravi is covering a table top in mosaic tiles that each measure one square inch. The table is in the shape of a rectangle with length 1.5 feet and width 2 feet. How many tiles does Ravi need?

4. A company makes rectangular banners using sheets of vinyl that are 6 feet long and 4 feet wide. A customer has requested a special order—a banner in the shape of a parallelogram with an area of 48 square feet. Explain how the company can create this banner.

5. In the diagram below, the dashed trapezoid is a copy of the solid-lined trapezoid. Explain how the diagram shows that the area of the solid-lined trapezoid is $\frac{1}{2}h(b_1 + b_2)$.

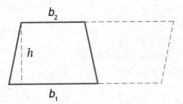

LESSON 7.2 Compute Circumference and Area of Circles

LESSON OBJECTIVES

- Compute the circumference and area of a circle
- Find the radius or diameter of a circle when given the area or circumference

CORE SKILLS & PRACTICES

- Perform Operations
- Calculate Area

Key Terms

area
the number of non-overlapping square units needed to exactly cover the entire inside of a two-dimensional figure

circle
a closed figure with all of its points the same distance from a fixed point called the center

circumference
the distance around the outside of a circle

Vocabulary

diameter
any segment that passes through the center of the circle and whose endpoints are on the circle.

pi (π)
an irrational number approximately equal to 3.14

radius
any segment within a circle whose endpoints are the center and a point on the circle

Key Concept

You can use formulas to find the circumference and area of circles.

Circumference

The **circumference** of a circle is the distance around the outside of the circle. To find circumference, we can use measurements across and around the outside of the circle.

Given the Radius

A **circle** is a closed figure in a plane in which every point is equidistant to a fixed point, called the center of the circle. The **radius** of a circle is any segment whose endpoints are the center of the circle and a point on the circle. Pi (π) is an irrational number that represents the ratio of the circumference of a circle to its diameter. The value of π is approximately 3.14.

$$C = 2\pi r$$

The relationship between the radius r of a circle and its circumference C is described by the formula $C = 2\pi r$.

Example 1: Use Radius to Find Circumference

Find the circumference of the circle. Use 3.14 for π.

Step 1 The radius of the circle is 5.
Substitute 5 for r in the formula.

Step 2 Multiply.

The circumference is 10π centimeters, or approximately 31.4 centimeters.

$$C = 2\pi r$$
$$= 2\pi(5)$$
$$= 10\pi$$
$$\approx 10 \times 3.14$$
$$\approx 31.4 \text{ cm}$$

Given the Diameter

The **diameter** of a circle is any segment that passes through the center of the circle and whose endpoints are on the circle. The diameter is twice the length of the radius: $d = 2r$.

The relationship between the diameter d of a circle and its circumference C is described by the formula $C = \pi d$.

$C = \pi d$

Example 2: Use Diameter to Find Circumference

Find the circumference of the circle. Use 3.14 for π.

8 ft

Step 1 The diameter of the circle is 8. Substitute 8 for d in the formula.

Step 2 Multiply.

The circumference is 8π feet, or approximately 25.1 feet.

$$C = \pi d$$
$$= \pi(8)$$
$$= 8\pi$$
$$\approx 8 \times 3.14$$
$$\approx 25.12 \text{ ft}$$

Think about Math

Directions: Answer the following question.

1. The length of the rectangle is the same as the diameter of the circle. What is the approximate circumference of the circle?

    14 in.

    A.   22 in.
    B.   44 in.
    C.   88 in.
    D.   615 in.

To find the area of a circle, you must square the radius. One way to do this is to multiply the radius by itself. On a calculator, you can use the x^2 function. For example, to find the value of 13^2, press   .

What is the value of 18^2?

CORE SKILL

Calculate Area

To find the area of a circle, you must first know the formula for area of a circle, $A = \pi r^2$. You must understand that r represents the radius, and you must be able to identify the correct value to substitute for r. Finally, you must be able to perform the operations in the formula to calculate the correct answer.

At many pizza restaurants, the diameter of a large pizza is 16 inches. Explain how to find the area of a large pizza. Then calculate the area.

Area

Given the Radius

Remember, the **area** of a figure is the number of non-overlapping square units it takes to exactly cover the inside of the figure. The formula for the area of a circle is $A = \pi r^2$.

$A = \pi r^2$

Example 3: Use Radius to Find Area

Find the area of the circle. Use 3.14 for π.

Step 1 The radius of the circle is 7. Substitute 7 for r in the area formula.

Step 2 Simplify.

The area is 49π ft^2, or approximately 153.9 ft^2.

$$A = \pi r^2$$
$$= \pi (7)^2$$
$$= 49\pi$$
$$\approx 49 \times 3.14$$
$$\approx 153.86 \text{ ft}^2$$

Given the Diameter

If the diameter is given instead of the radius, remember that $d = 2r$. So, before using the formula for area of a circle, divide the diameter by 2 to find the radius.

Example 4: Use Diameter to Find Area

Find the area of the circle. Use 3.14 for π.

Step 1 Find the radius by dividing the diameter by 2.

Step 2 Substitute 6 for r in the area formula.

Step 3 Simplify.

The area is 36π cm^2, or approximately 113 cm^2.

$12 \div 2 = 6$

$$A = \pi r^2$$
$$= \pi (6)^2$$
$$= 36\pi$$
$$\approx 36 \times 3.14$$
$$\approx 113.04 \text{ cm}^2$$

Think about Math

Directions: Answer the following questions.

1. What is the area of a circle whose diameter is 24 inches?
   A. about 37.7 in.^{2}
   B. about 75.4 in.^{2}
   C. about 452.16 in.^{2}
   D. about 1,808.64 in.^{2}

2. Find the area of a circle whose radius is 9 meters. Then round the area to the nearest whole number.

Find Radius or Diameter

If you know the circumference or area of a circle, you can determine the radius and the diameter.

Given Circumference

If you are given the circumference of a circle, you can determine the radius and diameter by substituting the given value into the appropriate formula and solving the resulting equation.

> **Example 5: Use Circumference to Find Radius and Diameter**
>
> Find the diameter and radius of a circle whose circumference is 62.8 cm. Use 3.14 for π.
>
> **Step 1** Substitute 62.8 into the formula for circumference.
>
> $$C = \pi d$$
> $$62.8 = \pi d$$
>
> **Step 2** Solve for d.
>
> The diameter is approximately 20 cm.
>
> $$\frac{62.8}{\pi} = d$$
> $$\frac{62.8}{3.14}$$
> $$20 \approx d$$
>
> **Step 3** Find the radius by dividing the diameter by 2.
>
> $$\frac{20}{2} = d$$
>
> The radius is approximately 10 cm.
>
> $$r \approx 10 \text{ cm}$$

Given Area

If you are given the area of a circle, you can determine the radius by substituting the given value into the formula and solving for r. To then find the diameter, multiply r by 2.

> **Example 6: Use Area to Find Radius and Diameter**
>
> Find the radius and diameter of a circle whose area is 379.94 yd². Use 3.14 for π.
>
> **Step 1** Substitute 379.94 into the formula for area.
>
> $$A = \pi r^2$$
> $$379.94 = \pi r^2$$
>
> **Step 2** Solve for r. Divide both sides by 3.14. Then take the square root of both sides.
>
> The radius is approximately 11 yd.
>
> $$\frac{379.94}{3.14} \approx r^2$$
> $$121 \approx r^2$$
> $$11 \approx r$$
>
> **Step 3** To find the diameter, multiply the radius by 2.
>
> The diameter is approximately 22 yd.
>
> $$d = 2r$$
> $$d \approx 2 \times 11$$
> $$d \approx 22 \text{ yd}$$

Think about Math

Directions: Answer the following questions.

1. What is the approximate radius of a circle whose circumference is 138.2 cm?

    A. 3.7 cm   B. 22 cm   C. 44 cm   D. 88 cm

2. What is the approximate diameter of a circle whose area is 5,026.5 ft²?

Plan and Organize

Before a construction project begins, there is a planning phase. During this phase, all of the specifications are calculated to account for space at the site of construction.

A circular fountain with a circumference of 47 feet is going to be built at a park. Describe how the planners can determine the diameter of the fountain. What is the diameter?

Vocabulary Review

Directions: Write the missing term in the blank.

area   circle   circumference
diameter   pi (π)   radius

1. The _____ of a circle is any segment whose endpoints are the center of the circle and a point on the circle.

2. The number of non-overlapping square units needed to cover the entire inside of a figure

    is the _____ .

3. Every point on a(n) _____ is the same distance from a fixed point called the center.

4. _____ is an irrational number that represents the ratio of a circle's circumference to its diameter.

5. The _____ of a circle is the distance around the outside of the circle.

6. The _____ of a circle is any segment whose endpoints are on the circle and that passes through the center of the circle.

Skill Review

Directions: Read each problem and complete the task.

1. Jessica is designing a mural to be painted on a large wall. Part of the mural will be inside a circle that Jessica will spray paint on the wall. The diameter of the circle will be 60 feet. She estimates that she can cover 30 feet with one can of spray paint. How many cans of spray paint will she need to paint the circumference of the circle?

    A. 2
    B. 6
    C. 7
    D. 9

2. What is the approximate radius of a circle whose area is 1,385 ft²?

    A. 21 ft
    B. 42 ft
    C. 84 ft
    D. 441 ft

3. Explain how to find the circumference of a circle if you are given the area.

4. What is the area of a circle whose diameter is 18 cm?

    A. 18π cm²
    B. 36π cm²
    C. 81π cm²
    D. 324π cm²

Compute Circumference and Area of Circles

5. The wheels on Dion's bicycle each measure 26 inches in diameter. The wheels on Tae's bicycle each measure 20 inches in diameter. Which is the best estimate of how much farther Dion's bicycle will travel in one complete rotation of the wheels than Tae's? *(Hint: In one rotation, a bicycle will travel a distance equal to the circumference of one wheel.)*

A. about 20 inches
B. about 40 inches
C. about 60 inches
D. about 80 inches

6. Before songs could be downloaded from the Internet, people listened to music on vinyl record albums. People also listen to music on compact discs (CDs). Both vinyl records and CDs are in the shape of a circle. There are several sizes of vinyl records; one common size is about 12 inches in diameter. A CD has a diameter of about 4.5 inches. Which statement or statements about these two objects are correct?

A. The radius of the record album is greater than the circumference of the CD.
B. The area of the CD is about half the area of the record album.
C. The area of the record album is about 7 times the area of the CD.
D. The circumference of the record album is more than 3 times the circumference of the CD.

Skill Practice

Directions: Read each problem and complete the task.

1. What is the approximate circumference of the circle shown?

15 ft

A. 47.1 ft
B. 94.2 ft
C. 706.5 ft
D. 2,826 ft

2. What is the approximate area of a circle whose circumference is 157 meters?

A. 78.5 m²
B. 493.2 m²
C. 1,961.5 m²
D. 7,854 m²

3. Carol said that the diameter of the circle is 3 cm. Which best describes Carol's mistake?

$A = 28.26$ cm²

A. Carol used the formula for circumference instead of the formula for area.
B. Carol found the radius of the circle instead of the diameter.
C. Carol calculated the diameter by dividing the radius by 2 instead of multiplying by 2.
D. Carol did not make a mistake. The correct diameter is 3 cm.

4. Explain how to calculate the diameter of a circle when you know its area.

5. Complete the table to find the circumference and area of a circle with the given radius. Leave your answers in terms of π.

Radius	Circumference	Area
2		
4		
8		
16		

Look for patterns in the table to make conjectures: How does the circumference change when the radius is doubled? How does the area change when the radius is doubled?

6. Olga wants to buy a premade pie crust to bake a pie. She knows that the area of the bottom of her circular pie pan is 50 in². At the grocery store, there are three different sizes of pie crusts. The small pie crust is for pans with 6-inch diameters, the medium pie crust is for pans with 8-inch diameters, and the large pie crust is for pans with 10-inch diameters. Which pie crust should Olga buy? Explain.

LESSON OBJECTIVES

- Find the volume and surface area of three-dimensional objects
- Find a missing dimension when given the volume of three-dimensional objects

CORE SKILLS & PRACTICES

- Calculate Volume
- Calculate Surface Area

Key Terms

surface area
the sum of the areas of all the faces of a three-dimensional figure

volume
a measure of the amount of space enclosed by a three-dimensional figure

Vocabulary

cylinder
a solid formed by two bases that are parallel, congruent circles

prism
a solid with two bases that are congruent, parallel shapes and rectangular lateral faces that connect the bases

pyramid
a solid in which all of the faces, except for the base, intersect at a point called the vertex

sphere
a solid formed by the set of all points that are a given distance from a point called the center

Key Concept

The volume of a three-dimensional object is the number of cubic units it takes to fill the object. The surface area of a three-dimensional object is the number of square units it takes to cover all sides of the object.

Rectangular Prisms

When people give gifts, they sometimes wrap a box in decorative paper to hide the gift. The surface area of the box determines the minimum amount of wrapping paper needed to wrap the gift.

Volume

A **prism** is a solid with two bases that are congruent, parallel shapes and lateral faces that intersect with the bases. In a rectangular prism, the bases are rectangles.

Volume is a measure of the amount of space enclosed by a three-dimensional object, such as a prism. Volume is measured in cubic units. The formula for the volume of a rectangular prism is $V = \ell w h$, where ℓ represents the length, w represents the width, and h represents the height.

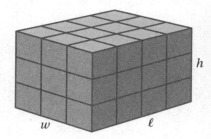

$$V = \ell w h$$

Example 1: Volume of a Rectangular Prism

Anastasia is buying a gift box with length 11 inches, width 6 inches, and height 4 inches. What is the volume of the box?

Step 1 Substitute the given values into the formula. $V = \ell w h = 11(6)(4)$

Step 2 Multiply. $= 66(4)$
     $= 264$

The volume of the box is 264 in³. We read this "264 cubic inches."

Surface Area

The **surface area** of a solid is the sum of the areas of all faces of the solid. To determine the surface area of a rectangular prism, you must find the sum of all six faces—the top, bottom, front, back, left, and right.

Each face of a rectangular prism is a rectangle. Therefore, the area of each face is found by multiplying the length by the width. The surface area is the sum of all six areas.

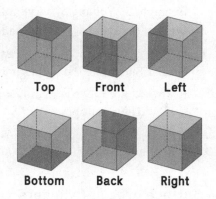

Top   Front   Left

Bottom   Back   Right

Example 2: Surface Area of a Rectangular Prism

A storage container measures 3 feet wide by 4 feet long by 2 feet tall. What is the surface area of the storage container?

2 ft

4 ft   3 ft

Step 1 Find the area of each face. Multiply the length by the width.

Top: $3(4) = 12$ Bottom: $3(4) = 12$ Front: $3(2) = 6$
Back: $3(2) = 6$ Left: $2(4) = 8$ Right: $2(4) = 8$

Step 2 Add the areas. $12 + 12 + 6 + 6 + 8 + 8 = 52$

The surface area is 52 ft^2. We read this "52 square feet."

Cylinders and Prisms

At the grocery store, you will see that juice containers come in many different shapes and sizes. The amount of liquid the container can hold, or the volume, varies depending on the size of the container's base and on its height. The volume can greatly affect the unit price you pay for the container of juice.

Volume of Cylinders

A **cylinder** is a solid that consists of two bases that are parallel congruent circles and have a curved surface connecting the bases. The formula for the volume of a cylinder is $V = \pi r^2 h$.

Volume of a Cylinder

$$V = \pi r^2 h$$

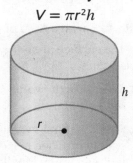

h

r

Calculate Volume

The formulas to find the volumes of rectangular prisms and cylinders are very similar. In both, the volume is equal to the area of the base B multiplied by the height h: $V = Bh$.

- The base of a rectangular prism is a rectangle. The area of the rectangular base is $B = \ell w$, so the formula for the volume of a rectangular prism is $V = Bh = \ell wh$.

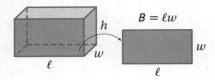

- The base of a cylinder is a circle. The area of the circular base is $B = \pi r^2$, so the formula for the volume of a cylinder is $V = Bh = \pi r^2 h$.

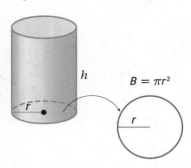

To calculate the volume of any prism, find the area of the base and then multiply by the height of the prism. A triangular prism has a base that is a triangle. Find the volume of the triangular prism below.

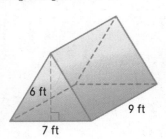

Example 3: Volume of a Cylinder

What is the volume of a cylinder whose height is 12 cm and whose radius is 8 cm?

Step 1 Substitute the values into the formula. $\quad V = \pi r^2 h = \pi \times 8^2 \times 12$

Step 2 Simplify. First evaluate the exponent. Then multiply.
$$= \pi \times 64 \times 12$$
$$= 768\pi$$
$$\approx 2412.7$$

The volume of the cylinder is about 2412.7 cm^3.

Surface Area of Cylinders

A cylinder has three surfaces: two circular bases and a curved lateral surface that connects the bases. To find the surface area of a cylinder, find the area of each surface and then add the areas together.

The bases are identical circles whose area can be expressed as $A = \pi r^2$. The curved lateral surface is actually a rectangle whose length is the circumference of the base ($C = 2\pi r$) and whose width is the height of the cylinder. The area of this rectangle, then, is $A = \ell w = 2\pi rh$.

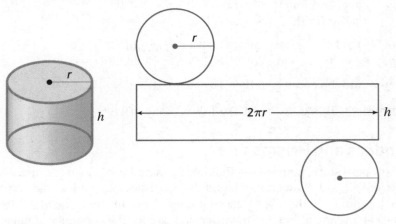

Surface Area = Base + Base + Lateral Surface
$$\pi r^2 + \pi r^2 + 2\pi rh = 2\pi r^2 + 2\pi rh$$

Example 4: Surface Area of a Cylinder

What is the surface area of a cylinder whose height is 12 cm and whose radius is 8 cm?

Step 1 Substitute the values into the formula.
$$SA = 2\pi r^2 + 2\pi rh$$
$$= 2\pi \times 8^2 + 2\pi \times 8 \times 12$$

Step 2 Simplify. First evaluate the exponent. Then perform all multiplication. Finally, add.
$$= 2\pi \times 64 + 2\pi \times 96$$
$$= 128\pi + 192\pi$$
$$= 320\pi$$
$$\approx 1005.3$$

The surface area is about 1005.3 cm^2.

Surface Area of Prisms

Like a cylinder, a prism has two identical bases. Instead of a curved surface, however, a prism has 4-sided lateral faces that connect the bases. To find the surface area, you must add the areas of the bases and all lateral faces.

To find the lateral area of a prism L, multiply the perimeter of the base, P, by the height of the prism, h. A prism's surface area equals the lateral area plus two times the area of the base.

Prism
$L = Ph$
$SA = L + 2B$

Example 5: Surface Area of a Non-Rectangular Prism

What is the surface area of the prism shown?

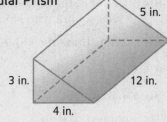

5 in.
3 in.
12 in.
4 in.

Step 1 Find the perimeter of the base.
The base is a triangle.
$$P = 3 + 4 + 5 = 12$$

Step 2 Find the lateral area by multiplying the perimeter by the height.
$$L = Ph = 12 \times 12 = 144$$

Step 3 Find B, the area of the base. It is a triangle with base 4 inches and height 3 inches. Use the formula for area of a triangle.
$$B = \tfrac{1}{2}bh$$
$$= \tfrac{1}{2} \times 4 \times 3$$
$$= 2 \times 3 = 6$$

Step 4 Add the lateral area and the area of the bases to find the total surface area.
$$SA = L + 2B$$
$$= 144 + 2 \times 6$$
$$= 144 + 12$$
$$= 156$$

The surface area is 156 in.²

Think about Math

Directions: Answer the following question.

1. Shayla is decorating the outside of a cylindrical oatmeal box. The box is 24 cm high with a radius of 7 cm. About how much surface area does Shayla have to cover?

Pyramids, Cones, and Spheres

The balls used in many different sports are in the shape of a sphere.

Right Pyramids

A **pyramid** is a solid whose faces, except for the base, intersect at a point called the vertex.

The formula for volume of a right pyramid is $V = \tfrac{1}{3}Bh$, where B is the area of the base and h is the height.

The formula for the surface area of a right pyramid is $SA = \tfrac{1}{2}P\ell + B$, where P is the perimeter of the base, ℓ is the slant height, and B is the area of the base.

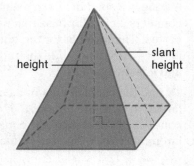

height
slant height

Environmental Literacy

The Environmental Protection Agency (EPA) is responsible for monitoring the effects of business and industry on the environment. Many industries generate waste products that are harmful to the environment if they are not stored and disposed of properly. In some cases, businesses may be required to record the volume of waste they produce and report this information to the EPA.

ABC Company stores its waste in cylindrical barrels that have a radius of 1.5 feet and a height of 4.5 feet. Last month, the company filled 30 barrels with waste. What was the total volume of waste that ABC Company produced last month?

Calculate Surface Area

A net is a two-dimensional diagram of a three-dimensional figure that can be folded to form the figure. For example, the net below can be folded to form the pyramid. You can use a net to find surface area.

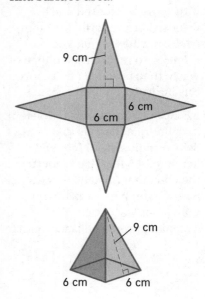

To find the surface area of this pyramid, find the area of each two-dimensional shape in the net. The net of this pyramid has a square with side length 6 cm. It also has four triangles with base 6 cm and height 9 cm. The surface area is equal to the total area of these five figures.

Surface Area = area of square + 4 × area of each triangle

Find the area of the square and the area of each triangle. Then find the surface area of the pyramid.

Example 6: Volume of a Right Pyramid

What is the volume of the pyramid shown?

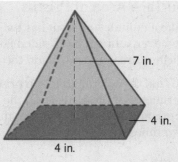

Step 1 Find B, the area of the base. The base is a square with side length 4 inches. $B = 4 \times 4 = 16$

Step 2 Substitute the values into the formula. $V = \frac{1}{3}Bh$

$$= \frac{1}{3} \times 16 \times 7$$

Step 3 Multiply. $= \frac{112}{3} = 37\frac{1}{3}$

The volume is $37\frac{1}{3}$ in³.

Example 7: Surface Area of a Right Pyramid

What is the surface area of the pyramid shown?

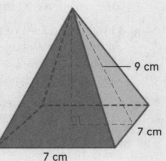

Step 1 Find P, the perimeter of the base. The base is a square with side length 7 cm. $P = 7 + 7 + 7 + 7$
$= 4 \times 7 = 28$

Step 2 Find B, the area of the base. $B = 7 \times 7 = 49$

Step 3 Substitute the values into the formula. $SA = \frac{1}{2}P\ell + B$
$= \frac{1}{2} \times 28 \times 9 + 49$

Step 4 Simplify. First perform multiplication from left to right. Then add. $= \frac{1}{2} \times 28 \times 9 + 49$
$= 14 \times 9 + 49$

The surface area is 175 cm². $= 126 + 49 = 175$

Right Cones

A cone is a solid in which a curved surface connects a circular base to a point called the vertex. The formula for the volume of a cone is $V = \frac{1}{3}\pi r^2 h$, where r is the radius of the base and h is the height.

To find the surface area of a cone, add the lateral area to the area of the base of the cone. The lateral area L is equal to $\frac{1}{2}$ times the circumference of the base times the slant height, ℓ.

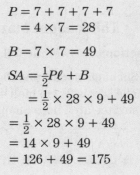

$$L = \frac{1}{2} \times 2\pi r \times \ell = \pi r\ell \qquad SA = \pi r\ell + \pi r^2$$

Example 8: Volume and Surface Area of a Cone

Josh has a cone-shaped paper cup with a radius of 6 cm, a height of 8 cm, and a slant height of 10 cm. The cup has a circular lid. Find the volume and surface area of Josh's cup.

Step 1 To find the volume, substitute the given values into the formula for volume of a cone.

Simplify. First evaluate the exponent. Then multiply.

The volume is about 301.6 cm^3.

$V = \frac{1}{3}\pi r^2 h$
$= \frac{1}{3}\pi \times 6^2 \times 8$
$= \frac{1}{3}\pi \times 36 \times 8$
$= 96\pi$
≈ 301.6

Step 2 To find the surface area, substitute the given values into the formula for surface area of a cone.

Simplify. First evaluate the exponent. Then multiply, and then add.

The surface area is about 301.6 cm^2.

$SA = \pi r \ell + \pi r^2$
$= \pi \times 6 \times 10 + \pi \times 6^2$
$= \pi \times 6 \times 10 + \pi \times 36$
$= 60\pi + 36\pi$
$= 96\pi$
≈ 301.6

CALCULATOR SKILL

π is an irrational number. This means that there is an endless number of digits after the decimal point, and these digits never repeat in any pattern. For this reason, the exact value of π cannot be used in any calculations. Instead, we must use an estimate. Common estimates of π are 3.14 and $\frac{22}{7}$.

A calculator can provide a better estimate of π. Scientific and graphing calculators generally have a button for π. (It may be the second function of another button.) You can use this button, $\boxed{\pi}$, whenever a calculation involves π.

Spheres

A sphere is a solid formed by the set of all points that are the same distance from a point called the center of the sphere. Because a sphere is circular in shape, the formulas for volume and surface area involve π. For both formulas, the only value you need is the radius of the sphere, r.

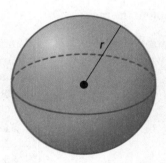

$$V = \frac{4}{3}\pi r^3 \qquad SA = 4\pi r^2$$

Example 9: Volume and Surface Area of a Sphere

Find the volume and surface area of the sphere shown.

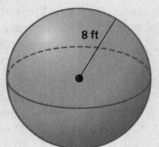

8 ft

Step 1 To find the volume, substitute the value of r into the formula for volume of a sphere.

Simplify. First evaluate the exponent. Then multiply.

The volume is about 2144.7 ft^3.

$V = \frac{4}{3}\pi r^3$
$= \frac{4}{3}\pi \times 8^3$
$= \frac{4}{3}\pi \times 512$
≈ 2144.7

Step 2 To find the surface area, substitute the value of r into the formula for surface area of a sphere.

Simplify. First evaluate the exponent. Then multiply.

The surface area is about 804.2 ft^2.

$SA = 4\pi r^2$
$= 4\pi \times 8^2$
$= 4\pi \times 64$
$= 256\pi$
≈ 804.2

Vocabulary Review

Directions: Write the missing term in the blank.

cylinder   prism   pyramid
sphere   surface area   volume

1. _____ is a measure of the amount of space enclosed by a three-dimensional figure.

2. A _____ is a solid that contains two parallel, congruent bases and rectangular lateral faces that intersect with the bases.

3. To find the _____ of a prism or pyramid, add the areas of all faces.

4. A solid with two bases that are parallel, congruent circles is a _____.

5. A _____ is a three-dimensional figure consisting of all points that are the same distance from a fixed point called the center.

6. In a _____, all of the lateral faces intersect at a point called the vertex.

Skill Review

Directions: Read each problem and complete the task.

1. What is the surface area of a box whose base measures 11 inches by 6 inches and whose height is 4 inches?

   A. 134 in.^{2}
   B. 264 in.^{2}
   C. 268 in.^{2}
   D. 528 in.^{2}

2. A traffic cone of solid plastic has radius 4 inches and slant height 25 inches. Which expression can be used to determine how much surface area needs to be painted orange, including the base?

   A. $\frac{1}{3}\pi \times 4 \times 25$

   B. $\pi \times 4 \times 25 + \pi \times 4^2$

   C. $\pi \times 4^2 \times 25$

   D. $\pi \times 25^2 \times 4$

3. Abner and his friends rolled a giant snowball with a radius of 9 inches. What was the volume of snow in the snowball?

   A. about 254 in.^{3}
   B. about 1018 in.^{3}
   C. about 2290 in.^{3}
   D. about 3054 in.^{3}

4. Lucinda wants to make a pair of silver earrings that are shaped like cylinders. The radius will be 4 mm and the height will be 11 mm. Each cubic millimeter of silver costs $0.21. What will be the total cost of the silver to make these earrings?

   A. $116.11
   B. $232.23
   C. $552.92
   D. $1105.84

5. A tent at a county fair is in the shape of a pyramid. The base of the tent is a square with each side length 24 meters, and the tent encloses a space of 960 m^3. How tall is the central tent pole?

   A. 2 meters
   B. 5 meters
   C. 13 meters
   D. 40 meters

6. You want to make the canvas tent shown. The tent will be pitched on top of grass, so no canvas is needed for a floor. How much canvas will you need to make the tent?

   A. 336 ft^2
   B. 528 ft^2
   C. 576 ft^2
   D. 692 ft^2

10 ft   6 ft   12 ft   16 ft

Skill Practice

Directions: Read each problem and complete the task.

1. Fernanda made a square pyramid out of sand at the beach. The base of the pyramid was 4 feet on each side. The lateral surface area of the pyramid was 24 ft^2. What was the slant height?

   A. 1 foot

   B. $2\frac{1}{4}$ feet

   C. 3 feet

   D. $4\frac{1}{2}$ feet

2. What is the volume of the figure below?

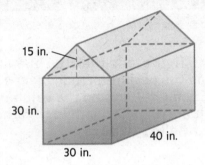

3. A paper drinking cup is in the shape of a cone with an open top. The cone has a diameter of 6 cm, and the height is 10 cm. Which is the best estimate of the amount of water that the cup will hold?

   A. 90 cm^3

   B. 270 cm^3

   C. 360 cm^3

   D. 1080 cm^3

4. A beach ball has a radius of 6 inches. Which expression gives the surface area of the beach ball?

   A. $4\pi \times 3^2$

   B. $4\pi \times 6^2$

   C. $4\pi \times 12^2$

   D. $4\pi \times 6^3$

5. A water tank is in the shape of a rectangular prism. The tank is 40 feet long and 20 feet wide. One cubic foot of water is equivalent to $7\frac{1}{2}$ gallons, and the tank holds 30,000 gallons of water. What is the height of the tank?

   A. 5 ft

   B. 22 ft

   C. 38 ft

   D. 281 ft

6. Stefano used a 3-D printer to make a solid plastic sphere with a radius of 3.5 cm. The cost of the plastic was $0.07/cm^3. What was the total cost of the plastic used to make the sphere?

   A. $10.78

   B. $12.57

   C. $153.94

   D. $179.59

7. Carmen is told that the diameter of a container is 6 cm, and that its volume is 577 cm^3. She calculates the height of the container to be about 20.4 cm. What assumption has Carmen made about the container?

8. To find the surface area of a cylinder, Wallace first adds the radius and the height. He then multiplies his answer by $2\pi r$. Is Wallace's method correct? Explain why or why not.

LESSON OBJECTIVES

- Calculate perimeter and area of 2-dimensional composite figures
- Calculate surface area and volume of 3-dimensional composite figures

CORE SKILLS & PRACTICES

- Calculate Area
- Make Sense of Problems

Key Terms

composite figure
a figure that is made up of two or more shapes

composite solid
an object that is made up of two or more solids

Vocabulary

2-dimensional
a flat shape having only two dimensions, often length and width

3-dimensional
an object consisting of 3 dimensions, usually length, width, and height

hemisphere
half of a sphere

Key Concept

To find the area of a composite figure, add the area of each figure in the composite. To find the perimeter, add pieces of the perimeter of each figure. Similarly, to find the volume of a composite solid, add the volume of each solid. To find the surface area, add parts of each solid's surface area.

2-Dimensional Figures

Shapes made up of two or more simple shapes are all around. For example, an area on a basketball court called the key is a rectangle and a semicircle.

Compute Perimeter

A **2-dimensional** figure is flat shape that has only two dimensions. A **composite figure** is made of two or more 2-dimensional shapes. The perimeter of a composite figure is the distance around the outside.

Example 1: Perimeter of a Composite Figure

Find the perimeter of each figure.

a.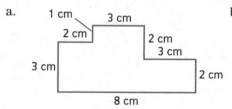

b.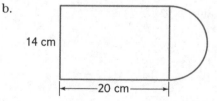

a. When a composite figure is composed entirely of straight edges, add the side lengths to determine the perimeter.

$P = 3 + 2 + 1 + 3 + 2 + 3 + 2 + 8 = 24$ cm

b. When a composite figure includes a semicircle, add the lengths of the straight sides and the circumference of the semicircle.

$P = s + s + s + \frac{1}{2}\pi d$
$= 14 + 20 + 20 + \frac{1}{2} \times \pi \times 14$
≈ 76 cm

Compute Area

To find the area of a composite figure, determine what shapes make up the composite figure and the formulas for the area of each shape. The area of the composite figure is the sum of the areas of each shape.

Example 2: Area of a Composite Figure

Find the area of each figure.

a.
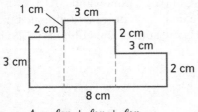

$$A = \ell w + \ell w + \ell w$$
$$= 3 \times 2 + 4 \times 3 + 2 \times 3$$
$$= 24 \text{ cm}^2$$

b.
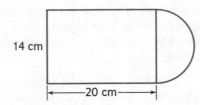

$$A = \ell w + \tfrac{1}{2}\pi r^2$$
$$A = (14)(20) + \tfrac{1}{2}\pi(7)^2$$
$$\approx 357 \text{ cm}^2$$

a. The composite figure is made up of three rectangles. Find the area of each rectangle. Add to find the total area.

$$A = \ell w + \ell w + \ell w$$
$$= 3 \times 2 + 4 \times 3 + 2 \times 3$$
$$= 24 \text{ cm}^2$$

b. The composite figure is made up of a rectangle and a semicircle. Find the area of each shape. Add to find the total area.

$$A = \ell w + \tfrac{1}{2}\pi r^2$$
$$A = (14)(20) + \tfrac{1}{2}\pi(7)^2$$
$$\approx 357 \text{ cm}^2$$

 ### Think about Math

Directions: Answer the following questions.

1. Farrah is going to run the trail shown. What is the distance that she will run? Round to the nearest yard.

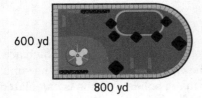

2. Ben is painting the interior wall of a room with the dimensions shown. What is the area of the wall to be painted?

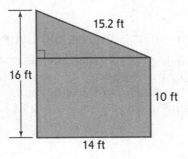

Calculate Area

Ceramic tiles are commonly used for kitchen and bathroom floors. Oftentimes, the tiles are sold by the square foot. So, it is necessary to calculate the area of the floor to be covered in order to determine how many tiles are needed.

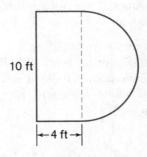

The figure represents a section of floor. What is the area of this floor section? Explain how you found your answer.

When farmers harvest grain, they must have a place to store the grain until they need to use it. Many farms have a building called a silo to store grain. When constructing a silo, the volume of the building is a measure of how much grain can be stored in the silo.

A farmer is constructing a grain silo for his farm with the dimensions shown. What is the approximate volume of the grain silo?

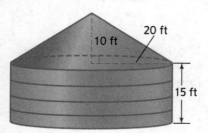

Volume of 3-Dimensional Solids

Many times containers are composed of multiple solids. For example, some grain silos are composed of a cylinder and a cone. The volume, or amount of grain that such a silo can contain, varies depending on the size of each solid that composes the building.

Volume of Prisms and Pyramids

A **3-dimensional** figure is an object consisting of 3 dimensions. A **composite solid** is an object that is made up of two or more 3-dimensional figures. The volume of a composite solid is the sum of the volumes of each individual solid.

Example 3: Volume of a Composite Solid

Find the volume of the figure.

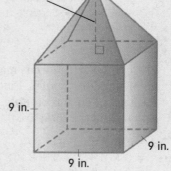

Step 1 Identify the solids that make up the figure and the correct formula for the volume of each solid. This composite solid is made up of a rectangular prism and a pyramid.

$$V_{Prism} = lwh$$

$$V_{Pyramid} = \frac{1}{3}lwh$$

Step 2 Substitute the values given into each formula and find the volume of each individual solid.

$$
\begin{aligned}
V_{Prism} &= lwh & V_{Pyramid} &= \frac{1}{3}lwh \\
&= 9 \times 9 \times 9 & &= \frac{1}{3} \times 9 \times 9 \times 6 \\
&= 729 \text{ in.}^3 & &= 162 \text{ in.}^3
\end{aligned}
$$

Step 3 Add the volumes of the individual solids to determine the volume of the composite solid.

$$
\begin{aligned}
V_{Prism} + V_{Pyramid} &= 729 + 162 \\
&= 891 \text{ in.}^3
\end{aligned}
$$

Volume of Spheres and Cylinders

A composite solid may contain cylinders, spheres, or half-spheres, which are called **hemispheres.** The total volume of the composite solid is still the sum of the volumes of the individual solids. For many solids, volume is calculated by multiplying the area of the base times the height of the solid.

Example 4: Volume of a Composite Solid

Find the volume of the figure.

Step 1 Identify the solids that make up the figure and the correct formula for the volume of each solid. This composite solid is made up of a cylinder and a hemisphere.

$$V_{Cylinder} = \pi r^2 h$$

$$V_{Hemisphere} = \frac{1}{2}\left(\frac{4}{3}\pi r^3\right)$$

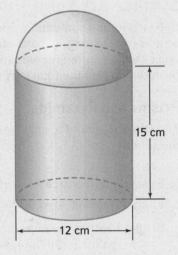

15 cm

12 cm

Step 2 Substitute the values given into each formula and find the volume of each individual solid.

$$V_{Cylinder} = \pi r^2 h$$
$$= \pi \times 6^2 \times 15$$
$$\approx 1{,}696.5 \text{ cm}^3$$

$$V_{Hemisphere} = \frac{1}{2}\left(\frac{4}{3}\pi r^3\right)$$
$$= \frac{1}{2}\left(\frac{4}{3}\pi(6)^3\right)$$
$$\approx 452.4 \text{ cm}^3$$

Step 3 Add the volumes of the individual solids to determine the volume of the composite solid.

$$V_{Cylinder} = V_{Hemisphere} \approx 1{,}696.5 + 452.4$$
$$= 2{,}148.9 \text{ cm}^3$$

 Think about Math

Directions: Answer the following questions.

1. A storage unit has the dimensions shown. What is the volume of the storage unit?

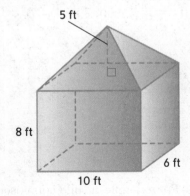

5 ft

8 ft

6 ft

10 ft

A. 480 ft³
B. 580 ft³
C. 4,800 ft³
D. 5,800 ft³

2. An observatory is in the shape of a cylinder with a radius of 10 meters and a height of 11 meters. The top of the observatory is a hemisphere. What is the approximate volume of the observatory?

A. 1,100 m³
B. 2,093 m³
C. 3,454 m³
D. 5,547 m³

CALCULATOR SKILL

To calculate the volume of a sphere or hemisphere, you will have to find the cube of the radius. One way to do this is to multiply the radius by itself 3 times. On the TI-30XS MultiView™ calculator, you can use the ⌃ button. For example, to find the value of 14^3, press

[1] [4] [⌃] [3] [enter].

Surface Area of 3-Dimensional Solids

Before painting something, you must determine the total area to be painted so you can ensure you have enough paint for the project. If the shape of the object to be painted is complex, the surface areas of the different parts can be added to find the total surface area.

Prisms and Pyramids

The surface area of a solid is the sum of the areas of all of the faces of the solid.

Example 5: Surface Area of a Composite Solid

Find the surface area of the figure.

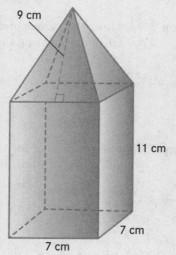

Step 1 Identify the faces that make up the figure and the correct formula for the area of each face. The faces of this composite solid are 5 rectangles and 4 triangles.

- Prism
  Front: ℓh_1
  Back: ℓh_1
  Left: ℓh_1
  Right: ℓh_1
  Bottom: ℓ^2

- Pyramid
  Each face: $\frac{1}{2}\ell h_2$

  All 4 faces: $4\left(\frac{1}{2}\ell h_2\right) = 2\ell h_2$

Step 2 Substitute the given values into the formula for the area of each face.

- Prism
  Front: $7 \times 11 = 77$ Back: $7 \times 11 = 77$ Left: $7 \times 11 = 77$
  Right: $7 \times 11 = 77$ Bottom: $7^2 = 49$

- Pyramid
  All 4 faces: $2 \times 7 \times 9 = 126$

Step 3 Add the individual areas to determine the surface area of the composite solid.

$SA = 77 + 77 + 77 + 77 + 49 + 126 = 483$ cm^2

Hemispheres and Cylinders

A composite solid may have curved surfaces. For example, a solid composed of a cylinder and a hemisphere has curved surfaces. The surface area of the composite figure is still the sum of the individual surface areas.

Compute Perimeter, Area, Surface Area, and Volume of Composite Figures

Example 6: Surface Area of a Composite Solid

Find the surface area of the figure.

Step 1 This solid is composed of a cylinder and a hemisphere. The surface area will be the surface area of the cylinder minus its top circle plus the curved surface area of the hemisphere. Identify the formula for each of these areas.

$$SA_{Cylinder} - A_{Circle} =$$
$$\pi dh + 2\pi r^2 - \pi r^2 = \pi dh + \pi r^2$$

Step 2 Substitute the given values into the formulas.

$$SA_{Cylinder} - A_{Circle} = \pi dh + \pi r^2$$
$$= \pi \times 12 \times 15 + \pi \times 6^2$$
$$\approx 678.6 \text{ cm}^2$$

$$SA_{Curved\ Hemisphere} = 2\pi r^2$$
$$= 2\pi \times 6^2$$
$$\approx 226.2 \text{ cm}^2$$

Step 3 Add the areas to determine the surface area of the composite solid.

$$678.6 + 226.2 \approx 904.8 \text{ cm2}$$

The surface area is about 904.8 cm^2.

15 cm

12 cm

Think about Math

Directions: Answer the following question using the composite solid.

1. The surface area of this solid is 612.3 yd^2. What is the approximate value of h?

   A. 6 yd
   B. 9 yd
   C. 12 yd
   D. 24 yd

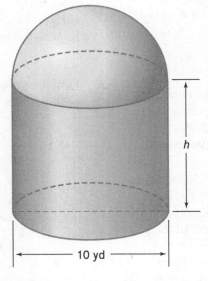

h

10 yd

CORE PRACTICE

Make Sense of Problems

An important step in making sense of problems is recognizing the prior knowledge required to solve the problem.

A tower consists of a cylinder and a cone, as shown below. The cylinder has radius $r = 20$ feet and height $h = 12$ feet. The cone has the same radius as the cylinder and slant height $\ell = 25$ feet. The formula for the surface area of the curved surface of a cone is $\pi r \ell$.

What additional formula that is not given in the problem is necessary to find the surface area of the tower? What is the surface area of the tower?

Compute Perimeter, Area, Surface Area, and Volume of Composite Figures

Vocabulary Review

Directions: Write the missing term in the blank.

composite figure	2-dimensional	hemisphere
3-dimensional	composite solid	

1. A _____ figure is a flat shape that has only two dimensions.

2. A _____ is half of a sphere.

3. A _____ is a 2-dimensional figure that is made up of two or more different shapes.

4. A _____ is a 3-dimensional object that is made up of two or more solids.

5. A _____ figure is an object that has 3 dimensions, usually length, width, and height.

Skill Review

Directions: Read each problem and complete the task.

A community vegetable garden is shown in the diagram. Use the diagram for 1 and 2.

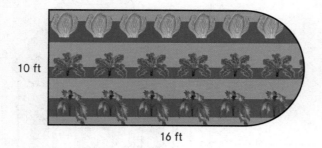

10 ft

16 ft

1. Fencing costs $12 per foot. To the nearest dollar, how much will it cost to put a fence around the entire garden?

   A. $500   B. $624
   C. $692   D. $809

2. One bag of organic fertilizer contains enough fertilizer to cover 375 ft². The head gardener wants to fertilize the entire garden three times during the summer. Which statement or statements are correct?

   A. One bag of fertilizer will last the gardener all summer.
   B. If the gardener buys 2 bags of fertilizer, he will have some left over at the end of the summer.
   C. There is more fertilizer than needed for the summer in 2 bags, so the gardener needs only 1 bag.
   D. By fertilizing only two times during the summer, the gardener can buy one less bag of fertilizer.

3. Roger built the toy house shown in the figure below. He now wants to paint the house, so he must determine the surface area to know how much paint to buy. What is the surface area of Roger's toy house?

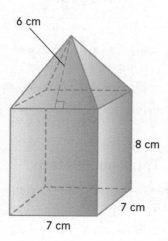

6 cm

8 cm

7 cm

7 cm

4. Find the volume and surface area of the figure. Round to the nearest whole number.

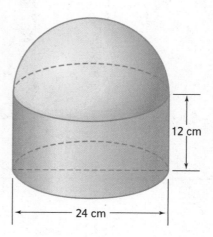

12 cm

24 cm

Compute Perimeter, Area, Surface Area, and Volume of Composite Figures

Skill Practice

Directions: Read each problem and complete the task.

Use the image below to answer questions 1 and 2.

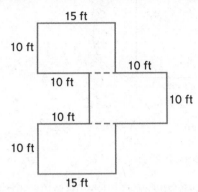

The diagram shows three adjoining rooms. Dashed lines represent doorways. An interior designer wants to put a decorative wallpaper border along the top wall of each room. The designer mistakenly did not include the wall space above the doorways when calculating the length of wallpaper border needed.

1. How much did the designer mistakenly calculate the length of the wallpaper border to be?

2. How much more border will she need?

3. Which is the best estimate of the volume of this figure?

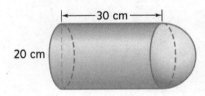

   A. 2,000 cm³
   B. 3,000 cm³
   C. 9,000 cm³
   D. 11,000 cm³

4. Scott found the volume of this figure as follows:

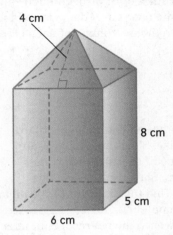

Volume of pyramid $= \frac{1}{3}\ell wh = \frac{1}{3} \cdot 6 \cdot 5 \cdot 4 = 40$ cm³

Volume of prism $= \ell wh = 6 \cdot 5 \cdot 8 = 240$ cm³

Total volume $= 40 + 240 = 280$ cm³

What error did Scott make?

5. The figure is composed of a rectangle and two semicircles. The area of the figure is 142 ft². What is the approximate length of the rectangle?

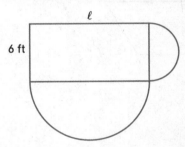

   A. 6 feet
   B. 12 feet
   C. 24 feet
   D. 136 feet

6. Draw and label a composite solid whose volume is 45 in³.

Directions: Choose the best answer to each question.

1. Angie is using cans that are shaped as a right circular cylinder for a craft project. Each can is 10 inches tall, and the base has a diameter of 4 inches. What is the approximate volume of each can?

    A. 40 in.³
    B. 125.7 in.³
    C. 160 in.³
    D. 502.4 in.³

2. Heather notices that a pillar in a courtyard has a circular base. She is able to measure the circumference of the pillar, which is 75.36 centimeters. She is able to use this measurement to find the diameter of the pillar, which is approximately _____ .
    Round your answer to the nearest centimeter.

3. Anthony draws a circle with a radius of 4.5 centimeters. He would like to glue a string around the circumference of the circle. So, the length of the string is approximately _____ .
    Use 3.14 for your approximation of pi.

4. The perimeter of the parallelogram below is 96 inches. What is the length of the side labeled with m?

    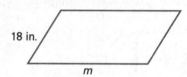

    A. 78 in.
    B. 60 in.
    C. 30 in.
    D. 15 in.

5. Using 3.14 as an approximation of pi, which is the approximate perimeter of the figure shown below?

    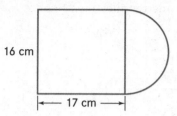

    A. 58 cm
    B. 75.12 cm
    C. 91.12 cm
    D. 100.24 cm

6. Jake drew an equilateral triangle with side lengths that are 18 centimeters. What is an approximation of the area of the triangle he drew?

    A. 15.59 cm²
    B. 140.3 cm²
    C. 162 cm²
    D. 181.12 cm²

7. The rectangular prism shown has a volume of 24 cubic centimeters. The measurement that is labeled p has a value of _____ .

    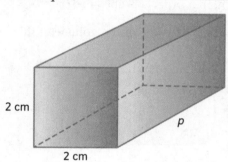

8. A circle has an approximate area of 379.94 square inches. Using 3.14 as an approximation of pi, what is the radius of the circle?

    A. 121 in.
    B. 19.49 in.
    C. 11 in.
    D. 6.21 in.

Directions: Choose the best answer to each question.

9. Phillip is going to paint the interior wall of a room with the dimensions shown. The area of the wall to be painted is _____.

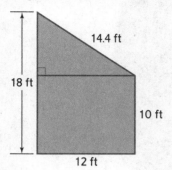

14.4 ft

18 ft

10 ft

12 ft

10. Which is the surface area of the solid shown?

A. 544 cm²
B. 608 cm²
C. 704 cm²
D. 6,400 cm²

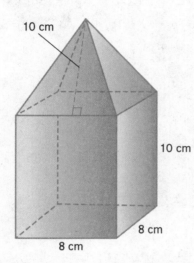

10 cm

10 cm

8 cm

8 cm

11. Jenna has a rectangular picture frame that is 6.5 inches wide and 4.75 inches tall. She is going to glue ribbon around the edge of the frame. She needs to know the perimeter to know the least amount of ribbon she'll need. What is the perimeter of the frame?

A. 30.875 in.
B. 22.5 in.
C. 17.75 in.
D. 11.25 in.

12. Jeff found a circle in a mural. He measures to find that the diameter is 16 inches. Which is an approximation of the area of the circle to the nearest square inch?

A. 64 in²
B. 201 in²
C. 256 in²
D. 804 in²

13. Which is the approximate volume of the solid? Use $\frac{22}{7}$ as the approximation of pi.

A. 2,034.72 cm³
B. 2,489.14 cm³
C. 2,502.72 cm³
D. 2,939.54 cm³

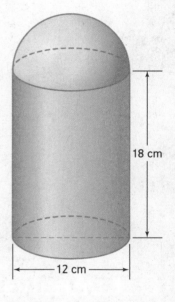

18 cm

12 cm

Check Your Understanding

On the following chart, circle the items you missed. The last column shows pages you can review to study the content covered in the question. Review those lessons in which you missed half or more of the questions.

Lesson	Item Number(s)			Review Page(s)
	Procedural	Conceptual	Problem Solving	
7.1 Compute Perimeter and Area of Polygons	4, 6		11	220–227
7.2 Compute Circumference and Area of Circles	8		2, 3, 12	228–233
7.3 Compute Surface Area and Volume	7		1	234–241
7.4 Compute Perimeter, Area, Surface Area, and Volume of Composite Figures	5, 10		9, 13	242–249

Chapter 8

Data Analysis

Collecting, analyzing, and displaying data is critical to numerous career fields. Scientists collecting information about a new chemical's boiling point, teachers analyzing students' test scores, and a politician looking at polling data, all have something in common. They are looking at data points and trying to see trends and draw conclusions about that data. This chapter discusses how to analyze data points and display the data in different types of graphs.

Lesson 8.1
Calculate Measures of Central Tendency

A plot summary sums up the major points of a movie. How could you summarize information in a data set? Measures like mean, median, and mode describe a data set. Learn how to calculate measures of central tendency for different sized data sets.

Lesson 8.2
Display Categorical Data

How do you organize your schedule? Do you spend more time with family or friends or at work? How could you see at a glance the way you spend your time? Learn how to display categorical data in bar graphs and circle graphs.

Lesson 8.3
Display One-Variable Data

You can calculate mean, median, and mode to summarize a data set. How can you display the data points visually? Learn how to display one-variable data in a dot plot, histogram, or box plot.

Lesson 8.4
Display Two-Variable Data

How did graphing an equation help you to understand the key features of the equation? How could a graph of data points help you to understand how the data is related? Learn how to display data in tables, scatter plots, and line graphs.

Goal Setting

When you watch the news or read the newspaper, how does data get displayed? What kinds of graphs and charts do you notice? What type of data are the charts describing? What point do you think the chart is trying to make? How does the chart reinforce the story it is displayed with?

How do you see charts and graphs being used on the Internet? If a graph or chart wasn't used, how would you list the data? Why does displaying the data graphically make more sense?

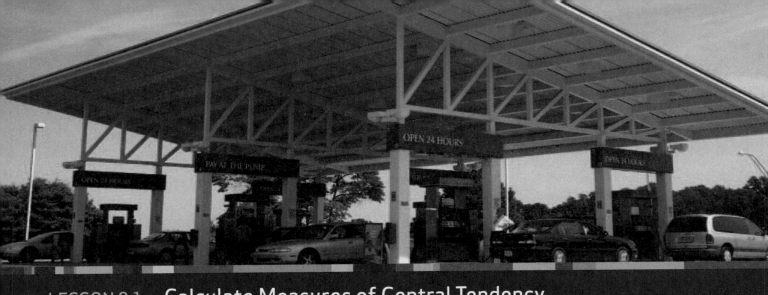

LESSON 8.1 Calculate Measures of Central Tendency

LESSON OBJECTIVES

- Calculate the mean, median, mode, and range of a data set
- Find a missing data item given the mean and other data
- Calculate a mean based on frequency counts
- Calculate a weighted average

CORE SKILLS & PRACTICES

- Interpret Data Displays

Key Terms

mean
average; the sum of all values in a data set divided by the number of values

median
the middle number of an ordered data set; in a data set with an even number of values, the average of the two middle values

mode
the value(s) that occur most often in a data set

Vocabulary

range
the difference between the greatest and least values in a data set

weighted average
an average of a data set in which some items carry more importance (weight) than others

Key Concept

A measure of central tendency is a number that can be used to summarize a group of numbers. Mean, median, and mode are measures of central tendency calculated in different ways.

Measures of Central Tendency

Mean, median, and mode are different measures of central tendency. They all attempt to show some kind of central point in the data. Some cars can calculate how many miles per gallon a driver gets based on his or her driving habits. This number is a mean.

Mean (Average)

The most frequently encountered measure of central tendency is the **mean**. The mean of a data set, also called the average, is the sum of all the data items, divided by the number of items.

$$\text{mean} = \frac{\text{sum of data items}}{\text{number of data items}}$$

Example 1: Finding the Mean

Let's say that the high temperatures on three days were 78°F, 81°F, and 87°F. What is the average temperature of the 3 days?

Step 1 Add the temperatures together.

$$78 + 81 + 87 = 246$$

Step 2 Divide 246 by the number 3 because there are 3 temperatures.

$$\frac{246}{3} = 82$$

The mean temperature for these three days is 82°F.

Median

The **median** of a group of numbers is the number that is in the middle when they are ordered from least to greatest value. In an odd-numbered group of items, the median is always the middle value.

Example 2: Odd Number of Items

In a group of five numbers, the median is the third-greatest value. So, we first order the numbers from least to greatest and then select the number in the middle.

21, 19, 37, 24, 31

Arrange in order: 19, 21, 24, 31, 37

The middle item, 24, is the median.

Example 3: Even Number of Items

If there is an even-numbered group of items, such as six numbers, there is no single number in the middle. In this case, the median is the mean of the two numbers closest to the middle.

8, 12, 19, 23, 30, 37

To find the median of an even-numbered set:

Step 1 Add the two numbers closest to the middle. In this example, they are 19 and 23.

$19 + 23 = 42$

Step 2 Divide the sum by 2. We divide by 2 because there are 2 numbers in the middle. The median is 21.

$\frac{42}{2} = 21$

Mode

The mode of a set of data items is the value that appears most often in the set.

45, 41, 47, 50, 41, 59, 47, 60, 41

41 is the mode of the data.

If two values appear more often than any others, and the values appear the same number of times, both values are the mode.

88, 79, 86, **85**, 79, 94, **85**, 77

Both 79 and 85 are modes of the data.

If each value appears only once in the set, the data set has no mode.

101, 98, 123, 85, 107, 91, 82

There is no mode of the data.

Interpret Data Displays

Being able to take a chart or table of data and determine a measure of central tendency can help you understand the data itself.

Bread Prices: 1 Loaf			
March	April	May	June
$2.70	$2.54	$2.78	$2.82

Look at the prices of bread presented in the table. What is the median price of a loaf of bread over the four-month period?

Range

In addition to the mean, median, and mode of a group of numbers, it is often helpful to know the **range**.

The **range** of a set of data items is the difference between the least and greatest values.

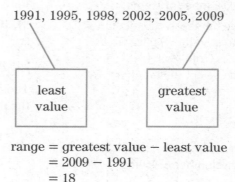

1991, 1995, 1998, 2002, 2005, 2009

least value   greatest value

$$\text{range} = \text{greatest value} - \text{least value}$$
$$= 2009 - 1991$$
$$= 18$$

The range of a group of numbers is not a measure of central tendency. The range shows the spread of the data. The greater the range, the greater the span of the data. If the range is smaller, then the span of data is smaller.

Think about Math

Directions: Use the following data:

13, 30, 30, 32, 40, 47

Match the measures on the left with the values on the right.

1. Mean   A. 30
2. Median   B. 31
3. Mode   C. 32
4. Range   D. 34

Finding a Missing Data Item

Sometimes an average is a goal you want to achieve, and you need to know what values will achieve the average you want. If you want to score at least a 90% in a class, you can determine what score you need on each individual test or project.

Setting an Average as a Goal

Example 4: Find the Percent Needed to Meet Goal

You want to earn an 88% in a class. On the first three tests you scored a 79%, 85%, and 90%. What percent would you need on the fourth test to earn an 88%? One way to find the answer is to write an equation.

Step 1 Write what you know as an equation, using n for the unknown score. The average will be the sum of the three scores plus n, divided by 4. Set the equation equal to 88.

$$\frac{79 + 85 + 90 + n}{4} = 88$$

Step 2 Add the known test scores together.

$$\frac{254 + n}{4} = 88$$

Step 3 Multiply both sides by the number of tests, in this case 4.

$$\left(\frac{254 + n}{4} = 88\right) \times 4$$
$$254 + n = 352$$

Step 4 Finally, subtract 254 from both sides of the equals sign. This will be the missing value. To earn an 88%, you would need at least a 98%.

Determining if an Average is Achievable

To determine if an average is achievable, follow the same steps as if you were trying to find a missing value.

Example 5: Find Out if Average is Achievable

Cara worked 7 hours on Monday, 6 on Tuesday, 5 on Wednesday, and 7 on Thursday. How many hours must she work on Friday to average 10 hours a day for the workweek?

Cara's Hours

Mon.	Tues.	Wed.	Thurs.	Fri.
7	6	5	7	?

Step 1 Write what you know as an equation, using n for Cara's hours on Friday. The average will be the sum of the hours from Monday to Thursday plus n, divided by 5. This will equal the average, 10.

$$\frac{7 + 6 + 5 + 7 + n}{5} = 10$$

Step 2 First, add the numbers on the left side of the equation.

$$\frac{25 + n}{5} = 10$$

Step 3 Next, multiply both sides by 5.

$$\left(\frac{25 + n}{5} = 10\right) \times 5$$
$$25 + n = 50$$

Step 4 Finally, get the variable by itself. To do this, subtract 25 from both sides of the equals sign. The result is 25. Cara cannot work 25 hours in a day. Therefore, she will not reach an average of 10 hours a day.

Plan and Organize

One thing that is important in running a business is the ability to determine trends in sales to prepare for the future. Tracking trends on the day, week, month and even year level can greatly reduce waste and save money towards other parts of a business.

Martin is a florist and is preparing for an upcoming holiday. In preparation, he bought 55 dozen roses. Because the flowers will wilt within a week of delivery to his shop, Martin must sell all of his roses within the week. Martin sold 5, 7, and 11 dozen roses on Monday, Tuesday, and Wednesday, respectively. How many roses must he sell, on average, on Thursday and Friday in order to sell all of his roses by the end of the week?

On the TI-30XS MultiView™ calculator, you can input data using the $\underset{\text{data}}{\overset{\text{stat}}{\boxed{\text{data}}}}$ button. After entering the data, one number per row, press $\boxed{\text{2nd}}$ $\underset{\text{data}}{\overset{\text{stat}}{\boxed{\text{data}}}}$, then $\boxed{1}$, displaying 1-Variable statistics. The screen will show the mean, median, minimum, and maximum (so you can find the range yourself). The mode, however, is not calculated.

Think about Math

Directions: Answer the following questions.

1. What must the golfer score to average 76 for the three rounds?

Golf Scores

Round 1	Round 2	Round 3
78	80	?

A. 68
B. 70
C. 79
D. There is no way.

2. How many points out of 100 must you score on the final test to average 96 for the three quizzes?

Test Scores

Quiz 1	Quiz 2	Quiz 3
91	89	?

A. 90
B. 96
C. 100
D. There is no way.

Weighted Averages

In some cases, certain values are given more weight than others. This happens often in classrooms. For example, when figuring semester grades, a teacher often gives more weight to the final exam than the first quiz of the semester. So, even if the score is the same on both tests, the final exam will have a larger impact on the semester grade. The average of a data set in which some values are given more weight than others is called a **weighted average.**

Finding a Mean Using Frequency Counts

When you want to find the average number of times a value occurs in a specific set of data, you can use frequency counts to find your information.

A class of students was asked how many television sets were in their house. The answers ranged between 1 and 4. What is the mean number of sets per house?

```
x
x
x x
x x x x
x x x x
─────────
1 2 3 4
```

First, determine the total number of televisions. Do this by multiplying the number of X's by the number of televisions for each column. Then, add the products together.

$$1 \times 5 = 5$$
$$2 \times 3 = 6$$
$$3 \times 2 = 6$$
$$4 \times 2 = 8$$
$$5 + 6 + 6 + 8 = 25$$

Next, count the total number of X's. Remember, each X represents a student surveyed. There are 12 students that were surveyed.

Finally, divide the total number of televisions (25) by the number of students surveyed (12). The mean number of televisions per house is 2.08.

$$\frac{25}{12} = 2.08$$

Finding a Weighted Average

Your teacher announces that your grade for the course will be determined by the average of your test scores, with the final exam counting double. Suppose you have test scores of 88, 72, and 85, and you score a 90 on the final exam. How will your test scores be averaged?

First, find the sum of weighted scores, multiply each item by its weight. The first three scores have a weight of 1; the last score has a weight of 2.

$$1(88) + 1(72) + 1(85) + 2(90) = 425$$

Next, find the total number of weighted items. Count each test according to its weight. Three items with a weight of 1 and one item with a weight of 2 gives 5 as the total number of weighted items.

$$425 \text{ total points}$$
$$1 + 1 + 1 + 2 = 5$$
$$5 \text{ weighted tests}$$

Finally, divide the weighted sum of the scores by the total number of weighted tests to find the mean.

You now know the weighted average of the test scores.

$$\frac{425}{5} = 85$$

◣ Think about Math

Directions: Answer the following questions.

1. A carwash station sold 80 regular carwashes at $8 and 20 premium carwashes at $10. What was the average price of a carwash sold?

   A. $8.00
   B. $8.40
   C. $9.00
   D. $10.00

2. What is the average of the test scores shown if the final test counts double?

Test 1	Test 2	Final
91	87	95

   A. 90
   B. 91
   C. 92
   D. 93

Understand Business Fundamentals

A business cannot operate without making a profit on each of its products. Some companies mark up the price of the products so that they can guarantee a certain percentage of profit, or a certain amount of profit.

Lisa sells two types of clothing: t-shirts and collared shirts. For each t-shirt she sells, she makes a profit of $4.50. For each collared shirt, she makes a profit of $7.50. If she sells 140 t-shirts and 60 collared shirts in a month, what is the average profit Lisa will make that month?

Vocabulary Review

Directions: Write the missing term in the blank.

average	mean	median
mode	range	weighted average

1. The most frequently appearing data value in a data set is called the _____.

2. When a data set has an odd number of values, the _____ is the middle value when the data set is ordered from least to greatest.

3. The difference between the greatest and least values in a data set is the _____ of the data set.

4. If you add all of the data values in a set and then divide by the total number of values in the set, you have calculated the _____ of the data set. This value is also called the _____.

5. A value based on a data set in which some values carry more weight than others is a(n) _____.

Skill Review

Directions: Read each problem and complete the task.

1. The rainfall in four successive months was 10.2 in., 7.7 in., 5.1 in., and 12.0 in. What was the mean monthly rainfall for the four-month period?

2. You have a data set of 8 values, arranged from least to greatest. Because there is an even number of values, there is no single value in the middle. Explain how to find the median of this data set.

3. What is the mode of the following set of numbers?

   8, 12, 8, 4, 9, 12, 15

   A. 8
   B. 12
   C. 8 and 12
   D. There is no mode.

4. What is the range of the following set of numbers?

   −7.1, 4.0, 8.5, −2.3, −9.1, 0.1

5. A web site had 12,000 hits the first day and 16,000 the second day. How many hits must it have on the third day to average 20,000 hits a day for the three days?

   A. 20,000
   B. 24,000
   C. 28,000
   D. 32,000

6. Your grade in a course is the average of two quizzes and a final exam, which counts double. So the instructor will add the grades for the two quizzes, twice the grade for the final exam, and divide by:

   A. 2
   B. 3
   C. 4
   D. 5

Calculate Measures of Central Tendency

Skill Practice

Directions: Read each problem and complete the task.

1. Listed below is the frequency chart for a set of data. What is the mode of the data?

```
    x
    x x
  x x x x
  x x x x x
 ─────────────
  0 1 2 3 4
```

2. Which of the following could be negative? Select all that apply.
   A. Mean
   B. Median
   C. Mode
   D. Range

3. Fill in the four data items based on the following information:

   Mean: 55
   Median: 51
   Mode: 48

   _____ _____ _____ _____

4. In a town of 100 residents, 99 have less than $1000 and one is a millionaire. Which measure of central tendency would best represent the net worth of a typical resident?
   A. Mean
   B. Median
   C. Mode
   D. None of the above

5. A bakery sold 400 bagels on Monday, 650 on Tuesday, and 350 on Wednesday. How many bagels, on average, must the bakery sell during the next two days to have average sales of 500 bagels for the five days?

6. Your grade in a course is determined by averaging two quizzes and a final exam, which counts double. Your grades on the two quizzes were 87 and 91. What grade must you get on the final exam to have a course average of 94?
   A. 91
   B. 94
   C. 97
   D. 99

7. Which set has the largest range?
   A. 25, 35, 45, 55
   B. 10, 20, 40, 80
   C. 1, 10, 100, 100
   D. 2, 20, 60, 100

8. A data set has values 5, 6, 8, 9, 10 with frequencies 2, 1, 4, 2, 3, respectively. What is the mean of the data set?
   A. 8
   B. 7.6
   C. 5
   D. 2.4

9. A certain web site is hoping to have at least 5,000 page views per week on average. So far, the web site has had 2,200 views last week and 3,100 views this week. What is the minimum number of views that the web site needs to receive this week to obtain its goal over a three-week period?

10. What is the median of the following set?

    1, 1, 2, 3, 5, 8, 13, 21, 34, 55, 89, 144

LESSON 8.2 Display Categorical Data

LESSON OBJECTIVES

- Interpret and display data in a bar graph
- Interpret and display data in a circle graph

CORE SKILLS & PRACTICES

- Interpret Data Displays
- Interpret Graphs

Key Terms

bar graph
a graph that uses the length of bars to represent data values

circle graph
a graph that uses sections of a circle to represent data values

Vocabulary

legend
A key printed on a graph or chart that shows the meanings of colors, symbols or markings used

Key Concept

Bar graphs and circle graphs are convenient ways of displaying data that fall into categories. Both types of graphs allow the viewer to see data at a glance. Bar graphs are appropriate to show the absolute size of various categories. Circle graphs show what percentage of the total is made up by the various categories.

Bar Graphs

Data is everywhere. Data helps you understand your performance in a video game as well as compares the amount of electricity used each month. Bar graphs compare the sizes or values of different categories in a way that can be seen at a glance.

Reading a Bar Graph

A **bar graph** is a data display that compares the relative size of data in different categories, such as the number of people who like one movie over another or the populations of local high schools.

A bar graph can use either vertical (up and down) or horizontal (side to side) bars to show data. One side, or axis, is labeled with numbers. The categories are listed along the other axis.

The length of the bar for each category corresponds to its number. If the end of a bar falls between two numbers, estimate where it lies between the two numbers.

School Band Members

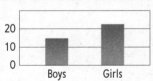

Purestock/SuperStock

Interpreting a Bar Graph

Graphs can contain a lot of information, but we have to know how to read them correctly. We want to use the given bar graph to find the minimum cost to purchase a Silvercuts franchise.

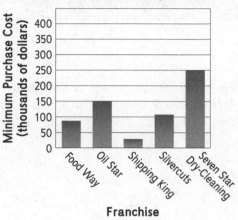

Minimum Cost to Purchase a Franchise for Selected Companies

Example 1: Interpreting Information

Step 1 Find the bar representing the purchase cost for Silvercuts. In the graph, the top of the bar for Silvercuts is a little higher than the $100,000 mark. It is about one-fifth the distance between 100 and 150. Since the distance between 100 and 150 is 50, we find one-fifth of 50 which is 10.

Step 2 The bar goes just past the 100 mark which means the 10 is added to the 100. You might estimate the length of the bar for Silvercuts to be 100 plus 10, or 110.

Step 3 Because the data are represented in thousands of dollars, a reading of 110 on the graph means the minimum cost to purchase a Silvercuts franchise is about $110,000.

Setting up a Bar Graph

Example 2: Creating a Bar Graph

Step 1 When creating a bar graph, the first choice to make is whether to use vertical or horizontal bars. Notice that the data are the same in both of these bar graphs.

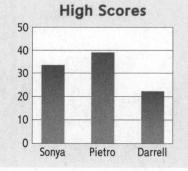

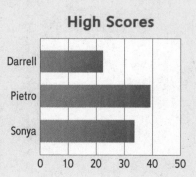

High Scores

Interpret Data Displays

To show similar information about two or more things, you can use a double bar graph. This graph shows money earned in sales at the two locations of Herman's Gift Shop during the last six months of the year. The key, or **legend**, shows the meaning of the graph's colors, symbols, or markings (in this case showing the difference between the two stores).

To find the months where a store earned at most $13,000, look for months where either the black bar or the white bar indicates $13,000 or less.

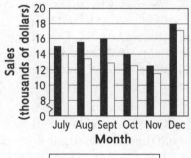

Sales at Herman's Gift Shops

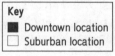

Key
- ■ Downtown location
- □ Suburban location

For September, the suburban sales appear to be about $13,000. For October, suburban sales are below $13,000. For November, both downtown and suburban sales are below $13,000. There were 3 months during which sales at either location of Herman's Gift Shops are at or below $13,000.

Now you try. The store managers receive a sales prize every time their location sells $15,000 or more in a month. Which managers received bonuses in which months?

CORE SKILL

Interpret Graphs

Sometimes a designer replaces the bars in a bar graph with a graphic. The result can be misleading.

Corn Production by County

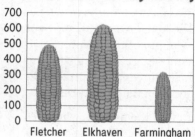

Corn production in Elkhaven County was nearly twice that in Farmingham County. However, using a two-dimensional graphic as a bar means that the ear of corn that is twice is tall is also twice as wide and has four times the area. That gives the impression that Elkhaven County produced four times as much corn as Farmingham County.

Look at the graph below. How many times as much snow fell in Brushwood as in Dover? How many times as much does the graphic make it appear?

February Snowfalls in Nearby Towns

Step 2 Next you must determine the range of the data to be displayed. Start at zero at one end of an axis. The greatest value at the other end of the axis should be at least as great as the highest value in your data set.

Pushups

Jared	Shawn	Ellen	Helga	Franco
21	9	14	25	15

(low)   (high)

Step 3 Now determine the scale of the axis. The greater the values of data, the greater the scale should be. For this graph a scale of 10 is used since the values of the data are fairly small.

Step 4 Between the least and the greatest values you need to insert lines at regular intervals. Then, on the other axis, label what each bar will represent (names) and include a label to tell what all the bars represent (pushups).

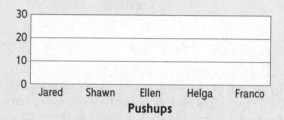

Step 5 Once you have placed the categories under the horizontal axis and the range of values along the vertical axis, draw bars to represent the data in each category.

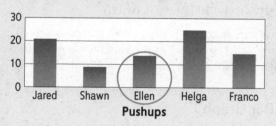

If a value lies between two lines on the vertical axis, estimate the proper distance according to the part that lies between the lines. For instance, a value of 14 should be less than halfway between the lines for 10 and 20.

Think about Math

Directions: Answer the following question.

1. What is the approximate difference in population between Dubney and Centerville?
   A. 25,000
   B. 28,000
   C. 31,000
   D. 35,000

Population of Nearby Towns

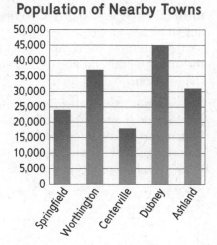

Circle Graphs

If you have ever completed a survey you have helped build a circle graph. Circle graphs are excellent ways to look at data of a whole. For example, if you determined that 60% of your friends prefer country music you could show this on a circle graph. A **circle graph** shows what portion of the whole, or 100%, each category takes up by dividing the whole into wedges of different sizes.

Reading a Circle Graph

In a circle graph, a circle is divided into wedges. Each wedge or section represents a fraction of the total—the larger the section, the greater the fraction of the whole represented. This type of graph allows the viewer to compare the sizes of the parts more easily.

Normally each wedge is marked with a percent mark showing what percent it is of the circle. If a wedge is too small, a callout is used, consisting of a line from inside the wedge to label the information on the outside of the wedge, like the label for U.S. Mail.

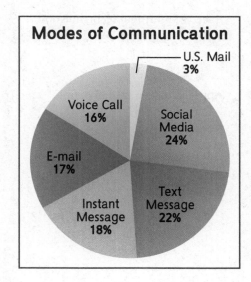

All the percents in a circle graph must add to 100%. If the percents add up to more than 100%, there is a mistake and you cannot draw the graph.

If the percents add up to less than 100%, you must add a category such as "Other" or "Undetermined" to hold the remainder.

TEST-TAKING SKILL

Circle Graphs

When reading a circle graph, look for an indication of what the whole represents, such as: "Survey of 500 Women" or "Total: 2000." If you need to find the number of units represented by a sector, multiply the total units by the percent that sector represents.

If the sectors aren't labeled with percents, you can estimate the size of the sector by comparing it with common fractions. For example, this sector is larger than a third $\left(33\frac{1}{3}\%\right)$ but smaller than a half (50%). You might estimate it as 40%.

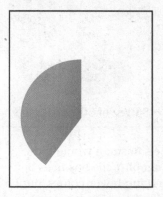

Based on this chart, about how many people interviewed rarely wear a hat?

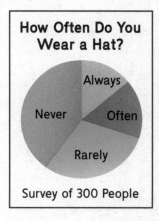

Find Information in Workplace Graphics

Circle graphs are often used in the workplace to summarize information. A car salesman creates a graph showing the percentage of customers who prefer new cars in each of the colors most commonly available.

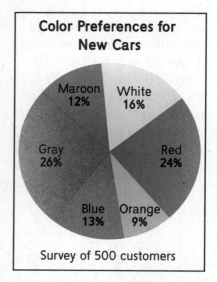

Color Preferences for New Cars

Maroon 12%
White 16%
Gray 26%
Red 24%
Blue 13%
Orange 9%

Survey of 500 customers

The salesman would like to simplify his business by carrying fewer colors, but he doesn't want to lose more than 20% of his potential customers. What would you advise him and why?

Interpreting a Circle Graph

Frequently the amount represented by the total circle is also given. To find the actual number represented by a section, multiply the total amount for the graph by the fraction or percent for that section.

For example, the circle graph shows people who immigrated to the United States from other countries in North America in 2001. About how many of those people came from Canada?

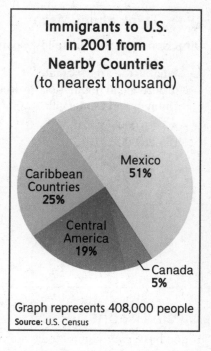

Immigrants to U.S. in 2001 from Nearby Countries (to nearest thousand)

Mexico 51%
Caribbean Countries 25%
Central America 19%
Canada 5%

Graph represents 408,000 people
Source: U.S. Census

Example 3: Interpreting Data

Step 1 Observe the percent of immigrants from Canada given in the graph. It states that approximately 5% of the 408,000 came from Canada.

Step 2 Next, convert 5% to a decimal, 0.05.

Step 3 Multiply 0.05 by 408,000.

$$\begin{array}{r} 408{,}000 \\ \times\ 0.05 \\ \hline 20{,}400.00 \end{array}$$

Approximately 20,400 people came from Canada.

Displaying Financial Data in a Circle Graph

A circle graph can also be used to show data displayed as cents per dollar.

When a circle graph is divided into percents the sections must total 100%. Likewise, when a circle graph representing parts of a dollar is divided by cents the sections must total $1.00.

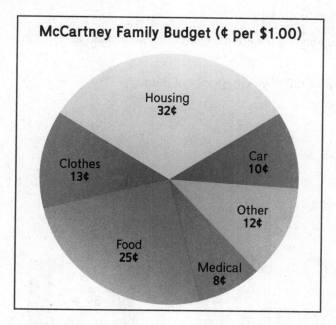

McCartney Family Budget (¢ per $1.00)

Housing 32¢
Clothes 13¢
Car 10¢
Other 12¢
Food 25¢
Medical 8¢

The circle graph shows that for each $1.00 the McCartney family spends, $0.25 goes to food. If you know the total monthly budget for the McCartney family, you can calculate how much money they spend on each component.

For example, if their monthly budget is $3,650 then you can calculate the amount spent on housing by multiplying the total by the cent per dollar.

$3,650 × $0.32 = $1,168

Think about Math

Directions: Choose the best answer to the questions.

Where Is Your Family's Car Parked?

Parking Lot
Garage
Street
Driveway

Hours of the Day

1. Which wedge of the circle graph is close to 25%?

   A. Garage   C. Street
   B. Driveway   D. Parking Lot

2. Use the information in this bar graph to create a circle graph.

CALCULATOR SKILL

Using a Calculator for Circle Graphs

Consider the following circle graph that describes the percentage of 200 students with different hair colors. Use a calculator to find the number of students that have each hair color.

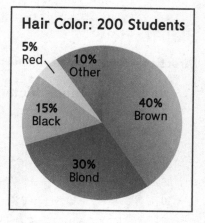

Hair Color: 200 Students

5% Red
10% Other
15% Black
40% Brown
30% Blond

Example 4: Using a Calculator

Step 1 Change the percent to a decimal. (Divide the percent number by 100.)

Step 2 Enter the number you want to find a percent of.

Step 3 Press the multiply button .

Step 4 Enter the decimal version of the percent.

Step 5 Press ENTER.

How many students have each hair color described in the graph?

Vocabulary Review

Directions: Write the missing term in the blank.

bar graph   **circle graph**   **graph legend**

1. You can show how parts of something are related to the whole with a _____.

2. A key called a _____ is used to identify the colors, symbols, and markings on a bar graph.

3. A _____ shows the relative size of different categories of data.

Skill Review

Directions: Read each problem and complete the task.

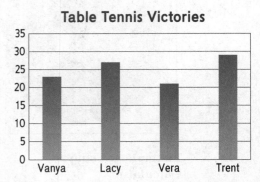

Table Tennis Victories

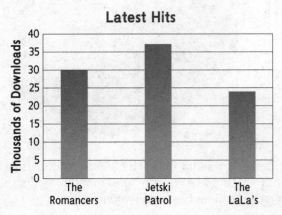

Latest Hits

1. Who had the second-highest score?

A. Vanya
B. Lacy
C. Vera
D. Trent

2. If Vanya and Trent played as a team and Lacy and Vera played as a team, which team won?

A. Vanya and Trent
B. Lacy and Vera
C. It was a tie.
D. It cannot be determined.

3. The downloads for The LaLa's are what portion of the downloads for the Romancers?

A. 60%
B. 75%
C. 80%
D. 125%

Survey of 200 Coffee Drinkers

Black	Sugar	Cream	Cream & Sugar
15%	10%	30%	45%

4. How many of the coffee drinkers surveyed like their coffee black?

5. How many people surveyed drink their coffee with sugar or cream but not both?

6. Make a circle graph of the data.

Skill Practice

Directions: Read each problem and complete the task.

1. You are making a bar graph to show the following data:

Favorite Pizza

Pepperoni	Pepper and Onion	Plain Cheese
37	24	19

What will be the minimum and maximum values on the vertical scale to the left? At what interval will you set ticks? At what interval will you set numbers? Explain your choices.

2. Observe the following data.

Money Raised by Candidates

Gomez	Nguyen	Bernacke
$290,000	$5,500	$4,300

Why will these data be hard to display as a bar graph?

3. In a survey of favorite colors, 45 out of 250 like the color green. What percent would be used to help make the sector of the circle graph represent green?

4. A sector in a circle graph is 37% of the whole. How many people out of 1,200 does this cover?

A. 370
B. 400
C. 444
D. 475

5. If you start at 12 o'clock and lay out a sector of 37.5%, where on the clock face would that sector end?

6. Which of these data sets would be displayed better as a bar graph and which as a circle graph?

A. Market share of cellphone brands
B. Amount spent by different cities on the arts
C. Internet usage in different countries
D. Percent of reasons given for calling 911

LESSON 8.3 Display One-Variable Data

LESSON OBJECTIVES

• Interpret and display data in a
dot plot, histogram, and box plot

CORE SKILLS & PRACTICES

• Interpret Data Displays
• Model with Mathematics

Key Terms

dot plot
a data display that uses a
number line and dots, or other
symbols, to show how often each
data value occurs in a data set

histogram
a display of data that have been
divided into intervals

box plot
a display that shows the range
and distribution of a data set

first quartile
the median of the lower half of a
data set

third quartile
the median of the upper half of a
data set

Vocabulary

distribution
a description of how the data
values in a set are spread out

median
the middle number of an ordered
data set; in a data set with an
even number of values, the
average of the two middle values

Key Concept

Dot plots, histograms, and box plots are different ways to
display one-variable data, data in which only one quantity is
measured. Each display highlights different characteristics
of the data set.

Dot Plots

Data is very important to businesses. If you have ever completed an online
survey from a receipt, your information has been placed on a dot plot. Dot
plots clearly display important pieces of information.

Reading Dot Plots

A **dot plot** is a data display that uses a number line to show how often each
data value occurs in a data set. In a dot plot, each data value is represented
by a dot or other symbol above the number line.

Example 1: Reading a Dot Plot

Ms. Morgan asked each student in her
homeroom how many pets he or she
has. She used the students' answers to
make the dot plot shown. How many
students in Ms. Morgan's homeroom
have two pets?

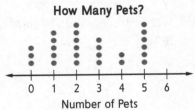

How Many Pets?

Number of Pets

Count the number of dots above 2 on
the number line. There are six dots above
the number 2, so six students have two pets.

Displaying Data in Dot Plots

It is relatively straightforward to make a dot plot from a set of data. Draw a
number line and place a dot above the line for each data value.

Example 2: Making a Dot Plot

Baseball Games Attended									
6	1	0	2	2	1	4	3	1	0
2	5	2	1	1	3	2	4	3	1

The table shows the number of baseball games that 20 of Gabriel's friends attended last season. Make a dot plot of the data.

Step 1 Sort the data values in order from least to greatest:

0, 0, 1, 1, 1, 1, 1, 1, 2, 2, 2, 2, 2, 3, 3, 3, 4, 4, 5, 6

Step 2 The least value is 0 and the greatest value is 6, so draw a number line from 0 to 6.

Step 3 Draw a dot above the number line for each data value. For example, two of Gabriel's friends responded 0, so there are 2 dots above 0.

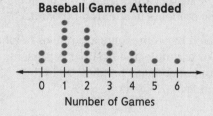

Baseball Games Attended

Number of Games

Step 4 Label the values on the number line and give the dot plot a title.

Think about Math

Answer the following questions based on the information in the dot plot.

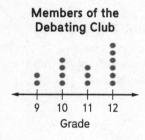

Members of the Debating Club

Grade

1. What percent of the debating club are in grade 11?
2. How many more debaters are in grade 12 than in grade 9?

Interpret Data Displays

A dot plot contains a lot of information about a data set. Because a dot plot shows every data value in a set, you can use a dot plot to find the range, mean, median, and mode of a data set. You can also use a dot plot to make comparisons within a data set.

A dental clinic surveyed several clients and asked them how many dental cleanings they have per year. The clients' responses are shown in the dot plot.

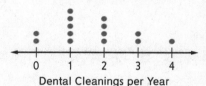

Dental Cleanings

Dental Cleanings per Year

Use the dot plot to compare the number of clients who have at least two cleanings per year with the total number of clients surveyed. What fraction of the clients surveyed have at least two cleanings per year?

Histograms

The census is taken every ten years. This important data tells a lot about the population of the United States. For example, a recent census confirmed that a large majority of the US population is less than 30 years of age. This is the type of information you would find in a histogram.

Reading Histograms

A histogram is a data display similar to a bar graph, but the data is grouped into intervals or ranges. Also, notice how the bars touch.

The histogram displays the heights of the students in a marching band.

Each bar represents a range of heights. For example, the first bar represents students whose heights are in the interval 51–55 inches.

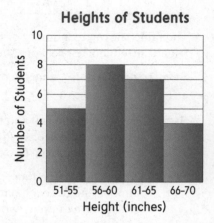

Heights of Students

Example 3: Reading a Histogram

How many of the students in the marching band are between 56 and 60 inches tall?

Step 1 Locate the bar labeled 56–60.

Step 2 Look for the height of the bar. The bar is at 8 students.
Therefore, there are 8 students between 56 and 60 inches tall.

Displaying Data in Histograms

To make a histogram from a given data set, you must decide how to divide the data into equally-sized intervals. Note that there may be more than one reasonable way to divide the data set.

Display One-Variable Data

Example 4: Making a Histogram

The ages in years of last month's applicants to the Ferndale Police Academy are as follows: 32, 24, 34, 22, 20, 26, 22, 27, 30, 24, 25, 38, 25, 21, 32, 24, 29, 23, 23, 27. Make a histogram of the data.

Step 1 Sort the data values in order from least to greatest:

20, 21, 22, 22, 23, 23, 24, 24, 24, 25, 25, 26, 27, 27, 29, 30, 32, 32, 34, 38

Step 2 Divide the data values into equally sized intervals of 5 years: 20–24, 25–29, 30–34, and 35–39 years of age. Use these intervals to draw the horizontal axis of the histogram.

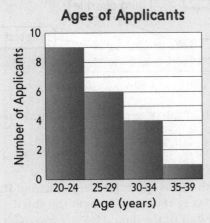

Ages of Applicants

This histogram shows that applicants to the police academy are skewed toward younger applicants.

Step 3 Draw a bar to show the number of applicants in each interval. For example, there are 9 applicants aged 20–24, so the height of the first bar is 9. Each bar should be of equal width. The bars should touch but not overlap.

Step 4 Label the axes and give the histogram a title.

Think about Math

The histogram shows the years of experience of teachers at Montvale High School. Answer the following questions based on the information in the histogram.

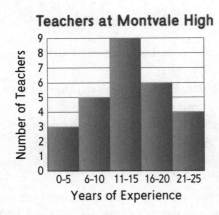

Teachers at Montvale High

1. How many teachers have between 0 and 10 years of experience?
2. What percent of teachers have 11 to 20 years of experience?

WORKPLACE SKILL

Make Decisions Based on Workplace Graphics

Sofia is considering a new seating plan for her restaurant. The restaurant has space for no more than 15 tables. Sofia has tables that seat two people, tables that seat four people, and tables that seat six people. She must decide among four different seating plan options:

Option 1: 10 tables of four and 5 tables of six

Option 2: 3 tables of two, 7 tables of four, and 5 tables of six

Option 3: 5 tables of two, 8 tables of four, and 2 tables of six

Option 4: 5 tables of two and 10 tables of four

Last night, the restaurant served a total of 30 tables during dinner hours. Sofia recorded the number of people at each table.

Number of People	Number of Tables
1	3
2	7
3	7
4	9
5	3
6	1

Make a histogram of Sofia's data. Then use the histogram to recommend one of the four seating plan options. Use your histogram to justify your recommendation.

When displaying data in a box plot, you need to find the median of the data set. When there is an odd number of data values, the median is the middle value when the data are ordered from least to greatest. When there is an even number of data values, the median is the mean of the two middle values. You may also have to calculate the mean of two middle values when finding the first or third quartile.

To find the mean of two numbers, such as 7 and 12.5, add the two numbers and then divide by 2. Be careful when using a calculator to perform these operations. If you enter 7 + 12.5/2, you will get an incorrect answer because the calculator follows the order of operations. It will divide 12.5 by 2 and then add 7.

One way to fix this is to use parentheses when entering the numbers into a calculator: (7 + 12.5)/2. Another method is to enter the sum 7 + 12.5, press (enter), and then divide the answer by 2.

Box Plots

One of the best things about the weather is all the rich data it provides. However, all that data can be challenging to manage and interpret. Box plots consolidate large amounts of data into a graphic.

Features of a Box Plot

The **distribution** of a data set refers to how the data values in the set are spread out. A **box plot**, sometimes called a box-and-whisker plot, is a data display that uses a box and two line segments (the "whiskers") to show the range and the distribution of a data set.

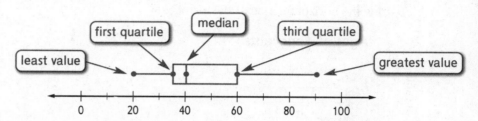

- The box represents the middle half of the data and the whiskers extend to the least and greatest data values.
- The box is divided by a vertical line at the **median**, the middle value when the data values are in order.
- The left side of the box occurs at the **first quartile**, the median of the lower half of the data. The right side of the box occurs at the **third quartile**, the median of the upper half of the data.
- The left whisker, then, represents the bottom quarter (25%) of the data, the box represents the middle half (50%), and the right whisker represents the top quarter (25%).

Reading Box Plots

The box plot shown displays data about the final sales prices of several used cars sold at a dealership last month.

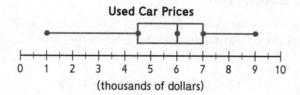

- You can see from the box plot that the median price of a used car last month was $6,000.
- The left whisker ends at 1. The lowest sales price for a used car last month was $1,000.
- The right whisker ends at 9. The highest price for a used car last month was $9,000.
- The box goes from 4.5 to 7. Half of the used cars sold last month had prices between $4,500 and $7,000.
- One-quarter (25%) of the used cars had prices between $1,000 and $4,500, and one-quarter (25%) had prices between $7,000 and $9,000.

Displaying Data in Box Plots

To create a box plot from a given data set, you must identify the least value, first quartile, median, third quartile, and greatest value. These five values will determine how you draw the box and whiskers.

Example 5: Making a Box Plot

The table shows Lisa's scores on her last nine spelling tests, in order from least to greatest. Make a box plot of the data.

Spelling Test Scores								
52	65	67	70	74	78	84	90	92

Step 1 Identify the least and greatest values and the median.

Step 2 Find the first and third quartiles.

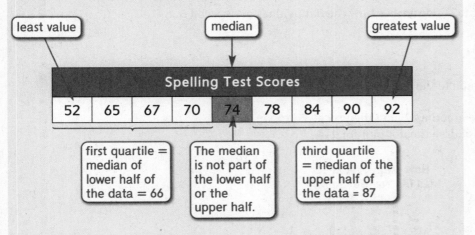

first quartile = median of lower half of the data = 66

The median is not part of the lower half or the upper half.

third quartile = median of the upper half of the data = 87

Step 3 All of the data values are between 50 and 100, so draw a number line from 50 to 100. Above the number line, draw a box from the first quartile to the third quartile. Draw a line inside the box at the median, 74. Draw the whiskers from the box to the least and greatest values. Add labels and a title to the box plot.

Lisa's Spelling Test Scores

Test Score

CORE PRACTICE

Model with Mathematics

Sometimes a data value is extremely different from the other data values in the set. A data value like this is called an *outlier*. In a box plot, an outlier is indicated with an asterisk. A whisker does not extend to the outlier, but instead stops at the least or greatest data value that is not an outlier. The outlier is not used to calculate the median, the first quartile, or the third quartile.

The weights in pounds of several pumpkins grown at Wilson Farm are given below:

20, 28, 25, 23, 32, 15, 22, 55, 17, 31, 21, 39

Identify the outlier. Then make a box plot of the data that shows the outlier.

Think about Math

Answer the following questions based on the information in the box plot.

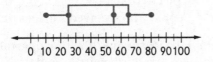

1. What are the least and greatest values of the data set?
2. What are the first and third quartiles?
3. What is the median?

Vocabulary Review

Directions: Write the missing term in the blank.

dot plot	histogram	box plot	first quartile
third quartile	distribution	median	

1. The median of the upper half of a data set is the _____.

2. A _____ is a display that shows the range and distribution of a data set.

3. A _____ is a display that shows how often each data value occurs.

4. The median of the lower half of a data set is the _____.

5. A _____ shows data that have been divided into intervals.

6. The middle of an ordered data set is the _____ of the data set.

7. The _____ of a data set describes how the data values are spread out.

Skill Review

Directions: Read each problem and complete the task.

Several people were surveyed about the amount of mail they received today.
Their responses are shown in the dot plot. Use the dot plot for 1–4.

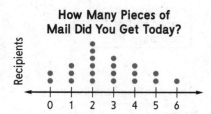

1. How many people received fewer than 3 pieces of mail?

   A. 4
   B. 11
   C. 13
   D. 15

2. What is the median of the data?

   A. 2
   B. 3
   C. 4
   D. 11

3. Write *less than*, *greater than*, or *equal to* in each blank.

   a. The range of the data is _____ 5.

   b. The number of people surveyed is

      _____ 25

   c. The mean of the data is _____ the median.

   d. The number of people who received at least

      4 pieces of mail is _____ the number of people who received 2 pieces of mail.

4. Find the least value, the first quartile, the median, the third quartile, and the greatest value. Use these values to make a box plot of the data.

5. The histogram shows data about the ages of runners participating in a race. How many runners are in the race?

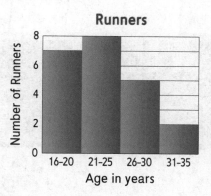

Runners

A. 21
B. 22
C. 23
D. 35

6. Sandy, Jackson, and Gina were looking at a histogram about annual incomes, and each had a different interpretation about the range of the histogram. Sandy says that you cannot determine the range of a data set from a histogram. Jackson disagrees. He says that the range of the data shown in the histogram is $99,999 because the greatest data value is $99,999 and the least data value is $0. Gina says the range is $79,999 because there are no data values in the first interval, so the least value is $20,000. Who is correct? Explain.

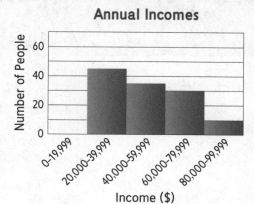

Annual Incomes

Skill Practice

Directions: Read each problem and complete the task.

1. Examine the histogram below. Which box plot could represent the same data as the histogram?

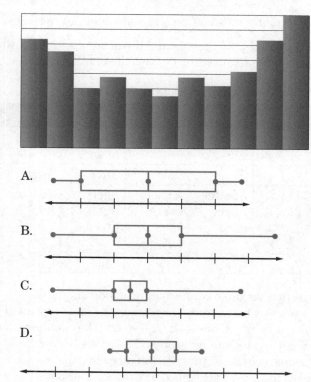

2. Which box plot shows data that are clustered around the median?

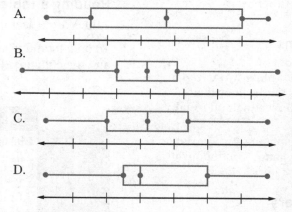

3. Ursula has data about the ages at which people graduate from college. The ages range from 16 to 65 years. Ursula decides to make a histogram and to divide the data into intervals of 2 years. Do you think this is a good way to divide the data? Explain.

4. Suppose you have a data set on the recent sales prices of single-family homes in your county. What would be the advantages of displaying the data in a histogram? What would be the advantages of a box plot?

LESSON OBJECTIVES

- Interpret and display two-variable data in tables
- Interpret and display two-variable data in scatter plots
- Interpret and display two-variable data in line graphs

CORE SKILLS & PRACTICES

- Build Lines of Reasoning
- Interpret Graphs

Key Terms

scatter plot
a graph that plots two-variable data items on the coordinate plane to show a general trend

line graph
a graph displaying two-variable data that change continuously over time

Vocabulary

positive trend
as one variable increases, the other variable tends to increase

negative trend
as one variable increases, the other variable tends to decrease

no trend
there is no pattern between two variables

Key Concept

Tables, scatter plots, and line graphs are all ways to show information that relates one thing to another, like temperature to time of day or height to weight. We call these displays of two-variable data, because there are two related items.

Tables

Tables can be used to organize information in two ways at once. Similar to a crossword puzzle, the rows organize the items in one way and the columns in another. Each item is placed in a cell that is in the right row and the right column for that item.

Reading a Table

Tables are a useful way to organize information and make it easier to read.

To understand what a table shows, read the title and the headings of the rows and the columns. They explain the relationships of the information shown.

Morning Work Schedule at Petra's Cafe

	Sun.	Tues.	Wed.	Thurs.	Fri.	Sat.
Cook	Marcus Allen	Marcus	Allen	Marcus	Marcus Allen	Allen
Bus staff	Stan			Stan	Stan	Stan
Waiters	Alicia Adolfo Connie Greta	Adolfo Lana Alicia	Adolfo Allen Lana	Alicia Greta Lana	Alicia Adolfo Connie	Connie Adolfo
Host	Lana				Lana	Lana

The title of the table tells us that the table contains information about the schedule for people who work at a restaurant. The rows of the table represent the different jobs the people have; cook, bus staff, and so on. The columns represent the days of the week the restaurant is open. The boxes within the table show the names of people scheduled for each job on each day. The shaded box shows the names of waiters scheduled to work on Tuesday.

Displaying Data in a Table

Put the following information in table form: Small cheese pizza, $7.50; medium cheese pizza, $8.50; large cheese pizza, $9.50. Small pepperoni pizza, $8.00; medium pepperoni pizza, $9.00; large pepperoni pizza, $10.00. Small pepper-and-onion pizza, $7.75; medium pepper-and-onion pizza, $8.75; large pepper-and-onion pizza, $9.75.

Example 1: Construct a Table

Step 1 Since all the prices are for pizza, we can put that information in the title of the table.

Step 2 Each price depends on the size and the topping. We will make a table with 3 columns for the 3 sizes and 3 rows for the 3 different toppings. There will be an additional row and column for labels.

Step 3 Finally, we put each item in the row and column where it belongs. The table is complete. Customers can easily find the price of the pizza they want by looking along the row with the topping they like and down the column with the size they want.

Pizza Prices

	Small	Medium	Large
Cheese	$7.50	$8.50	$9.50
Pepperoni	$8.00	$9.00	$10.00
Pepper and Onion	$7.75	$8.75	$9.75

Think about Math

Directions: Answer the following questions.

1. Use the following gas prices to make a table:
   Regular at ValuGas, $3.49; Premium at ValuGas, $3.79; Regular at Fuelfast, $3.59; Premium at Fuelfast, $3.69.

2. Luigi sells neckties in silk, cotton, and wool by 5 different designers. Describe a table that Luigi could use to show the price of each necktie.

WORKPLACE SKILL

Making a Business Decision Based on a Graphic

Evening Shift

	Tues.	Wed.	Thurs.	Fri.
Cook	Dan S.	Bailey	Dan S.	Dan S.
Bus staff			Roger	Roger
Waiters	Juana Maria	Juana Maria	Ali Mary	Ali Juana
Host				Maria

Petra organizes her workers by shift, job, and day of the week. This table shows Petra's current schedule for weekday evenings. If Petra wants to add a bus staff person so that every day is staffed, how much more will that cost her per week than what she's paying now? Bus staff persons make $9 per hour and work 6-hour shifts.

When determining the scale of an axis for a scatter plot or line graph, a rule of thumb is to use no more than 10 tick marks to span the range of the data. Calculate the range of the data by subtracting the least value from the greatest value, and then divide by 10. This number tells you how much each tick mark will represent. (Note that you may sometimes want to round this number.)

Particularly if the numbers involve decimals, you may find it useful to use a calculator. As always, keep in mind that a calculator will follow the order of operations. If you enter *greatest value–least value*/2, you will get an incorrect answer. You can use parentheses, (*greatest value–least value*)/2, or you can press (enter) or (◂ ▸) after performing the subtraction and then divide by 10.

Given a data set of class rank plotted against GPA with GPAs ranging from 1.75 to 3.4, use a calculator to find the range of GPAs and then choose an appropriate scale for tick marks.

Scatter Plots

Do people who own more books tend to make more money? Scatter plots can answer questions like that by showing whether a lot of individual cases add up to a trend.

Reading a Scatter Plot

A **scatter plot** is a graph showing two values for each item, one value on the horizontal axis and one value on the vertical axis. Each item is shown as a point on the coordinate plane.

The scatter plot shows the heights in inches of 10 men at a gym and the weights in pounds they were able to lift. For each dot, you can read the height of a man by looking at the horizontal scale and the amount he could lift by looking across to the vertical scale. Point A, for instance, shows that one of the men was 70 inches tall and could lift 160 pounds.

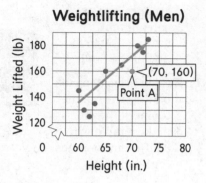

Interpreting Trends

When data is plotted on a scatter plot, it becomes possible to see whether the points form a trend, or a pattern showing a relation between the two values.

For instance, we see in the graph above that the values for weight lifted tend to increase as the values for height increase; the points cluster around an upward line. There is a **positive trend** if, as one variable increases, the other variable tends to increase.

A scatter plot can also show a **negative trend** if one variable increases as the other decreases. For instance, as a car increases in age, its resale value decreases. The points cluster around a downward line.

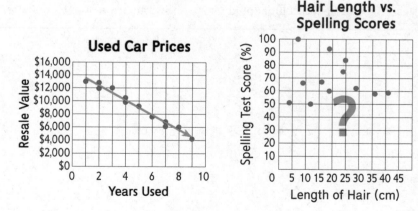

Some scatter plots show **no trend** if there is no relation between the two variables. The length of girls' hair has no relation to the girls' scores on a spelling test. The points don't cluster around an upward or downward line or show any pattern at all.

Display Two-Variable Data

Displaying Data in a Scatter Plot

The scatter plot, if it shows a trend, will show how the y-values are related to the x-values.

> ### Example 2: Build a Scatter Plot
>
> The plot will show whether the speed of the young runners is influenced by their age.
>
> | Age (yr) | 4 | 5 | 7 | 8 | 10 | 11 | 14 | 15 | 15 | 17 |
> |----------|---|---|---|---|----|----|----|----|----|----|
> | Speed (mph) | 1 | 3 | 4 | 4.5 | 5.5 | 5 | 6 | 7 | 8.5 | 9 |
>
> **Step 1** First, choose one category to be the x-values and one to be the y-values. For each axis, choose a range that will hold the data comfortably. For ages we choose 0–20 years. For speeds we choose 0–10 mph.
>
> **Step 2** Add enough lines and number labels to the axes to be able to accurately place the points, but still be readable. For the y-axis, we will set a whole-number label at every line. For the x-axis we will set an even-numbered label for every other line on the axis.
>
>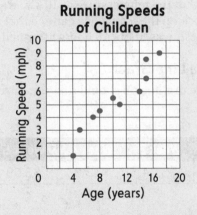
>
> **Running Speeds of Children**
>
> **Step 3** Plot each data item as a point in the coordinate plane, using the age as an x-value and the speed as a y-value. Not surprisingly, the finished plot shows a positive trend: as the children's age increases, the speed also increases.

Think about Math

Directions: Answer the following questions.

1. What would you expect to see in a scatter plot correlating the daily pollen count with the price of silver?

   A. A positive trend
   B. A negative trend
   C. No trend

2. What would you expect to see in a scatter plot correlating the monthly rainfall with the incidence of forest fires?

   A. A positive trend
   B. A negative trend
   C. No trend

Build Lines of Reasoning

The manager of a customer service desk wants to know what makes a good customer service person. He has customer feedback forms about each of his workers showing customer satisfaction as a percent.

He has several theories about employee performance. One is that perhaps better-educated workers can guide customers more skillfully to the right outcome.

He decides to make a scatter plot. The x-axis will show the years of education of the workers and the y-axis will be the percent of satisfaction recorded by the customers.

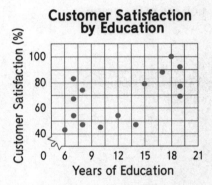

Customer Satisfaction by Education

Workers with more education were not significantly better and there is not an apparent trend. Apparently, at this company, other qualities matter more.

What are some other qualities of employees that you think would show a positive trend with customer satisfaction?

Perhaps the statistic most often viewed as a line graph in the United States is the Dow Jones Industrial Average, calculated from the buying and selling of 30 publicly traded stocks. The Dow, as it is often called, is looked to as a major index of the health of the U.S. economy. On trading days you can watch the Dow being charted in real time. Sites where you can find it include yahoo.com, marketwatch.com, money.cnn.com, and others.

Find a current line graph of the Dow Jones Industrial Average. What is the most recent reading? When was it last updated? Describe any trends you see in the graph.

Line Graphs

The price of a share of stock, the outside temperature, and the number of visits to a website are all examples of continuous data that can go up and down at any time. Line graphs are a way of displaying continuous data and showing the upward and downward trends.

Reading a Line Graph

A **line graph** shows continuously changing data, especially data that change over time. A line graph is constructed by first sampling the data at intervals and plotting the results as points on a graph. Then lines are added, connecting the points to show that change is continuous.

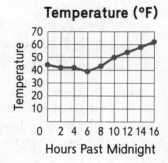

Temperature (°F)

The x-axis is often measured in time units, whether seconds or years. The y-axis shows the other variable—temperature, gas prices, or something else that is always changing.

We can read the time and temperature values of any point on the line using the horizontal and vertical scales. By looking up from the 10-hour mark on the x-axis, we can see that the temperature at 10 A.M. was 50°F. By looking across from the 60° mark on the y-axis, we can see that the time when the temperature reached 60°F was about 16 hours after midnight, or 4 P.M.

Displaying Data in a Line Graph

To display data in a line graph, you should first organize the data in a table.

Gas Prices

Week 1	Week 2	Week 3	Week 4	Week 5
$3.39	$3.52	$3.84	$3.78	$3.69

Example 3: Build a Line Graph

Step 1 For each axis, choose a unit and a scale that display the data sensibly. Here the x-axis will be measured in weeks and the y-axis will show dollars. Since the data show relatively small changes, we indicate a jump between $0 and $3.00 with a break symbol (the zigzag line) in the y-axis.

Step 2 Add enough lines and number labels on the axes to make the graph readable, but not so many as to make the graph cluttered. On the x-axis, we will put a line and a label for each week. On the y-axis, we will put lines at intervals of $0.10 and labels at intervals of $0.20.

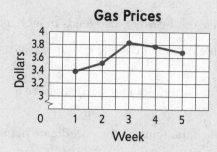

Gas Prices

Step 3 Plot each item above the label for its time value. Plot the other value according to the vertical scale. Connect the points with lines from each to the next. The line graph is complete. We can see that during the weeks shown the price of gas first rose and then fell.

Think about Math

Directions: Answer the following questions.

1. How much per week, on average, did the stock price lose between Week 1 and Week 5?

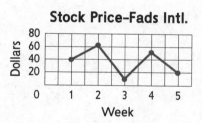

Stock Price–Fads Intl.

   A. $4.00
   B. $5.00
   C. $10.00
   D. $20.00

2. Erik kept track of the scores of his school's basketball team for a three-month season. Would a line graph be a good display for his data?

Interpret Graphs

Line graphs are open to manipulation. By including some data and not others or adjusting the axes, it is possible to change the general impressions made by the data.

Share Price Is Skyrocketing!

This line graph of the share price of stock for Blurbo International is made to look like there is a significant increase. To highlight the general rise in price, the vertical scale starts at $30 rather than at $0. The line graph is stretched vertically making the increase appear larger. The magnitude of the increase would appear smaller if the scale started at $0.

Suppose that in Weeks 6 through 8, Blurbo's share price was decreasing. How could a rival company use this to manipulate data from those eight weeks to give the impression that Blurbo International's stock price is plummeting?

Vocabulary Review

Directions: Identify the correct word to complete each definition.

scatter plot   **line graph**   **positive trend**
negative trend   **no trend**

1. A _____ shows that as one variable increases, the other also tends to increase.

2. A _____ is a graph displaying two-variable data that change continuously over one variable such as time.

3. A _____ scatter plot shows the result of a scatter plot that does not have a pattern between the two variables.

4. A _____ is a graph that plots two-variable data items on the coordinate plane to show a general trend.

5. A _____ shows that as one variable increases, the other tends to decrease.

Skill Review

Directions: Read each problem and complete the task.
Use the scatter plot to answer questions 1 and 2.

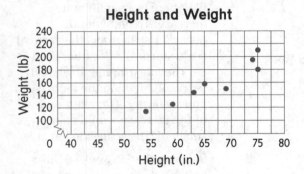

1. How tall was the person who weighed 180 lb?

2. How many people weighed between 140 and 160 lb?

3. Which brand in which size is the best value?

Dish Detergent Prices

	12 oz	24 oz
Dazzle	$1.79	$3.25
Sparkle	$1.69	$3.29
Glimmer	$1.74	$3.35

Display Two-Variable Data

4. Name two factors that would show a positive trend if graphed as a scatter plot.

5. Which of these would likely show a negative trend? Select all that apply.

   A. Number of hours spent playing video games related to grade point average
   B. Population of towns related to number of restaurants
   C. Mean temperatures related to distance north of the equator
   D. Age related to hours spent on the Internet

Skill Practice

Directions: Read each problem and complete the task.

1. The table shows distances in miles between cities. Copy the table and fill in the blank cells.

	Wharton	Essex	Gardner
Wharton		6	
Essex			12
Gardner	5		

2. A dealer sells cars, vans, and trucks of several different manufacturers in a number of colors. He wants to list his offerings in a table. How should he set it up?

3. How could you make the modest gains in the stock price of Everest Enterprises seem more impressive?

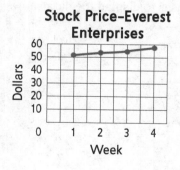

4. Which item's sales are most likely shown in the graph?

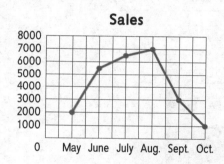

   A. Snow shovels
   B. Bathing suits
   C. Leaf blowers
   D. Parkas

Directions: Choose the answer to each question.

1. Which statement is true about the data set below?

10	11	12	12	12
12	10	11	11	14
15	10	15	16	15

   A. The mean is greater than the median.
   B. The mean is less than median.
   C. The mode is greater than the mean.
   D. The median is greater than the mode.

2. A sales manager collected data about how many years of sales experience each member of his team had. He compiled the data in a histogram. _____ members had less than 15 years of sales experience.

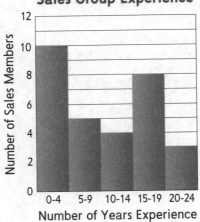

3. The menu prices are shown in the table below. How much would it cost to buy two small beef stew entrees and a large beans and rice entree?

Menu Prices			
Entre	**Small**	**Medium**	**Large**
Chicken and Vegetables	$4.00	$5.00	$8.00
Beans and Rice	$2.50	$3.50	$6.00
Beef Stew	$4.50	$6.00	$7.00

   A. $10.50
   B. $13.00
   C. $15.00
   D. $16.50

4. The North Store sold _____ less than the South Store during the four months shown.

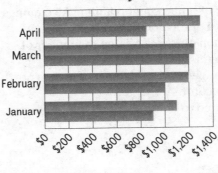

5. Which is the mean of this data?

   A. 24
   B. 22
   C. 20
   D. 5

6. Andrea's class grade is based on two test grades and a final that counts double. She earned a 95% and 86% on her first two tests and an 84% on the final. What is her average for the class?

   A. 86.6%
   B. 88.3%
   C. 87.25%
   D. 116.33%

7. Mr. Blackburn sells used cars. He has collected data about the age of a car and its selling price. The scatter plot he uses to display the data shows a _____.

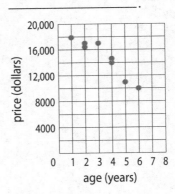

Data Analysis

Directions: Choose the best answer to each question.

8. The Buchanan Family uses this circle graph to represent their family budget. They spend $1,200 each month on housing, so their entire budget is _____ each month.

Budget

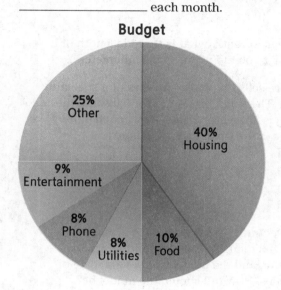

- 25% Other
- 40% Housing
- 9% Entertainment
- 8% Phone
- 8% Utilities
- 10% Food

9. The dot plot represents the number of high school students in Chorus. Which is true about the data set?

Chorus Members

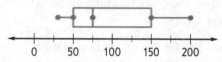

Grade

A. 35% of students are in the 10th grade.
B. There are 3 more 9th grade students than 12th grade students.
C. There are 20 students in Chorus.
D. 45% of the students are in the 10th and 11th grades.

10. Andrew wants to earn an 85% in a class. On the first four tests he scored a 92%, 82%, 75%, and 78%. What percent would he need to score on a fifth test to earn an 85%?

A. 93%
B. 38%
C. 82%
D. 98%

11. What is the difference between the highest temperature and the lowest temperature?

High Tempurature for 10 days

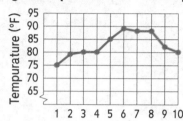

A. 5°
B. 9°
C. 14°
D. 75°

12. Which value is the median?

A. 25
B. 75
C. 100
D. 150

Check Your Understanding

On the following chart, circle the items you missed. The last column shows pages you can review to study the content covered in the question. Review those lessons in which you missed half or more of the questions.

Lesson	Item Number(s)			Review Page(s)
	Procedural	Conceptual	Problem Solving	
8.1 Calculate Measures of Central Tendency	6	1	10	254–261
8.2 Display Categorical Data	4		8	262–269
8.3 Display One-Variable Data	12	9	2, 5	270–277
8.4 Display Two-Variable Data	3	7	11	278–285

1. Which number has a 5 in the tenths place and a 2 in the hundreds place?

   A. 5.02
   B. 2.05
   C. 500.2
   D. 200.5

2. Jake needs to build a dock that will be $2\frac{3}{5}$ feet above the water at low tide. If the dock is 1 foot above the water at high tide, how high does the water rise from low tide to high tide?

   A. $1\frac{2}{5}$ feet
   B. $1\frac{3}{5}$ feet
   C. $3\frac{2}{5}$ feet
   D. $3\frac{3}{5}$ feet

3. Find the GCF of 16 and 36.

4. Which property is illustrated in the following equation?

   $7 \times 3 + 7 \times 2 = 7 \times (3 + 2)$

   A. Associative Property of Addition
   B. Associative Property of Multiplication
   C. Commutative Property of Multiplication
   D. Distributive Property

5. Charice invested $5,000 in a fund that returns 5% interest compounded yearly (at the end of the year) and makes no additional deposits or withdrawals. The total value of the fund, including accrued interest, at the end of n years is given by the equation $V = 5,000(1.05)^n$. How much accrued interest was earned during the second year?

6. What is the number 30,500,000,000 expressed in scientific notation?

   A. 3.05×10^9
   B. 30.5×10^9
   C. 3.05×10^{10}
   D. 30.5×10^{10}

7. To the nearest tenth, what is the side length of a cube with a volume of 857 cubic millimeters?

8. A pet store sells two different sizes of a specialty dog food. The small bag is $18 for 5 lbs and the large bag is $61.60 for 28 lbs. How much more expensive is the small bag per pound?

   A. $1.20
   B. $1.40
   C. $2.20
   D. $3.60

9. An architectural firm is creating a scale model of a new building. The actual building will be 240 feet tall, and the model will have a scale of 1:300 inches. How tall should the model be?

   ____ inches

10. In a survey of 800 students, 60% of students reported being involved in an after school sport. How many of the surveyed students does this represent?

    A. 160
    B. 280
    C. 424
    D. 480

11. A car dealership finances simple interest car loans at a fixed rate of 5.3% per year for 5 years. How much interest will a buyer pay if she borrows $17,999.00 for a new car?

A. $953.95
B. $4,769.74
C. $9,539.47
D. $23,301.84

12. How many possible ways can you choose a committee of 5 people out of a pool of 18 employees?

13. A number cube is rolled twice. Is this probability situation independent or dependent?

14. Which algebraic expression could be described in words as "2 times the difference of 7 minus a number"?

A. $2n - 7$
B. $\frac{2 \times 7}{n}$
C. $2(7 - n)$
D. $n - 2 \times 7$

15. Which inverse operation should you use to solve this one-step equation?

$81 + x = 108$

A. Addition
B. Subtraction
C. Multiplication
D. Division

16. The Gunsallus family is saving up to buy a new TV. The TV they would like to purchase costs $1,080, and they plan to save $135 per month. The equation $135m = 1,080$ represents this situation, where m is the number of months the family saves. Solve the equation to determine the number of months it will take for them to save up enough money for the TV.

_____ months

17. Is the number -7 in the solution set for $x - 5 \le -3$? Yes or no.

18. Solve the inequality $4(s - 3) > 3(s + 1)$.

$s >$ _____

19. Servers at a local café earn $3.75 per hour plus tips. On average, a server will earn 15% of her total sales in tips. Last Friday, one of the servers worked 5 hours and had total sales of $980. How much money did that server make last Friday?

A. $147
B. $149.81
C. $155.75
D. $165.75

20. Which of the following is a binomial of degree 3?

A. $x^3 + 3x - 2$
B. $x^2 - 3x$
C. $x^2 + 3x - 2$
D. $x^3 - 3$

21. What is the value of the polynomial expression below when $y = 2$?

$y^3 - 5y + 7y - y^2 + 9$

A. 9
B. 13
C. 17
D. 33

22. Which of the following is not a factor of the polynomial $2x^2 - 2x - 24$?

A. 2
B. $(x - 2)$
C. $(x + 3)$
D. $(x - 4)$

23. What is the leading coefficient of the polynomial $3x^2 + 5x + 2x^3 + 7$?

A. 2
B. 3
C. 5
D. 7

24. The distance d in feet that a dropped object falls in t seconds is given by the equation $d = 16t^2$. If an object is dropped from a height of 1,024 feet, how long will it take to reach the ground?

_____ seconds

25. Which shows the rational expression $\dfrac{x - 5}{x^2 - 4x - 5}$ correctly simplified with its restricted values?

A. $\dfrac{1}{x - 5}; x \neq -1$

B. $\dfrac{1}{x - 5}; x \neq -1; x \neq 5$

C. $\dfrac{1}{x + 1}; x \neq -1$

D. $\dfrac{1}{x + 1}; x \neq -1; x \neq 5$

26. Which quadrant of the coordinate plane contains points of the form $(+x, -y)$?

A. Quadrant I
B. Quadrant II
C. Quadrant III
D. Quadrant IV

27. A proportional relationship is modeled by $y = kx$. If $k = 6$, and (2, 12) is a point on the line, what is the slope of the line?

$m = $ _____

28. In what form is the equation $y = 5x + 2$?

A. Standard form of a linear equation
B. Slope-intercept form
C. Point-slope form
D. None of the above

29. The graph of a line passes through the points $(3, -2)$ and $(-1, 5)$. Graph the line and determine which of the following is true.

   A. The line crosses the negative x-axis and the negative y-axis.
   B. The line crosses the negative x-axis and the positive y-axis.
   C. The line crosses the positive x-axis and the negative y-axis.
   D. The line crosses the positive x-axis and the positive y-axis.

30. How many solutions does the system have?

$3x + 2y = 5$
$x + y = 2$

   A. no solutions
   B. one solution
   C. two solutions
   D. infinitely many solutions

31. Shawna is making trail mix for her hiking trip. She buys 20 total ounces of peanuts and raisins. She wants 3 times as many ounces of peanuts as raisins in her mix. How many ounces of raisins does she buy?

32. Which of the following is not a function?

   A. A horizontal line
   B. A vertical line
   C. A line passing through the point $(8, 2)$ and the origin
   D. A line parallel to the x-axis

33. What is the value of $f(x)$ when x is 0 in the piecewise function below?

$$f(x) = \begin{cases} x - 1 \text{ when } x < -1 \\ 0 \text{ when } x = -1 \\ -x + 1 \text{ when } x > -1 \end{cases}$$

34. Make a table of values of the function $y = 3 - x^2$. Which of the following is true about the function?

   A. 1st difference $= -2$, linear
   B. 1st difference $= -2$, quadratic
   C. 2nd difference $= -2$, linear
   D. 2nd difference $= -2$, quadratic

35. The height of a ball above the ground is represented on a graph where y is the height, and x is the time that has passed. If the ball is thrown straight up in the air, what is true of the graph of the ball's motion?

   A. There is one relative minimum
   B. There is one relative maximum
   C. The graph is symmetric across the x-axis
   D. The graph is always increasing

36. Which of the following descriptions do not apply to the graph of a function that is a parabola?

   A. The graph has a relative maximum or minimum.
   B. The end behavior is the same for both ends of the parabola
   C. The graph is periodic
   D. There is a line of symmetry on the graph

37. The hourly sales at two local convenience stores are modeled below with y representing total dollars sold, and x representing the number of minutes the store has been open. Which of the stores is selling at a faster rate?

Store A: $4y - 3x = 16$

Store B:

x	12	15	21
y	8	10	14

38. A triangle has side lengths x, $x + a$, and $x + b$. The perimeter of the triangle will be a polynomial of what degree?

39. A trapezoid has an area of 42 square inches and a height of 12 inches. If one base has a length of 6 inches, what is the length of the other base?

   A. 1
   B. 2
   C. 6
   D. 7

40. A contractor is installing a circular fish pond in a park. The circumference of the fish pond must be less than 100 feet. What is the largest possible diameter of the pool, if it cannot be a fractional length?

   A. 10
   B. 11
   C. 31
   D. 32

41. If a cone and a cylinder have the same base area and the same height, which will have a greater volume?

42. A sphere has a diameter of 6 inches. What is the surface area of the sphere?

   A. 36π
   B. 48π
   C. 108π
   D. 144π

43. A farmer is buying a cylindrical silo that has a hemispherical roof. If the radius of his silo is 5 feet and the height is 20 feet, what is the approximate volume of his silo? Round your answer to the nearest whole number.

44. Which of the following measures of central tendency might not exist for a data set?

   A. Mean
   B. Median
   C. Mode
   D. Range

45. A restaurant records the average wait times during their peak hour every night for a week. What is the median value of this data?

11 min, 5 min, 9 min, 8 min, 11 min, 19 min, 7 min

   A. 5 min
   B. 8 min
   C. 9 min
   D. 11 min

46. If you would like to see what portion of your yearly salary goes to housing expenses, would a bar graph or a circle graph more clearly display the data?

47. If you are given a box-and-whisker plot but not the data values it was based on, what measure of central tendency would you not be able to calculate?

  A. Mean
  B. Median
  C. Range
  D. Outliers

48. What is the first quartile and median value of the data set below?

17, 8, 4, 7, 9, 3, 15

  A. The first quartile is 4 and the median is 8.
  B. The first quartile is 7 and the median is 4.
  C. The first quartile is 8 and the median is 7.
  D. The first quartile is 7 and the median is 8.

49. Which of the following is not true about a histogram?

  A. The bars are all of equal width
  B. Each bar represents an equal-sized interval
  C. Each bar represents a different sized interval
  D. Each bar can be a different height

50. A scatter plot is used to compare the price of gas and the outside temperature. If the data shows no correlation, which of the following would describe the line of best fit?

  A. The line has a positive slope.
  B. The line is horizontal.
  C. The line is vertical.
  D. There is no line of best fit.

1. **D** The tenths place is the number to the right of the decimal point, and the hundreds place is three places to the left of the decimal point. Incorrect answers switch the 2 and the 5, or mistake tenths place for tens, and hundreds place for hundredths.

2. **B** The difference between high tide and low tide is the absolute value of the difference between the distance from dock at high tide and low tide. $\left|1 - 2\frac{3}{5}\right| = 1\frac{3}{5}$.
Incorrect answers may indicate errors in subtracting fractions, or adding the distances instead of subtracting.

3. **4** The factors of 16 are 1, 2, 4, 8, and 16. The factors of 36 are 1, 2, 3, 4, 6, 9, 12, 18, and 36. The greatest common factor is 4. Incorrect responses may choose other factors, or identify the least common multiple instead of the greatest common factor.

4. **D** The Distributive Property states that $a(b \times c) = a \times b + a \times c$. This property is illustrated in the equation. Incorrect responses may incorrectly choose the Associative Property based on the parentheses in the equation.

5. **$512.50** The total value of the fund after two years is $5,000(1.05)^2 = 5,512.50$. The accrued interest is the difference between the total value and the original investment: $5,512.50 - 5,000 = 512.50$. Incorrect responses may find the total value of the fund and not the accrued interest.

6. **C** The number is 240 feet tall. Multiply by 12 to translate to inches. $240 \times 12 = 2,880$. For every 300 inches, the model will be 1 inch high. The model is $2,880 \div 300 = 9.6$ inches high. Incorrect responses may not convert to inches before applying the scale factor.

7. **9.5** The cube root of 857 is about 9.498614, rounded to the nearest tenth is 9.5. Incorrect responses may calculate permutations instead of combinations or make an error calculating with factorials.

8. **B** The unit price for each size is the price divided by the number of pounds. The unit price of the small bag is $18 \div 5 = \$3.60$, the unit price of the large bag is $61.60 \div 28 = \$2.20$. The difference between the two is $3.60 - \$2.20 = \$1.40$. Incorrect answers may indicate errors in the difference.

9. **9.6 inches** The building is 240 feet tall. Multiply by 12 to translate to inches. $240 \times 12 = 2,880$. For every 300 inches, the model will be 1 inch high. The model is $2,880 \div 300 = 9.6$ inches high. Incorrect responses may not convert to inches before applying the scale factor.

10. **D** To calculate the number of students, convert the percentage to a decimal and multiply. The number of surveyed students who are involved in an after school sport is $800 \times 0.60 = 480$. Incorrect responses may convert the percentage to a decimal incorrectly or make an error when multiplying.

11. **B** Simple interest is calculated with the formula $I = P \times r \times t = 17{,}999 \times 0.053 \times 5 = \$4{,}769.74$. Incorrect responses may convert the percentage to a decimal incorrectly or make an error while multiplying.

12. The correct answer is 8,568. When choosing a group of 5 order doesn't matter, so it is calculated using combinations. $C(18, 5) = \frac{P(18,\, 5)}{5!} = \frac{18!}{(18 - 5)!5!} = \frac{18 \times 17 \times 16 \times 15 \times 14}{5 \times 4 \times 3 \times 2 \times 1} = 8{,}568$. Incorrect responses may calculate permutations instead of combinations or make an error calculating with factorials.

13. **independent** The number rolled the first time does not impact the result of the second number-cube roll, therefore the probability situation is independent. Incorrect responses may result from confusion over independent and dependent probability situations.

14. **C** Look for the key words "times" and "difference" to translate the words into an algebraic expression, $2(7 - n)$. The other answer options misinterpret the words that are translated into mathematical operations.

15. **B** The one-step equation involves addition, so the inverse operation needed to solve is subtraction. The other answer options may identify the operation in the equation, or mistakenly identify the inverse of addition.

16. **8 months** The equation $135m = 1{,}080$ is a one-step equation involving multiplication, so you can solve by using the inverse operation and dividing each side by 135. Incorrect responses may use the wrong inverse operation or make an arithmetic error when dividing.

17. **yes** To check if the number -7 is in the solution set, substitute it into the inequality and check if it is valid. $-7 - 5 = -12 \le -3$. Incorrectly answering no may result from an error in calculation or misinterpreting the inequality symbol.

18. $s > 15$ To solve the inequality, multiply out each side and then get the variable to one side. Incorrect responses result from calculation errors or forgetting to move all of the variables to one side of the inequality.

19. **D** The server earned \$3.75 for each of the 5 hours worked, $3.75 \times 5 = 18.75$. The server earned 15% of her total sales in tips. $0.15 \times 980 = 147$. Add the totals together and you get the total earnings of \$166.75. Incorrect responses may find only the tip amount, or incorrectly take 15% of the server's hourly wages.

20. **D** A binomial is a polynomial with two terms, and degree 3 means the highest power of the variable is 3, so $x^3 - 3$ is the only polynomial that meets the description. Incorrect responses may confuse binomials and trinomials, or mistake the degree for the number of terms.

21. **C** The value of the polynomial expression can be found by substituting 2 and evaluating,
$$2^3 - 5(2) + 7(2) - (2)^2 + 9$$
$$= 8 - 10 + 14 - 4 + 9 = 17$$
Incorrect responses may make errors in calculating powers of 2 or evaluating the expression.

22. **B** Factor the polynomial completely to determine its factors, $2x^2 - 2x - 24 = 2(x + 3)(x - 4)$. The only choice that is not a factor is $(x - 2)$. Incorrect responses made an error while factoring the polynomial.

23. **A** The leading coefficient is the coefficient of the highest term of the polynomial, regardless of its position in the expression. In this polynomial the highest term is $2x^3$ so the leading coefficient is 2. Incorrect responses may result from choosing the coefficient of the first term.

24. **8** To solve this problem substitute 1024 for d and solve for t, $1,024 = 16t^2$; $64 = t^2$; $t = 8$. Incorrect responses may forget to divide by 16 first, or may find the negative square root, which does not make sense in this real world scenario.

25. **D** When the rational expression is simplified, you can divide $(x - 5)$ from the numerator and the denominator; however, that needs to be included in the restricted values. So the expression can be reduced to $\frac{1}{x + 1}$ but both -1 and 5 are restricted values that would make the denominator of the expression 0. Incorrect responses make an error when reducing the expression or forget to include 5 as a restricted value.

26. **D** The quadrant that contains positive x-values and negative y-values is Quadrant IV. Incorrect responses may confuse the location of the quadrants or the x- and y-axes.

27. **6** The slope of a proportional relationship is the same as the value of k. Incorrect responses may use the x-value or y-value of the point on the line as the slope instead of the value of k.

28. **B** The equation is in slope intercept form, $y = mx + b$, where the slope is 5 and the y-intercept is 2. Incorrect responses may misidentify the form of the line.

29. **D** When you graph the line you clearly see that it crosses the positive x- axis and the positive y-axis. Incorrect responses may come from not graphing the line first, or incorrectly graphing the points.

30. **B** The system is not dependent, so there are not infinite solutions, and the system is not inconsistent so there is a solution. Two lines can only intersect at 0, 1, or infinitely many points, so the correct answer is B. Incorrect responses may incorrectly identify the system as inconsistent or dependent.

31. **5 ounces** Set up the problem as a system of equations where x is the number of ounces of peanuts she buys, and y is the number of ounces of raisins. The system is $x + y = 20$, and $x = 3y$. The system can be solved using graphing, or substitution to find the answer $(15, 5)$. Incorrect responses may set up the wrong system to solve, or switch the variables and solve for the wrong value.

32. **B** A graph must pass the vertical line test to be a function. All of the lines pass this test except for a vertical line. Incorrect responses may mistake the line parallel to the x-axis as a vertical line, or confuse the definition of a function.

33. **1** To evaluate a piecewise function, determine which criteria the value of x meets and then evaluate that piece of the function. When $x = 0$, it is > -1 so, $-x + 1 = 0 + 1 = 1$. Incorrect responses may evaluate the wrong piece of the function or misinterpret the criteria for x.

34. **D** The table of values for $y = 3 - x^2$ has no common first difference, and has a common second difference of -2. Therefore the function is quadratic. Incorrect responses may make an error when calculating, or misinterpret the meaning of the second difference.

35. **B** The graph of the ball's motion will increase, hit a relative maximum and then decrease until hitting the ground. The other descriptions are not possible for the graph. There can be no negative x-values for time so the graph can't be symmetric about the x-axis, and the graph can't keep increasing because gravity will bring the ball down.

36. **C** A parabola will always have a relative maximum or minimum, and then end behavior will either increase or decrease indefinitely but it will be the same for both ends of the parabola and there will always be a line of symmetry. Incorrect responses might confuse the key features of the graph.

37. **Store A** To find the rate at which the stores are selling, find the slope of each line. Store A's slope can be found by converting the equation into point-slope form to find $m = 0.75$. Store B's slope can be found by using the slope formula and two points to find $m = \frac{2}{3}$. Incorrect responses may make errors in calculating the slope, or not recognize slope as being the same as the rate.

38. **1** The perimeter of a triangle is found by adding the sides together, and all of the sides are of degree 1, so the sum will also be of degree 1. Incorrect responses may confuse perimeter and area, or add the degrees of the sides.

39. **A** The formula for the area of a trapezoid is $A = \frac{1}{2}h(b_1 + b_2)$. Substitute all the known values and solve for the length of the other base, $42 = \frac{1}{2}(12)(6 + b_2)$ and so $b_2 = 1$. Incorrect responses may forget the $\frac{1}{2}$ or divide the area by the length instead of using the trapezoid formula.

40. **C** The circumference of a circle is calculated by multiplying the diameter by π. To find the largest diameter divide 100 by π and find the next smallest whole number, 31. Incorrect responses may find the square root of 100 or incorrectly round their answer.

41. **cylinder** The volume of the cylinder is base area time height, and the volume of the cone is $\frac{1}{3}$ base area times height, so the cone will be $\frac{1}{3}$ the volume of the cylinder. Incorrect responses may confuse the volume formulas of cones and cylinders.

42. **A** The formula for the surface area of a sphere is $SA = 4\pi r^2$, and the diameter of the sphere is 6, so the radius is 3 and the surface area is 36π. Incorrect responses may use the diameter instead of the radius, or cube the radius instead of square it.

43. **1833 ft³** The volume of his silo is the volume of the cylinder, $V = \pi r^2 h = \pi(5)^2(20) = 1,571$, plus the volume of the hemisphere section, $V = \frac{2}{3}\pi r^3 = \frac{2}{3}\pi(5)^3 \approx 262$. Add the two totals, $1,571 + 262$, to find the total volume is approximately $1,833 \text{ ft}^3$. Incorrect responses may use the volume of a sphere instead of a hemisphere, or make a calculation error.

44. **C** A data set may not have a mode if there are no repeated values. All of the other measures of central tendency can be calculated on any data set. Incorrect responses may confuse the measures of central tendency.

45. **C** The median value is the middle value when all of the data set is arranged in order. The incorrect answers may find the middle value before the data is arranged, or may find the mode or mean instead of the median.

46. **circle graph** A circle graph shows the data as parts of a whole so would be more useful in this scenario. A bar graph would be more useful to show year-over-year housing expenses.

47. **A** The box plot will show the median as a line in the box, and the outliers will be indicated by an asterisk. The range can be calculated by subtracting the least value from the greatest value. However you will not be able to determine the mean just from the box plot.

48. **A** In order to find the values, arrange the data set from least to greatest and then find the middle value, 7, and then the middle value of the lower half of the data, 4. Incorrect responses may include forgetting to arrange the data before finding the values.

49. **C** A histogram's bars need to all represent equal-sized intervals and should be the same width; however, the height can vary depending on the number of data points in that interval. Incorrect responses may confuse bar graphs and histograms.

50. **D** If there is no correlation of the data in the scatter plot, then there is no line that can be drawn that fits the data points. Incorrect answers may interpret a horizontal or vertical line as indicating no correlation.

Check Your Understanding

On the following chart, circle any items you missed. This helps you determine which areas you need to study the most. If you missed many of the questions that correspond to a certain skill, you should pay special attention to that skill as you work through this book.

Lesson	Item Number(s)			Review Page(s)
	Procedural	Conceptual	Problem Solving	
Chapter 1: Number Sense and Operations	3, 6, 7	1, 4	2, 5	44–45
Chapter 2: Ratio, Proportion, and Probability	10, 12	13	8, 9, 11	78–79
Chapter 3: Linear Equations and Inequalities	17, 18	14, 15	16, 19	114–115
Chapter 4: Polynomials and Rational Expressions	21, 22, 25	20, 23	24	146–147
Chapter 5: Linear Equations in the Coordinate Plane	27, 29	26, 28	30, 31	180–181
Chapter 6: Functions	33, 34	32,36	35, 37	216–217
Chapter 7: Geometry and Measurement	39, 42	41	38, 40, 43	250–251
Chapter 8: Data Analysis	46, 48	44, 47, 49, 50	45	286–287

Lesson 1.1

Workplace Skill, page 13

$27.50

Think about Math, page 14

1. C

2. A, B, and C

Core Skill, page 15

$4.8 < 5.25 < 5.375$

Core Practice, page 16

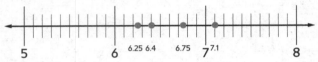

$6.25 < 6.4 < 6.75 < 7.1$

Think about Math, page 16

1. greater than 4.65

2. less than 4.65

3. greater than 4.65

4. less than 4.65

Think about Math, page 17

1. C

2. C

21st Century Skill, page 17

Pools A and D have acceptable pH levels.

Vocabulary Review, page 18

1. order

2. numerator

3. rational number

4. denominator

5. absolute value

6. integers

Skill Review, page 18–19

1. D

2. $3.65 < 3\frac{11}{16} < 4.1$

3. Possible answer: Rational numbers can be written as the ratio of two integers (e.g., $\frac{7}{1}$ or 7) while irrational numbers (e.g., pi) cannot be written as a ratio or fraction. The fraction $\frac{22}{7}$ is only an approximation of pi.

4. A

5. A

6. C

Skill Practice, page 19

1. D

2. C

3. $\frac{9}{5} < \frac{5}{2}$

4. C and D

5. Possible answer: Absolute value is always positive or zero because it tells a number's distance from zero, so it cannot be a negative number.

6. B

Lesson 1.2

Think about Math, page 21

1. D

2. B

Core Skill, page 21

12 bracelets

Think about Math, page 23

1. $(25 \times 4) \times 7 = 700$

2. $9 \times 12 - 9 \times 3 = 108 - 27 = 81$

3. $\left(\frac{1}{4} \times 5\right) \times 20 = \left(\frac{1}{4}\right) \times (5 \times 20) = \frac{1}{4} \times 100 = 25$

21st Century Skill, page 23

1. expression $= 365(875.22 - 750.79)$, $45,416.95$

Core Practice, page 24

Answer: $(3 + 4)^2 - 5 \times 7 + 11$

Think about Math, page 25

1. 60

2. 44

3. undefined

Calculator Skill, page 25

Answers will vary: for example, taking square roots of negative numbers

Vocabulary Review, page 26

1. least common multiple

2. addend

3. greatest common factor

4. undefined

5. order of operations

6. factors

Skill Review, page 26

1. B

2. A

3. C

4. $2 \times 3^2 \times 5$, or $2 \times 3 \times 3 \times 5$

5. 5 boxes

6. B

Skill Practice, page 27

1. 5 square inch grid squares

2. 5 and -5

3. A

4. B

5. Distributive Property

6. April 22

7. C

8. Distributive Property; Distributive Property

9. $2^3 \times 11$

Lesson 1.3

Think about Math, page 29

1. Expression: $3.75 \times 15.5^2 + 50$
   Cost: $951

Core Skill, page 29

0.7×12^3

Core Practice, page 30

Sample answer: The exponent rule $a^m \times a^n = a^{m+n}$ should hold true for all values of a, m, and n. So $1 = a^0 = a^{n-n} = a^n \times a^{-n}$ is true and so $a^{-n} = \frac{1}{a^n}$.

21st Century Skill, page 33

1.75×10^8

Think about Math, page 33

1. 1.118×10^8

2. 5.58×10^7

3. -6.46×10^{-2}

4. 3.625×10^3

Vocabulary Review, page 34

1. b

2. d

3. e

4. c

5. f

6. a

Skill Review, page 34–35

1. A

2. Power of a Quotient Property

3. D

4. $2,519.42

5. A

6. A

Skill Practice, page 35

1. D

2. $\dfrac{5^2 \times 2^6 \times 5^6}{10^6} = \dfrac{5^2 \times 2^6 \times 5^6}{(2 \times 5)^6}$

$\quad = \dfrac{5^2 \times 2^6 \times 5^6}{2^6 \times 5^6}$ Power of a Product

$\quad = \dfrac{5^2 \times 2^6 \times 5^6}{2^6 \times 5^6}$

$\quad = \dfrac{5^{2+6}}{5^6} = \dfrac{5^8}{5^6}$ Product of Powers

$\quad = 5^{8-6} = 5^2$ Quotient of Powers

$\quad = 25$

3. $3.42 \times 20^2 + 100$

4. B

5. $(2.5 \times 10^5) \times (7.6 \times 10^{-4}) = (2.5 \times 7.6) \times (10^5 \times 10^{-4})$

$\qquad\qquad\qquad\qquad = 19 \times 10^{5+(-4)}$

$\qquad\qquad\qquad\qquad = 19 \times 10^1$

$\qquad\qquad\qquad\qquad = 190 = 1.9 \times 10^2$

6. C

Lesson 1.4

Core Skill, page 37
24 inches by 24 inches

Think about Math, page 39
B

Core Practice, page 40
17.3 miles

Think about Math, page 41
1. B
2. A

Vocabulary Review, page 42
1. index
2. prime factorization
3. rational exponent
4. cube root
5. irrational number
6. square root

Skill Review, page 42–43
1. Division Property of Radicals
2. B
3. 13.82
4. 56.6 yd
5. B
6. D

Skill Practice, page 43
1. C
2. $\dfrac{\sqrt{9} \times \sqrt[3]{9} \times \sqrt[3]{81}}{\sqrt{3} \times \sqrt{27}} = \dfrac{\sqrt{9} \times \sqrt[3]{9 \times 81}}{\sqrt{3 \times 27}}$ Mult. Prop. of Radicals

      $= \dfrac{\sqrt{9} \times \sqrt[3]{729}}{\sqrt{81}}$

      $= \dfrac{3 \times 9}{9}$ Evaluate the roots.

      $= 3$

3. Multiplication Property of Radicals
4. D
5. 10 feet
6. A

Chapter Review

Review, page 44–45

1. D
2. A
3. D
4. B
5. D
6. D
7. D
8. A
9. C
10. A
11. B
12. undefined
13. Commutative
14. 10
15. square root
16. A
17. A
18. rational
19. $648.44
20. 512

Lesson 2.1

Core Skill, page 49

the 50-lb bag

Think about Math, page 49

1. $1.85

2. $0.188 or 19 cents

3. $0.04

Think about Math, page 50

1. B

2. C

Core Skill, page 50

$159.50

21st Century Skill, page 51

960 Feet

Think about Math, page 51

1. 32 inches

2. 4

Vocabulary Review, page 52

1. ratio

2. similar

3. unit rate

4. proportion

5. equivalent

6. scale factor

Skill Review, page 52

1. B

2. C

3. 12 books

4. B

Skill Practice, page 53

1. A ratio is a comparison of two numbers. For example, 3 out of 4 people can be written as the ratio $\frac{3}{4}$. A proportion is a statement that two ratios are equivalent, such as $\frac{3}{4} = \frac{9}{12}$.

2. B

3. B

4. 10 feet

5. The unit rate for the 12-ounce box of cereal is $0.24 per ounce. The unit rate for the 16-ounce box of cereal is $0.22 per ounce, so the 16-ounce box is a better value because it has a lower unit rate.

6. C

7. B

8. 9 feet; $\frac{1}{2}$

Lesson 2.2

Core Practice, page 56

about 85; possible benchmark: 48% is close to 50% or $\frac{1}{2}$

Think about Math, page 56

1. 9
2. 630
3. 135
4. 225
5. 80
6. 675

Core Skill, page 57

The cost increased from April to August, so the percent change cannot be negative. The customer subtracted the amounts in the wrong order. The correct percent change is about 51%.

Think about Math, page 58

1. D
2. C

21st Century Skill, page 59

Jin

Think about Math, page 59

1. A
2. B

Vocabulary Review, page 60

1. percent
2. discount
3. principal
4. Simple interest
5. interest rate
6. benchmark

Skill Review, page 60–61

1. 0.34, 34%, $\frac{34}{100}$, or $\frac{17}{50}$
2. B
3. C
4. about 8.3% increase
5. $6.82; about 15%
6. 330 shoppers
7. 150 shoppers

Skill Practice, page 61

1. 20% increase; Explanations should include subtracting $50 from $60 and dividing the difference by $50, and then changing the decimal to a percent.
2. No; Elena will pay $415.80 interest on the 3-year loan and $484 interest on the 4-year loan.
3. $23
4. B
5. Yes; the cost of the monitor
6. C

Lesson 2.3

Core Skill, page 64

$4 \times 5 \times 4 = 80$

Think about Math, page 64

1. C
2. B

Core Practice, page 65

1,320

Think about Math, page 65

1. C
2. B

Think about Math, page 67

1. A
2. C

Vocabulary Review, page 68

1. factorial
2. experiment
3. outcome
4. combination
5. tree diagram
6. permutation

Skill Review, page 68–69

1. B
2. C
3. D
4. A
5. D
6. C

Skill Practice, page 69

1. 8 options
2. No; Possible answer: order does not matter, so you should find the Combination of outcomes. Andrew found the Permutation of outcomes. There are 220 different outcomes.
3. 24 combinations. Possible explanation: I used the Counting Principle and multiplied $4 \times 3 \times 2$.
4. Possible answers: 14!; 87,178,291,200
5. Possible explanation: When you think about the formula in the lesson $P(n, k) / k!$, replace the permutation with its formula and you end up with: $(n!/(n - k)!) \div k!$ which is the same as $n!/(n - k)! \times 1/k!$. This simplifies to the formula shown here.

Lesson 2.4

Core Skill, page 73

Answers: blue – 2/9; white – 4/9; gray – 1/3; not blue – 7/9; not white – 5/9; not gray – 2/3

Think about Math, page 73

1. 12

2. 30

Calculator Skill, page 74

1. $\left(\frac{1}{2}\right)^{20} \approx 0.00000095$

2. $(0.7)^7 \approx 8.2\%$

21st Century Skill, page 75

0.87

Vocabulary Review, page 76

1. compound event

2. Probability

3. tree diagram

4. independent event

5. complement

6. dependent event

Skill Review, page 76–77

1. $\frac{1}{2}$

2. $\frac{1}{4}$

3. $\frac{1}{5}$

4. $\frac{1}{10}$

5. 5 times

6. $\frac{3}{10}$

7. Drawings will vary. The probability is 10.7%.

8. $\frac{2}{5}$

Skill Practice, page 77

1. Possible description: This is a dependent probability situation since she will not replace the shirt from the previous day back into her choices for the next day

2. Drawings will vary.

3. blue shirt

4. Drawings should include 12 sections, 4 sections red, 1 section blue, 3 sections green, 2 sections yellow, 2 sections purple.

5. Possible answer: For Monday through Thursday keep track of the colors of each car, find the probability that the next car parked is red. Then multiply that number by 25.

6. 2 people

7. A

8. B

Chapter Review

1. D
2. B
3. certain
4. 16
5. C
6. proportional
7. C
8. B
9. C
10. $8 \times 7 \times 6 \times 5 \times 4 \times 3 \times 2 \times 1$
11. A
12. 26%
13. B
14. 10
15. B
16. C
17. B
18. number cube twice
19. C
20. similiar

Lesson 3.1

Workplace Skill, page 83

$0.10s + 250$

Think about Math, page 83

1. $n - 5$
2. $-7 + 2n$
3. $10(2 + n)$

Core Skill, page 85

$21n + 8$

Think about Math, page 86

1. D
2. A

Core Skill, page 87

42

Calculator Skill, page 87

-5.9445

Think about Math, page 87

1. B, D
2. C

Vocabulary Review, page 88

1. coefficient
2. distribute
3. constant
4. evaluate
5. variable
6. algebraic expression

Skill Review, page 88

1. 2n – 7
2. $-3n$
3. $5n - 9$
4. $3w$
5. $-20p + 4$
6. 27
7. A, B
8. C

Skill Practice, page 89

1. Sample answer: Susan was correct. The expression is $2n - (-5)$. When the expression is evaluated at $n = -5$, the value of the expression is -5.
2. $15p + 200$
3. $-4x + 10$
4. $18,000
5. Sample answer: The new gym is more expensive. After one year, the new gym membership cost is $198 compared to the local gym membership cost of $180.
6. A, B, D

Lesson 3.2

Core Skill, page 91

$n = 3$

$r = 2$

$q = 6$

$g = 3$

Think about Math, page 93

1. $n - 10 = 62; n = 72$
2. $n + 20 = 30; n = 10$
3. $3n = 12; n = 4$
4. $-7 + n = 2; n = 9$
5. $\frac{n}{4} = -5; n = -20$

Calculator Skill, page 94

No

Think about Math, page 95

1. $x = -12$
2. $r = -6$
3. $y = 4$

Core Skill, page 95

$w = 5$

Vocabulary Review, page 96

1. expression
2. reciprocal
3. inverse operation
4. equation
5. variable
6. solution of an equation

Skill Review, page 96

1. expression; equation; expression; expression; equation; expression; equation; expression
2. $2n - 4 = 7$
3. $6n + 2 = 6$
4. $x = 4$
5. $x = -8$
6. $x = 36$
7. $x = 3$
8. $y = 7x$; 56 minutes
9. possible equation $y = 22x + 50$; 3 hours
10. B
11. A

Skill Practice, page 97

1. B
2. B
3. $x = 4$
4. 6 months
5. $25.50
6. In the last step, Jermaine multiplied the right side by 2 instead of dividing by 2. The correct solution is $r = -\frac{3}{2}$.
7. Greater than 100; to solve the equation, add c to both sides. Because c is a positive number, $100 + c$ is greater than 100.

Lesson 3.3

Core Skill, page 99

520

Think about Math, page 100

1. $t < 75$

2. yes

Core Skill, page 101

$n < -\frac{1}{2}$, -10 is a solution because -10 is less than $-\frac{1}{2}$

Calculator Skill, page 102

$x \leq 1.65$

21st Century Skill, page 103

20 months

Think about Math, page 103

1. $b < -6.5$

2. $q \geq -13.5$

3. $q > -2$

4. $x > -\frac{1}{3}$

Vocabulary Review, page 104

1. equation

2. inequality sign

3. Inverse operations

4. solution of an inequality

5. variable

6. inequality

Skill Review, page 104

1. $n + 4 < 6$

2. $3n \geq n + 1$

3. $x > -6$

4. $x \leq -\frac{1}{2}$

5. $q > -3$

6. yes; $x \leq 0$

7. yes; $x \geq 9$

8. yes; $x \geq -4$

Skill Practice, page 105

1. $p \geq \$25.60$

2. $n < -1$

3. no; $x \geq 0.44$ (rounded)

4. $400 - 25w \geq 175$, $w \leq 9$. Martin can spend $25 for 9 or less weeks

5. Answer will vary. Sample answer: the person has $2,000 in her savings account and is saving $30 per month. She is saving to purchase a car that will cost at least $9,000.

6. Michael incorrectly turned the symbol. Because the variable r did not have a negative coefficient, the symbol did not need to be reversed. Correct solution: $r < -\frac{2}{3}$

7. Inequalities may vary. Sample answer: $x < 0.3(150)$; $x < 45$; A student can answer no more than 45 questions incorrectly to pass the state nursing exam this year.

8. a. $<$; $<$

b. $>$; $>$

9. $x > \frac{1}{3}$; The number line should have an open circle at $\frac{1}{3}$ and an arrow pointing to the right from that circle to show that all values greater than $\frac{1}{3}$ are solutions.

10. Inequalities may vary. Sample answer: $44 - 6h \geq 20$; $h \leq 4$; Emily can lower the temperature for no more than 4 hours.

11. Answers will vary. Sample answer: You must reverse the inequality symbol to make the solution inequality true.

12. Answers will vary. Sample answer: We do not know the value of the variable b—it could be positive or negative. As a result, the solution instead could be $x > 5$.

Lesson 3.4

Core Skill, page 107

$11h - 5w$; $435; $430

21st Century Skill, page 108

$308.75 = P(0.065)(5)$; $P = $950

Think about Math, page 109

1. C
2. A

Core Skill, page 110

$28,000 + c \leq 80,000$; $c \leq 52,000$; allowable cargo weights are less than or equal to 52,000 pounds.

Think about Math, page 111

1. C
2. B

Vocabulary Review, page 112

1. inverse operations
2. algebraic expression
3. equation
4. inequality

Skill Review, page 112–113

1. a. $10b + 5$; b. $85; c. 3
2. C
3. 11 weeks
4. $h \geq 3$
5. $s \leq 2,000$
6. $h > 16$
7. 5 years

Skill Practice, page 113

1. A
2. A
3. $700 - 3m \geq 100$; no more than $200 per month
4. B
5. In the simple interest formula, Cal substituted 11,200 for the interest. However, the amount of interest is $11,200 - 10,000 = 1,200$. The correct interest rate is 4%.
6. $x + (x + 8) = 52$; 22 inches

Chapter Review

Review, page 114–115

1. Possible answer: 8 less than 5 times a number
2. D
3. D
4. $n > 8$
5. 194
6. D
7. B
8. A
9. Possible Answer: 85 divided by a number is equal to 17

10. D
11. $0.2x + 3 = 15$
12. B
13. B
14. C
15. A
16. $10x \geq 70$

Lesson 4.1

Core Practice, page 119

$-5m^3 - 5m^2 + 16m + 5$

Think about Math, page 119

1. B
2. D

Core Skill, page 120

18 feet

21st Century Skill, page 121

$-0.0002x^2 + 1.5x - 175$

Vocabulary Review, page 122

1. b
2. d
3. e
4. c
5. a

Skill Review, page 122–123

1. D

2. Sample answer: For polynomials written in standard form, the exponents are in order from greatest to least, left to right. Since the degree of a polynomial is the greatest exponent in the polynomial, the exponent of the first term of a polynomial written in standard form is the degree of the polynomial. For example, for the polynomial $2x^5 + 3x^4 - 7x$ written in standard form, the degree, 5, is the exponent of the first term

3. False; the product of two monomials is a monomial. Examples: $2(x^7) = 2x^7$; $3x^2(x) = 3x^3$; $2x^2(3x^3) = 6x^5$

4. A
5. B

Skill Practice, page 123

1. D
2. C
3. Yes; Sample explanation: If the terms with an exponent of 4 in both polynomials have the same coefficient, the terms with an exponent of 3 in both polynomials have the same coefficient, and the terms with an exponent of 2 in both polynomials do not have the same coefficient, then the difference of the two polynomials will have degree 2. For example, $5x^4 + 2x^3 - 3x^2 + x$ and $5x^4 + 2x^3 + 6x^2 - 4$ are two polynomials of degree 4, and their difference, $3x^2 + x - 4$ has degree 2.
4. 2
5. C
6. C
7. Sometimes; if the monomials are not like terms, the sum will be a binomial: $(2x^3) + (x^2) = 2x^3 + x^2$; if the monomials are like terms, the sum is a monomial: $(2x^3) + (4x^3) = 6x^3$
8. A

Lesson 4.2

Think about Math, page 126

1. $2x^3(7 + 2x^6)$

2. $x^2y(2x^5 - 3y^2)$

3. $2x^2y^2(2x + x^2y^2 - 3y)$

Core Practice, page 126

$(x + 4)(x - 3)$

Core Practice, page 127

$4x^2 + 12x - 40$

Vocabulary Review, page 128

1. c

2. a

3. f

4. e

5. b

6. d

Skill Review, page 128

1. B

2. A

3. $3x^2 - 8x + 4 = 3x^2 - 6x - 2x + 4$
$$= (3x^2 - 6x) + (-2x + 4)$$
$$= 3x(x - 2) - 2(x - 2)$$
$$= (3x - 2)(x - 2)$$

4. D

5. C

Skill Practice, page 129

1. B

2. Disagree; Possible explanation: A linear factor has general form $ax + b$. Multiplying four such linear factors would give a term that includes x^3 and therefore be a degree 3 polynomial; but the degree of the polynomial shown is 2.

3. C

4. $4x^3 + 2x^2y - 2xy^2 = 2x(2x^2 + xy - y^2)$
$$= 2x[(2x^2 + 2xy - xy + (-y^2)]$$
$$= 2x[(2x^2 + 2xy) - (xy + y^2)]$$
$$= 2x[2x(x + y) - y(x + y)]$$
$$= 2x(2x - y)(x + y)$$

5. A

6. B

7. False; factoring a quadratic expression could result in the product of a linear monomial and a linear binomial. Example: $x^2 + x = x(x + 1)$

8. D

Lesson 4.3

Think about Math, page 131
1. $[-2, 2]$
2. $[-7, 3]$

Test-Taking Skill, page 132
5 feet; 3 feet

Think about Math, page 133
1. $[-12, 2]$
2. $[-6, 0]$
3. $[-6, 2]$

Core Skill, page 134
Equations 1 & 2

Core Skill, page 135
6 seconds

Think about Math, page 135
1. $a = 1, b = 2, c = -8$
2. 36
3. 2
4. 2, -4

Vocabulary Review, page 136
1. Completing the square; perfect-square trinomial
2. solving by inspection
3. quadratic formula; discriminant

Skill Review, page 136
1. $[-7, 3]$
2. $[3, 12]$
3. $[-9, 8]$
4. $[-14, 6]$
5. B
6. C
7. C
8. 2
9. $[-\frac{2}{3}, -1]$

Skill Practice, page 137
1. $[2, 3]$
2. $[-4, 1]$
3. $[-4]$
4. C
5. a. $r = \sqrt{\frac{A}{\pi}}$

   b. No; the negative square root does not make sense because a radius cannot be negative.

   c. about 2 feet
6. A
7. The length and width of the rectangle; 35 feet and 15 feet
8. 35 sandwiches
9. 9; Possible explanation: For the equation to have one real solution, $b^2 - 4ac = 0$. Substitute $b = 6$ and $a = 1$ into this equation and solve for c.
10. Zach did not write the equation in the form $ax^2 + bx + c = 0$ before identifying a, b, and c. To find the correct values, subtract $4x$ from both sides and add 3 to both sides. This gives the equivalent equation $3x^2 - 6x + 4 = 0$; $a = 3$, $b = -6$, and $c = 4$. For this equation, $b^2 - 4ac = -12$, so there are no real solutions.

Lesson 4.4

Core Skill, page 139

-2

Think about Math, page 139

1. $\frac{x^2}{x+5}$; $x \neq 3$, $x \neq -5$

2. $\frac{x}{x+2}$; $x \neq -2$

3. $\frac{x-6}{x-4}$; $x \neq -3$, $x \neq 4$

Core Skill, page 140

$\frac{22}{27} \times \frac{63}{66} = \frac{2 \times 11}{3 \times 3 \times 3} \times \frac{3 \times 3 \times 7}{2 \times 3 \times 11} = \frac{7}{9}$

Think about Math, page 141

1. $\frac{3}{x^2 - 2x - 24}$

2. $\frac{x^2 - 4x}{12}$

Workplace Skill, page 142

$\frac{2x+5}{x^2+5x}$

Vocabulary Review, page 144

1. polynomial
2. reciprocals
3. rational expression
4. LCD
5. Restricted values
6. prime number

Skill Review, page 144–145

1. $\frac{r+1}{r-4}$, $\frac{4}{3x}$, $\frac{n^2-81}{n-1}$

2. a. $x \neq 1$

   b. $x \neq 0$

   c. $x \neq -3$, $x \neq 4$

3. D

4. $\frac{2x+2}{5x}$

5. $\frac{7r+7}{8r+1}$

6. $\frac{4n-4}{n-3}$

7. C

Skill Practice, page 145

1. B

2. The student made an error combining like terms in Step 3. The correct answer is $\frac{8x+14}{x^2+2x}$.

3. Possible answer: $\frac{2x-2}{x+4} + \frac{-x+1}{x+4}$

4. $x + 2$

5. $\frac{a+b}{2}$

6. a. $m + 3$

   b. $\frac{1}{m+3}$

   c. $\frac{2m+3}{m^2+3m}$

   d. $\frac{1}{2}$; 2 hours

Chapter Review

1. $2x^3 + 6x^2 + 5x + 15$
2. Zero Product Principle
3. C
4. C
5. B
6. -8 and 2
7. $4x^2y$
8. D
9. D
10. B
11. $\frac{x-3}{x}$
12. C
13. C
14. leading coefficient
15. A
16. $\frac{x^2 + 9x - 18}{(x+3)(x-3)}$
17. C

CHAPTER 5 Answer Key

Lesson 5.1

Core Practice, page 151

Quadrant II: negative x-coordinates, positive y-coordinates; Quadrant III: all negative coordinates; Quadrant IV: positive x-coordinates, negative y-coordinates

Think about Math, page 152

1. A, C
2. B, C

21st Century Skill, page 154

Person B

Think about Math, page 154

Line B, Line A, Line C

Core Skill, page 155

3 miles

Vocabulary Review, page 156

1. unit rate
2. coordinate plane
3. quadrant
4. proportional relationship
5. slope
6. ordered pair

Skill Review, page 156–157

1. Quadrant I: (11, 2); Quadrant II: (−1, 8), (−10, 12); Quadrant III: (−20, −1); Quadrant IV: (1, −15), (7, −18)]

2. Sample answer: The solutions of the equation are ordered pairs that represent points on the line. Every solution is represented by a point on the line, and all points on the line represent solutions. For example, the point (2, 1) lies on the graph shown; this means that (2, 1) is a solution of the line's equation.

3. C
4. A
5. $10.50

Skill Practice, page 157

1. Answers will vary. Possible response: (−2, −4); x and y coordinates must both be negative
2. A, C, D
3. Answer: Lincoln divided the run by the rise, instead of the rise by the run; 2
4. Point graphed and labeled at (1, 2)
5. Answer: Store A. Possible response: You can use the graph to determine that 1 pound of apples costs $1.50. The table shows that 1 pound of apples at Store B costs $1.70. Because 1 pound of apples costs less at Store A, 5 pounds of apples will cost less at Store A.

Lesson 5.2

Core Skill, page 159

$y = 2x + 10$

Think about Math, page 160

1. $y - 2 = -3x + 3$
2. $A = 3, B = 1, C = 5$
3. $y = -3x + 5$

Core Practice, page 161

$y = -\frac{5}{2}x + 110; 60$

Think about Math, page 162

1. 20
2. $y = -3x + 7$

Workplace Skill, page 163

The slope for Factory 1 is steeper than the slope for Factory 2. Factory 1 will produce the most items in an 8-hour workday because it has a steeper slope.

Think about Math, page 163

$(-2, 2)$, $(-4, 1)$, and $(4, 5)$ are on the line.

Vocabulary Review, page 164

1. y-intercept
2. point-slope form
3. slope
4. standard form of a linear equation
5. coefficient
6. slope-intercept form

Skill Review, page 164–165

1. D
2. A
3. D
4. $y = 2x + 2$
5. $y - 3 = -\frac{4}{3}x - \frac{16}{3}$
6. Possible points: $(0.5, 10)$, $(1, 15)$, $(1.5, 20)$, $(2, 25)$

Skill Practice, page 165

1. The equation is $y = 2x + 8$. For selling 20 T-shirts, she earns $48 in commission.
2. The slope is $\frac{4}{5}$. The equation of the line in slope-intercept form is $y = \frac{4}{5}x$.
3. The equation of the line is $y = 10x + 10$. The cost to rent a boat for 8 hours is $90.
4. B

Lesson 5.3

Core Skill, page 167

$x = -6$

Think about Math, page 167

1. 1
2. -3
3. -7

graph a line through points (1, 1) and (2, -3)

Calculator Skill, page 167

The answer is -17 using both methods; however, when stored as a variable I don't have to enter parentheses around the negative number.

Core Skill, page 168

Equations will vary.

Test-Taking Skill, page 169

No, the slope of the equation is negative, so it should be sloping the other direction. The y-intercept in the equation is 1, but the y-intercept of the line on the graph is 0.

Think about Math, page 169

1. D
2. $4.50

Vocabulary Review, page 170

1. slope
2. y-intercept
3. x-value
4. ordered pair
5. slope-intercept form
6. y-value

Skill Review, page 170–171

1. $-2, 1, 4$
2. B
3. a. graph a line through points (0, 20) and (1, 80)

   b. A
4. $x = -4$
5. 4

Skill Practice, page 171

1. a. cats cost $68 for a week, dogs cost $96 for a week

   b. A
2. a. Graph a line through points (1, 120) and (2, 240).

   b. 25 minutes
3. a. Graph a line through points (0, -2) and (1, -5)

   b. 9 units

Lesson 5.4

Core Skill, page 173

$x + y = 200$ and $10x + 15y = 2,600$

Think about Math, page 174

D

Workplace Skill, page 175

1. Company A charges $60 and Company B charges $55.

Think about Math, page 175

1. 75 boxes

Core Skill, page 177

elimination, graphing, substitution

Think about Math, page 177

1. 100 square and 50 rectangular blocks

Vocabulary Review, page 178

1. independent system
2. substitution method
3. dependent system
4. system of linear equations
5. inconsistent system
6. elimination method

Skill Review, page 178-179

1. independent
2. A, C
3. a. $x + y = 5$ and $x - y = 1$
   b. $x = 3, y = 2$
4. A
5. B
6. a. 50 prints
   b. Photographer B

Skill Practice, page 179

1. C
2. 5 classes
3. a. Set a line going through points (1, 29) and (2, 32), and another line going through points (1, 27) and (2, 32)
   b. $32
4. a. B
   b. Plan B

Chapter Review

Review, page 180–181

1. B
2. D
3. Answers will vary. Possible answer: (3, 9)
4. (0, 2)
5. C
6. A
7. C
8. $y = 5x + 3$
9. -2.25 and 2.25, or $-\sqrt{5}, \sqrt{5}$
10. $x > 0$
11. relative maximum
12. A
13. A
14. D
15. B
16. Answers will vary. Possible answer: $y = 5x + 2$

Lesson 6.1

Core Practice, page 185

Top graph: no; bottom graph: yes

Think about Math, page 186

1. No; there are many vertical lines that will intersect the graph in more than one place.

2. Yes; each domain value has exactly one range value.

21st Century Skill, page 187

$288.00

Think about Math, page 187

1. 18, −3, −24

2. 10, 2, $\frac{5}{2}$

Core Skill, page 188

256; 240; 192; 112; 0; the height of the object after 0, 1, 2, 3, and 4 seconds

Think about Math, page 189

4; 2; 2; 4; 6

Vocabulary Review, page 190

1. range

2. function

3. quadratic function

4. domain

5. linear function

6. one-to-one function

Skill Review, page 190–191

1. A, B, and D represent functions; A and D are one-to-one.

2. D

3. a. $6.50

b. $26; $32.50; $39; $45.50

4. A

5. 0; 2; 2; 0; −4

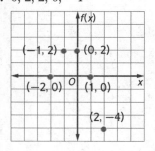

6.

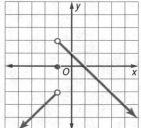

Skill Practice, page 191

1. Table C is not a function because the domain value −1 is paired with two different range elements.

2. Possible answer: If there is a vertical line that intersects the graph at more than two points, then these points have the same x-coordinate but different y-coordinates. This means that a value from the domain has more than one range value, so the graph cannot represent a function. If there is no such vertical line, then the graph does represent a function.

3. B

4. a. $f(x) = 8x$

b. 16 miles

c. Yes; each input (length of time) will result in a different output (distance).

d. 7.5 minutes, or 7 minutes and 30 seconds

e. Linear; it is in the form $f(x) = mx + b$ with $m = 8$ and $b = 0$.

5. 0; The ball's height after 5 seconds is 0, so the ball reached the ground after 5 seconds.

6. The relationship is not a function because the domain value $x = 0$ has two range values, 1 and −1; possible answer: either change $x \geq 0$ to $x > 0$ or change $x \leq 0$ to $x < 0$ (or both).

Lesson 6.2

Think about Math, page 193

1. -3

2. 1

Test-Taking Skill, page 193

Yes; the first and third consecutive differences

Core Practice, page 197

No; Mario subtracted the values of $f(x)$ in the wrong order in some cases.

Vocabulary Review, page 198

1. linear function

2. common differences

3. quadratic function

4. coordinate

5. consecutive difference

Skill Review, page 198

1. B

2. B

3. Answers will vary; check students' work.

Skill Practice, page 199

1. Heidi; the x-values in Neal's table do not change by 1.

2. $f(x) = x^4$ has common fourth consecutive differences; check students' tables.

3. Answers will vary; check students' work.

4. Answers will vary; check students' work.

5. Answers will vary; check students' work.

Lesson 6.3

Core Practice, page 201

A quadratic graph will have one y-intercept and may have no, one, or two x-intercepts.

Core Skill, page 204

Possible answer: The x-intercept is 3 and the y-intercept is 6.

Calculator Skill, page 205

$1.5, -\frac{1}{3}$

Think about Math, page 205

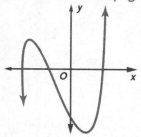

Vocabulary Review, page 206

1. y-intercept
2. relative maximum/minimum
3. x-intercept
4. line symmetry
5. End behavior
6. rotational symmetry

Skill Review, page 206–207

1. A; Possible explanation: Substitute 0 for x to find that the y-intercept is 3. Similarly, substitute 0 for y to find that the x-intercept is 3. Choose the graph with these intercepts.

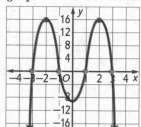

2.

3. C

Skill Practice, page 207

1. x-intercepts: -1, 1, 3

   y-intercept: 3

   positive: $-1 < x < 1$ and $x > 3$

   negative: $x < -1$ and $1 < x < 3$

   increasing: $x < 0$ and $x > 2$

   relative maximum: 3; relative minimum: -3

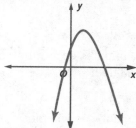

2. a. Possible answer:

   b. Not possible; because of the shape of a quadratic graph, it must have either a relative maximum or a relative minimum.]

   c. Not possible; all quadratic graphs are symmetrical about a vertical line.

3. Answers will vary. Check students' work.

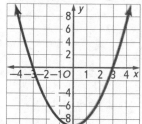

4.

   Possible description: Use the function rule to find that the y-intercept is –9 and the x-intercepts are –3 and 3. Plot the points $(0, -9)$, $(-3, 0)$, and $(0, 3)$. Because the graph is quadratic, it is symmetrical about a vertical line; this line must be the y-axis because of the location of the x-intercepts.

5. Yes; if a line has both an x-intercept and a y-intercept, those intercepts correspond to two points, and two points determine a unique line. If a line has only one intercept, then the line is vertical or horizontal through the point that corresponds to the intercept.

Lesson 6.4

Core Skill, page 210

$600

In one week or 7 days, Dean earns 7($75) = $525.

No; Dean spends $75 more per week than he earns.

Think about Math, page 210

B, A and C

21st Century Skill, page 212

at least about 24 minutes or 0.4 hour; at least about 14 minutes or 0.23 hour

Core Practice, page 213

Factor the function to see whether it had 1 and 5 as its x-intercepts.

Vocabulary Review, page 214

1. slope; y-intercepts
2. quadratic function
3. proportional relationship, slope

Skill Review, page 214

1. Jonas is correct that the table does not tell when Rocket A reached the ground. However, the table does show that Rocket A was still in the air after 2.5 seconds, and the graph shows that Rocket B was in the air for less than 2 seconds. Therefore, Rocket B reached the ground first.
2. A
3. C
4. Boris earns $8 more per day.

Skill Practice, page 215

1. B, D
2. Function B, Function A, Function C
3. Chicago; $4.00
4. Dallas; $1.80
5. B

Chapter Review

1. A
2. quadratic
3. D
4. Possible answer: kicking a ball in the air
5. domain
6. −4 and 5
7. $-2 < x < 3$
8. relative maximum
9. Ellen
10. B
11. C
12. D
13. B

Lesson 7.1

Core Skill, page 221

84 m^2; round 11.2 to 11 and round 7.5 to 8.
11 × 8 = 88, so 84 m^2 is a reasonable answer.

Think about Math, page 221

1. A

2 D

Test-Taking Skill, page 223

84 cm^2

Think about Math, page 224

1. A

Core Skill, page 225

60 cm^2

Think about Math, page 225

1. A

2. B

Vocabulary Review, page 226

1. trapezoid

2. area

3. polygon

4. parallelogram

5. perimeter

6. hypotenuse

Skill Review, page 226-227

1. B

2. C

3. The perimeter of a polygon is the distance around the outside of a polygon. It is measured in linear units. The area of a polygon is the number of square units inside the polygon. For example, the perimeter of a rectangle with length 8 cm and width 3 cm is 22 cm, while its area is 24 cm^2.

4. B

5. 10 pendants

6. Keisha; Howie found the area by multiplying $2 \times 1\frac{1}{4}$ instead of $2 \times \frac{3}{4}$.

Skill Practice, page 227

1. C

2. D

3. 432 tiles

4. Possible answer: The area of one sheet of vinyl is 6 × 4 = 24 square feet, so two sheets of vinyl are needed. Cut one sheet of vinyl in half along the diagonal, creating two right triangles. Attach the triangles to the second sheet of vinyl, one on each side, to form a parallelogram whose area is the same as two rectangles, or 48 ft^2.

5. Possible answer: The two trapezoids together form a parallelogram with height h and base $b_1 + b_2$. The area of this parallelogram is $h(b_1 + b_2)$. Therefore the area of the trapezoid is $\frac{1}{2}$ this amount, or $\frac{1}{2}h(b_1 + b_2)$.

Lesson 7.2

Core Practice, page 229

$C = \pi d$; because you are given the diameter, substitute the diameter (250 ft) for d in this formula.

about 785 ft

Think about Math, page 229

1. B

Calculator Skill, page 230

324

Core Skill, page 230

You must first find the radius by dividing the diameter by 2. The radius is 8 in. Substitute this value into the area formula and simplify. The area is 64π in^2, or about 200.96 in^2.

Think about Math, page 230

1. C

2. 254 m^2

Workplace Skill, page 231

Use the formula for circumference, $C = \pi d$. Substitute 47 for C and solve for d by dividing 47 by 3.14. The diameter is about 15 feet.

Think about Math, page 231

1. B

2. 80 feet

Vocabulary Review, page 232

1. radius

2. area

3. circle

4. Pi (π)

5. circumference

6. diameter

Skill Review, page 232–233

1. C

2. A

3. Sample answer: Divide the area by π, take the square root to find the radius, and then multiply by 2π.

4. C

5. A

6. C

Skill Practice, page 233

1. B

2. C

3. B

4. Sample answer: First, substitute the given area into the formula $A = \pi r^2$ and solve for r. Then, because diameter is twice the radius, multiply r by 2.

5. Circumference: 4π, 8π, 16π, 32π,

   Area: 4π, 16π, 64π, 256π

   When the radius is doubled, the circumference is doubled. When the radius is doubled, the area is multiplied by 4.

6. Medium; the radius of her pie pan is a little less than 4 inches, so the diameter is a little less than 8 inches.

Lesson 7.3

Core Skill, page 236

189 ft^3

21st Century Skill, page 237

about 954 ft^3

Think about Math, page 237

1. 1364 cm^2

Core Skill, page 238

Area of square: 36 cm^2; area of each triangle: 27 cm^2; surface area: 144 cm^2

Vocabulary Review, page 240

1. Volume
2. prism
3. surface area
4. cylinder
5. sphere
6. pyramid

Skill Review, page 240

1. C
2. B
3. D
4. B
5. B
6. A

Skill Practice, page 241

1. C
2. 45,000 in.^{3}
3. A
4. B
5. A
6. B
7. Carmen has assumed that the container is a cylinder.
8. Yes; the formula for surface area of a cylinder is $SA = 2\pi r^2 + 2\pi rh$. You can use the Distributive Property to rewrite the right side as $2\pi r(r + h)$, which is equivalent to Wallace's method.

Lesson 7.4

Core Skill, page 243

The floor is composed of a rectangle and a semicircle. Using the formulas for the areas of a rectangle and a semicircle, the area of the composite shape is 79.3 ft²

Think about Math, page 243

1. about 3,142 yd
2. 182 ft²

21st Century Skill, page 244

23,038 ft³

Calculator Skill, page 245

2,744

Think about Math, page 245

1. B
2. D

Core Practice, page 247

The formula for surface area of a cylinder, $2\pi rh$; 3,079 ft²

Think about Math, page 247

1. C

Vocabulary Review, page 248

1. 2-dimensional
2. hemisphere
3. composite figure
4. composite solid
5. 3-dimensional

Skill Review, page 248

1. C
2. B
3. 357 cm²
4. Volume = 9,048 cm³; surface area = 2,261 cm²

Skill Practice, page 249

1. 130 feet
2. 10 feet
3. D
4. Scott used the slant height in the formula instead of the height.
5. B
6. Answers will vary. Possible answer:

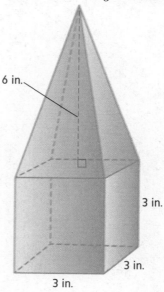

Chapter Review

Review, page 250–251

1. B
2. 24 centimeters
3. 28.26 centimeters
4. C
5. B
6. B
7. 6 cm
8. C
9. 168 ft^2
10. A
11. B
12. B
13. B

Lesson 8.1

Core Skill, page 256

$2.74

Think about Math, page 256

1. C
2. B
3. A
4. D

Workplace Skill, page 257

16 dozen roses each day

Think about Math, page 258

1. B
2. D

Workplace Skill, page 259

$5.40 per shirt

Think about Math, page 259

1. B
2. C

Vocabulary Review, page 260

1. mode
2. median
3. range
4. mean; average OR average; mean
5. weighted average

Skill Review, page 260

1. 8.75 in.
2. Possible answer: average the two items closest to the center, in this case the 4th and 5th
3. C
4. 17.6
5. D
6. C

Skill Practice, page 261

1. 1
2. A, B, C
3. 48 48 54 70
4. B
5. 550
6. D
7. C
8. A
9. 9,700
10. 10.5

Lesson 8.2

Core Skill, page 263

Downtown: July, August, September, and December; Suburban: December

Core Skill, page 264

3 times more and the graphic makes it appear like 9 times more

Think about Math, page 265

1. B

Test-Taking Skill, page 266

about 90

Workplace Skill, page 266

Sample answer: Eliminate orange. It's the least popular color. Eliminating maroon as well might lose more than 20% of customers, so orange is the only color to eliminate.

Calculator Skill, page 267

brown: 80; blond: 60; black: 30; red: 10; other: 20

Think about Math, page 267

1. B

2.
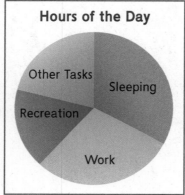

Vocabulary Review, page 268

1. circle graph
2. legend
3. bar graph

Skill Review, page 268–269

1. B
2. A
3. C
4. 30
5. 80
6.
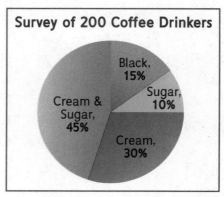

Skill Practice, page 269

1. Possible answer: 0 to 40, with ticks every 5 units and number labels every 10 units. This is enough to show all the data and see the lengths of the bars reasonably well without having so many labels it is difficult to read.

2. Gomez is so far ahead of the others that their bars will be virtually invisible next to his.

3. 18%

4. C

5. halfway between 4 o'clock and 5 o'clock

6. a. Circle graph
   b. Bar graph
   c. Bar graph
   d. Circle graph

Lesson 8.3

Core Skill, page 271

$\frac{7}{14}$ or $\frac{1}{2}$

Think about Math, page 271

1. 20%

2. 4

Workplace Skill, page 273

Option 3; Possible justification: The histogram shows that about half of the parties are 3 or 4 people, so about half of the 15 tables should seat 4 people. The histogram also shows that there about twice as many parties of 1 or 2 people as there are parties of 5 or 6 people. This means that of the remaining half of the 15 tables, there should be about twice as many tables that seat 2 people than tables that seat 6 people. Option 3 best meets these conditions.

Think about Math, page 273

1. 8

2. about 55.6%

Core Practice, page 275

Pumpkin Weights

Weight (lb)

The outlier is 55.

Think about Math, page 275

1. least: 10; greatest: 80

2. first: 25; third: 65

3. 55

Vocabulary Review, page 276

1. third quartile

2. box plot

3. dot plot

4. first quartile

5. histogram

6. median

7. distribution

Skill Review, page 276–277

1. B

2. A

3. a. greater than

   b. less than

   c. greater than

   d. equal to

4. least value: 0; first quartile: 1.5; median: 2; third quartile: 4; greatest value: 6

How Many Pieces of Mail Did You Get Today?

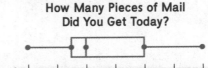

Number of Pieces

5. B

6. Sandy; you cannot determine the range from a histogram because a histogram shows only the intervals in which the least and greatest values lie, not their actual values.

Skill Practice, page 277

1. A

2. B

3. Possible answer: No; there will be too many intervals, and the histogram will be too wide. Intervals of 5 or 10 years would be more practical.

4. Possible answer: The histogram shows the distribution of sales prices in greater detail, while the box plot provides essential information in a compact form.

Lesson 8.4

Workplace Skill, page 279

$108

Think about Math, page 279

1. Answers will vary.

2. Answers vary. Possible answer: A table with 5 rows (where each row represents a designer) and 3 columns (where each column represents a material).

Calculator Skill, page 280

1.65, 0.2

Core Skill, page 281

sample answers: patience, compassion, good communication skills, cheerfulness

Think about Math, page 281

1. C

2. B

21st Century Skill, page 282

Answers will vary.

Core Skill, page 283

sample answer: Only plot the points for Weeks 6 through 8 where there is a decrease and use a vertical scale that has a range only slightly larger than the share prices.

Think about Math, page 283

1. D

2. sample answer: No, the scores don't represent continuous data. There is no score between games. The data should be graphed with a scatter plot.

Vocabulary Review, page 284

1. positive trend

2. line graph

3. no trend

4. scatter plot

5. negative trend

Skill Review, page 284–285

1. 75 in. or 6 ft 3 in.

2. 3

3. Dazzle in the 24-oz size.

4. Possible answer: years of education and annual income

5. A, C, D

Skill Practice, page 285

1.

	Wharton	Essex	Gardner
Wharton	0	6	5
Essex	6	0	12
Gardner	5	12	0

2. Possible answer: Put cars, vans, and trucks in different sections. In each section, list the manufacturers down the left-hand column and colors across the top row.

3. Possible answer: Stretch the vertical axis and start numbering at $50.

4. B

Chapter Review

1. A
2. 19
3. C
4. Possible Answer: $900
5. B
6. C
7. negative trend
8. $3,000
9. C
10. D
11. C
12. B

Name _____ **Date** _____ **Class** _____

Name _____ **Date** _____ **Class** _____

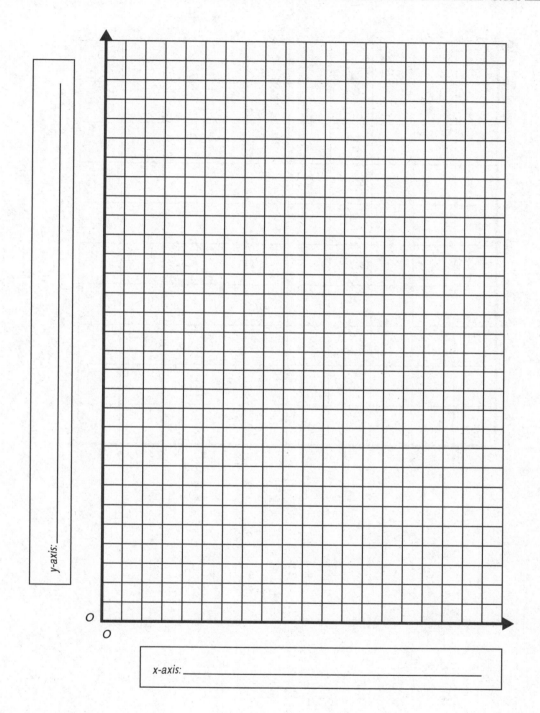

y-axis: _____

O

O

x-axis: _____

Name _____ **Date** _____ **Class** _____

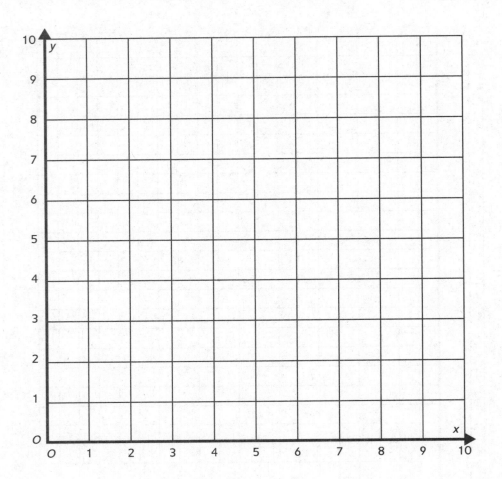

Name _____ **Date** _____ **Class** _____

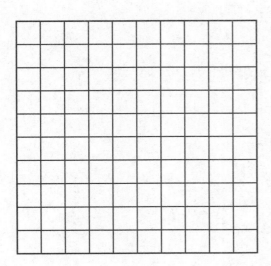

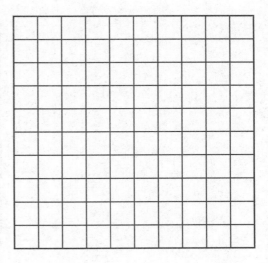

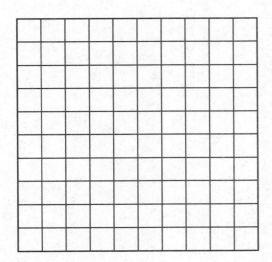

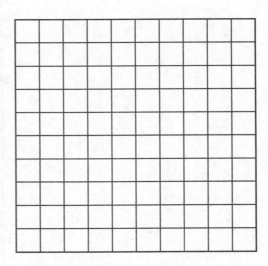

Name _____ **Date** _____ **Class** _____

Name _____ **Date** _____ **Class** _____

MATHEMATICAL FORMULAS

Area of a   square   Area = side^2

rectangle   Area = length × width

triangle   Area = $\frac{1}{2}$ × base × height

parallelogram   Area = base × height

trapezoid   Area = $\frac{1}{2}$ × (base$_1$ + base$_2$) × height

circle   Area = π × radius^2; π is approximately equal to 3.14

Perimeter of a   square   Perimeter = 4 × side

rectangle   Perimeter = 2 × length + 2 × width

triangle   Perimeter = side$_1$ + side$_2$ + side$_3$

Circumference of a   circle   Circumference = π × diameter; π is approximately equal to 3.14

Surface Area of a   rectangular/ right prism   Surface Area = 2(length × width) + 2(width × height) + 2(length × height)

cube   6 × side^2

square pyramid   Surface Area = ($\frac{1}{2}$ × perimeter of base × height of slant) + (base edge)^2

cylinder   Surface Area = (2 × π × radius × height) + (2 × π × radius^2); π is approximately equal to 3.14

cone   Surface Area = (π × radius × height of slant) + (π × radius^2)

sphere   Surface Area = 4 × π × radius^2

Volume of a   rectangular/ right prism   Volume = length × width × height

cube   Volume = edge^3

square pyramid   Volume = $\frac{1}{3}$ × (base edge)^2 × height

cylinder   Volume = π × radius^2 × height; π is approximately equal to 3.14

cone   Volume = $\frac{1}{3}$ × π × radius^2 × height

sphere   Volume = $\frac{4}{3}$ × π × radius^3

Coordinate Geometry	(x_1, y_1) and (x_2, y_2) are two points in a plane.
	slope of a line $= \dfrac{y_2 - y_1}{x_2 - x_1}$; (x_1, y_1) and (x_2, y_2) are two points on the line
	slope-intercept form of the equation of a line $y = mx + b$, when m is the slope of the line and b is the y-intercept
	point-slope form of the equation of a line $y - y_1 = m (x - x_1)$, when m is the slope of the line
Pythagorean Relationship	$a^2 + b^2 = c^2$; in a right triangle, a and b are legs, and c is the hypotenuse
Quadratic Equations	standard form of a quadratic equation $ax^2 + bx + c = 0$
	quadratic formula $x = \dfrac{-b \pm \sqrt{b^2 - 4ac}}{2a}$
Measures of Central Tendency	**mean** $= \dfrac{x_1 + x_2 + \dots + x_n}{n}$, where the x's are the values for which a mean is desired, and n is the total number of values for x
	median $=$ the middle value of an odd number of ordered scores, and the average of the two middle values of an even number of ordered scores
Simple Interest	interest $=$ principal $\times$ rate $\times$ time
Distance	distance $=$ rate $\times$ time

Order of Operations	The TI-30XS MultiView™ automatically evaluates numerical expressions using the Order of Operations based on how the expression is entered.	The correct answer is 23.

Example
$12 \div 2 \times 3 + 5 =$

Note that the 2 is **not** multiplied by the 3 before division occurs.

Decimals	To calculate with decimals, enter the whole number, then ⟨.⟩, then the fractional part.	The correct answer is 17.016.

Example
$11.526 + 5.89 - 0.4 =$

The decimal point helps line up the place value.

Fractions	To calculate with fractions, use the ⟨n/d⟩ button. The answer will automatically be in its simplest form.	The correct answer is $\frac{15}{28}$.

Example
$\frac{3}{7} \div \frac{4}{5} =$

⟨3⟩ ⟨n/d⟩ ⟨7⟩ ⟨enter⟩ ⟨÷⟩ ⟨4⟩ ⟨n/d⟩ ⟨5⟩ ⟨enter⟩

This key combination works if the calculator is in Classic mode or MathPrint™ mode.

Mixed Numbers	To calculate with mixed numbers, use the ⟨2nd⟩ ⟨n/d⟩ button. To see the fraction as an improper fraction, don't press the ⟨2nd⟩ ⟨x10ⁿ⟩ buttons in sequence below.	The correct answer is $39\frac{13}{15}$.

Example
$8\frac{2}{3} \times 4\frac{3}{5} =$

⟨8⟩ ⟨2nd⟩ ⟨n/d⟩ ⟨2⟩ ⟨▼⟩ ⟨3⟩ ⟨enter⟩ ⟨×⟩ ⟨4⟩ ⟨2nd⟩
⟨n/d⟩ ⟨3⟩ ⟨▼⟩ ⟨5⟩ ⟨enter⟩ ⟨2nd⟩ ⟨x10ⁿ⟩ ⟨enter⟩

This key combination only works if the calculator is in MathPrint™ mode.

Percentages	To calculate with percentages, enter the percent number, then ⟨2nd⟩ ⟨(⟩.	The correct answer is 360.

Example
$72\% \times 500 =$

Powers & Roots

To calculate with powers and roots, use the (x²) and (^) buttons for powers and the (2nd) (x²) and (2nd) (^) buttons for roots.

Example
$21^2 =$

[2] [1] (x²) (enter)

The correct answer is 441.

Example
$2^8 =$

[2] (^) [8] (enter)

The correct answer is 256.

Example
$\sqrt{729} =$

(2nd) (x²) [7] [2] [9] (enter)

The correct answer is 27.

Example
$\sqrt[5]{16807} =$

[5] (2nd) (^) [1] [6] [8] [0] [7] (enter)

The correct answer is 7.

You can use the (2nd) (x²) and (2nd) (^) buttons to also compute squares and square roots.

Scientific Notation

To calculate in scientific notation, use the (x10ⁿ) button as well as make sure your calculator is in Scientific notation in the (mode) menu.

The correct answer is 1.2011×10^5.

Example
$6.81 \times 10^4 + 5.201 \times 10^4 =$

[6] [.] [8] [1] (x10ⁿ) [4] (enter) (+)

[5] [.] [2] [0] [1] (x10ⁿ) [4] (enter)

When you are done using scientific notation, make sure to change back to Normal in the (mode) menu.

Toggle

In MathPrint™ mode, you can use the toggle button (◄►) to switch back and forth from exact answers (fractions, roots, π, etc.) and decimal approximations.

The correct answer is 0.428571429.

Example
$\frac{3}{7} =$

[3] (n/d) [7] (enter) (◄►)

If an exact answer is not required, you can press the toggle button (◄►) immediately to get a decimal approximation from an exact answer without reentering the expression.

GLOSSARY

A

absolute value the distance a number is from zero

addend a number that is added to another number

algebraic expression a mathematical statement containing letters and numbers organized as terms but with no equal sign

area the number of non-overlapping square units needed to exactly cover the entire inside of a two-dimensional figure

average the value found by adding all numbers in a data set and dividing by the total number of data in the set

B

bar graph a graph that uses the length of bars to represent data values

benchmark a point of reference from which other measurements or estimates can be made

box plot a display that shows the range and distribution of a data set

C

circle a closed figure with all of its points the same distance from a fixed point called the center

circle graph a graph that uses sections of a circle to represent data values

circumference the distance around the outside of the circle

coefficient a number that is multiplied by a variable

combination a selection of objects or values in which order is unimportant

common difference the amount that is the same between all consecutive differences

complement an event that shows all the ways that an event cannot happen

completing the square a technique of manipulating quadratic equations so that they can be solved by taking the square root of both sides

composite figure a figure that is made up of two or more shapes

composite solid an object that is made up of more than one type of solid

compound event an event formed by two or more simple events

consecutive difference the difference between the next and current terms in a table

constant an expression that stays the same

coordinate the pairs (x, y) graphed on a plane

coordinate plane a grid formed by the intersection of a horizontal number line and a vertical number line

cube a number raised to the third power

cube root a number that, when cubed, equals a given number

cylinder a solid formed by two bases that are parallel, congruent circles

D

degree of a polynomial the largest power of the variable

denominator the bottom number of a fraction that represents the total number of parts contained in the whole of a fraction

dependent event a second event whose probability depends upon a first event

dependent system a system that has an infinite number of solutions

diameter any segment that passes through the center of the circle and whose endpoints are on the circle

discount a decrease or reduction in price

discriminant the part of the quadratic formula that is under the square root

distribute to use multiplication over addition or subtraction

distribution a description of how the data values in a set are spread out

domain the set of inputs of a function

dot plot a display that shows how often each data value occurs

E

elimination method a method of solving a system of equations by adding or subtracting equations to eliminate one of the variables

end behavior the appearance of a graph as it extends in both directions away from zero

equation a mathematical statement that two expressions are equal

equivalent equal; having the same value

evaluate to substitute values for variables

experiment an activity or situation in which the results are uncertain

expression a mathematical statement that contains numbers, operations, and/or variables but no equal sign

F

factor a number that is multiplied by another number

factorial the product of a series of all descending consecutive positive integers from a given starting point

first quartile the median of the lower half of a data set

function a rule that assigns exactly one output to each input

G

greatest common factor (GCF) the greatest factor that is shared between the numbers

H

hemisphere half of a sphere

histogram a display that shows data that have been divided into intervals

hypotenuse the longest side of a right triangle, which is opposite the right angle

I

inconsistent system a system that has no solutions

independent event a second event whose probability does not depend upon a first event

independent system a system that has exactly one solution

index the small number next to a radical sign that indicates the degree of the root

inequality a mathematical statement showing that two quantities are not equal

inequality signs symbols used to show the relationship between the expressions in an inequality ($<$, $>$, $\leq$, or $\geq$)

integers the set of whole numbers and their opposites

interest rate the amount that is earned or charged during a certain amount of time

inverse operations operations that undo each other

irrational number the set of numbers that cannot be expressed as the ratio of two integers

L

leading coefficient the coefficient accompanying the first term in a polynomial that has been written in standard form

least common denominator (LCD) the least common multiple of two or more denominators

least common multiple the least multiple that is shared between the numbers

legend a key printed on a graph or chart that shows the meanings of colors, symbols, or markings used

line graph a graph displaying two-variable data that change continuously over time

line symmetry a figure for which there is a line that divides the figure into two halves that are mirror images of each other

linear function a function that can be written in the form $f(x) = mx + b$, where m and b are constants, whose graph is a non-vertical line

M

mean the sum of all values in a data set divided by the number of values

median the middle value of an ordered data set; in a data set with an even number of values, the average of the two middle values

mode the value(s) that occur most often in a data set

monomial a polynomial with one term, such as 10, $2x$, and $3xy$.

N

negative trend as one variable increases, the other variable tends to decrease

no trend there is no pattern between two variables

numerator the top number in a fraction that represents the part of the whole the fraction is describing

O

one-to-one function a function for which every value in the range has exactly one element assigned to it from the domain

opposite polynomial the polynomial with all of its signs changed to their opposites

order to place in the proper sequence

order of operations the rules for the order in which calculations should be done when evaluating an expression

ordered pair a pair of numbers (x, y) that is used to describe the location of a point in the coordinate plane

outcome a result of an experiment or activity that involves uncertainty

P

parallelogram a four-sided polygon whose opposite sides are parallel

percent a ratio of a number to 100

perfect square trinomial a quadratic expression that can be written as a perfect square of a linear expression

perimeter the distance around the outside of a polygon

permutation a selection of objects or values in which order is important

pi an irrational number approximately equal to 3.14 that represents the ratio of a circle's circumference to its diameter

point-slope form an equation that allows points on a line to be calculated if one point and the slope are known

polygon a closed figure in a plane that is formed by three or more segments

polynomial an algebraic expression consisting of one or more terms in which each term is a number or a product of numbers and variables with whole-number exponents

positive trend as one variable increases, the other variable tends to increase

prime factorization shows a number written as the product of its prime factors

prime number a whole number greater than 1 whose only two factors are 1 and itself

principal an amount of money invested or borrowed

prism a solid with two bases that are congruent, parallel shapes and rectangular lateral faces that connect the bases

probability the study of how likely it is for an event to occur

proportion an equation stating that two ratios are equal

proportional relationship an equation of the form $y = kx$ for some nonzero k

pyramid a solid with all of its faces, except for the base, that intersect at a point called the vertex

Q

quadrant one of the four regions of the coordinate plane formed by the intersection of the x- and y-axes

quadratic function a polynomial that has 2 as its highest power of x

R

radius any segment within a circle whose endpoints are the center of the circle and a point on the circle

range the set of outputs of a function

range the difference between the greatest and least values in a data set

ratio a comparison of two values

rational exponent an exponent that is a rational number

rational expression a ratio of two polynomials

rational number the set of numbers that can be expressed as the ratio of two integers

reciprocals two numbers or expressions whose product is 1

relative maximum/minimum the y-coordinate of any point that is the highest/lowest point for some section of the graph

restricted value (of a rational expression) a value of the variable for which the denominator of the rational expression is equal to 0

rotational symmetry a figure that can be rotated less than 360 degrees around a point to coincide with itself

S

scale factor a ratio of corresponding parts of similar figures

scatter plot a graph that plots two-variable data items on the coordinate plane to show a general trend

scientific notation a system of writing a number as the product of a decimal and a power of 10

similar having the same shape, but not necessarily the same size

simple interest the amount of interest charged or earned after an interest rate is applied to the principal

slope the ratio of rise to run

slope-intercept form $y = mx + b$, where m is the slope of the line and b is the y-intercept

solution of an equation a value for the variable that makes the equation true

solution of an inequality the numbers that, when substituted for the variable in an inequality, make the inequality statement true

solving by inspection determining the solution(s) of an equation simply by looking at the equation

sphere a solid formed by the set of all points that are a given distance from the center

square a number raised to the second power

square root a number that, when squared, equals a given number

standard form the form of a polynomial that shows the terms listed from left to right with the powers of the variables from greatest to least

standard form of a linear equation $Ax + By = C$, where A is a whole number and A and B cannot be 0

standard notation the way in which a number is typically written, using place value

substitute to replace a variable in an expression with a numerical value

substitution method a method of solving a system of equations by solving one equation for one variable and substituting the resulting expression into the other equation

surface area the sum of the areas of all the faces of a three-dimensional figure

system of linear equations a set of two or more linear equations with two or more variables

T

third quartile the median of the upper half of a data set

trapezoid a four-sided polygon with exactly one pair of parallel opposite sides

tree diagram a branching diagram that shows possible outcomes of an experiment

3-dimensional an object consisting of 3 dimensions, usually length, width, and height

2-dimensional a flat shape having only two dimensions, often length and width

U

undefined an expression that cannot be evaluated

unit rate a ratio that compares a quantity to a single unit

V

variable a symbol used to represent an unknown value

volume a measure of the amount of space enclosed by a three-dimensional figure

W

weighted average an average of a data set in which some items carry more importance (weight) than others

X

x-intercept the *x*-coordinate of a point where a graph crosses the *x*-axis

x-value the horizontal value in an ordered pair

Y

y-intercept the *y*-coordinate of the point where a graph crosses the *y*-axis

y-value the vertical value in an ordered pair

INDEX